APPLIED OPTICS
and
OPTICAL ENGINEERING

VOLUME V

Optical Instruments

Part II

Contributors to This Volume

Robert M. Corbin

M. S. Dickson

Walter G. Driscoll

Harry K. Hammond, III

D. Harkness

John H. Hett

Rudolf Kingslake

Henry A. Knoll

A. B. Meinel

Robert J. Meltzer

Francis B. Patrick

David Richardson

APPLIED OPTICS
and
OPTICAL ENGINEERING

EDITED BY

RUDOLF KINGSLAKE

Eastman Kodak Company
Rochester, New York
(Retired)

VOLUME V

Optical Instruments

Part II

ACADEMIC PRESS, New York and London 1969

ACADEMIC PRESS, INC.
111 Fifth Avenue, New York, New York 10003

United Kingdom Edition published by
ACADEMIC PRESS, INC. (LONDON) LTD.
Berkeley Square House, London W1X 6BA

LIBRARY OF CONGRESS CATALOG CARD NUMBER 65–17761

PRINTED IN THE UNITED STATES OF AMERICA

List of Contributors

Numbers in parentheses indicate the pages on which the authors' contributions begin.

ROBERT M. CORBIN, *Rochester, New York* (305)

M. S. DICKSON, *W. and L. E. Gurley, Troy, New York* (231)

WALTER G. DRISCOLL, *St. Vincent Hospital, Worcester, Massachusetts* (85)

HARRY K. HAMMOND, III, *Optics Metrology Branch, National Bureau of Standards, Washington, D.C.* (105)

D. HARKNESS, *W. and L. E. Gurley, Troy, New York* (231)

JOHN H. HETT, *Consultant, American Cystoscope Makers, Inc., Pelham Manor, New York* (251)

RUDOLF KINGSLAKE, *Eastman Kodak Company, Rochester, New York* (1)

HENRY A. KNOLL, *Bausch and Lomb, Inc., Rochester, New York* (281)

A. B. MEINEL, *University of Arizona, Tucson, Arizona* (133)

ROBERT J. MELTZER, *Dynamic Optics, Inc., Fairport, New York* (47)

FRANCIS B. PATRICK, *US Army, Frankford Arsenal, Philadelphia, Pennsylvania* (183)

DAVID RICHARDSON,* *Bausch and Lomb, Inc., Rochester, New York* (17)

* Deceased.

General Preface

It is only within recent years that a specific branch of engineering has arisen devoted to the theory, design, manufacture, testing, and use of optical instruments. There are at the present time hundreds of companies in this and other countries making thousands of optical devices in great variety, which are vitally important in industry and research, in medicine, in the armed forces, in space projects, and in everyday life, yet optical engineering as such is so young that only a very few institutions exist in the world today in which it is taught as a distinct and separate subject. Traditionally, the development of optical apparatus has been in the hands of a few expert craftsmen who have been more or less educated in engineering or science, and who have passed on their knowledge and skill to their successors. It was indeed quite revolutionary when Abbe in the late nineteenth century insisted that all optical instruments in the Zeiss Works be completely designed on paper before construction. Even when an optical device has been satisfactorily made it must be tested and later used with an awareness of its capabilities and limitations if it is to achieve its maximum usefulness.

Optical engineering is now so demanding of advanced knowledge that young men are entering this field from physics, from mechanical and electrical engineering, and other disciplines. They find that the standard physics and engineering textbooks are of little help. There is, therefore, great need for readily available information on the design, testing, and use of optical equipment, and for data on currently available apparatus. Technical handbooks have long been available in chemistry, electronics, and in branches of engineering, but in optics the needed data are scattered widely in many different places or are not available at all.

It is hoped that these volumes will provide information on many aspects of applied optics which may be obscure to an engineer called upon to design some piece of apparatus in which optical elements have an important place. No attempt has been made to provide a mere instruction manual, and the mechanical aspects of instrument design are omitted. Emphasis throughout has been placed on the principles of operation of the various devices described and on equipment existing at the present time.

The material offered has been divided arbitrarily into fifty chapters written by specialists in their fields. Some of the chapters are largely electrical in content, reflecting the present-day close connection between optics and electronics. The level of academic sophistication varies considerably from chapter to chapter; this is inevitable when many authors having quite different interests and backgrounds are involved. However, mathematical derivations have been held to a minimum as they are generally available elsewhere if required. Rapidly developing topics such as the laser have largely been omitted because anything written would probably be out of date by the time it was published, thus defeating the purpose of the book. Although microwaves and x rays are often treated by optical methods, these also have been omitted. However, the ultraviolet and infrared extensions of the visible spectrum, inseparable from "optics" as ordinarily understood, are included.

The fifty chapters have been divided into five groups, each occupying one volume in the series. Volume I deals with the production and modification of light and the formation of images. Volume II covers the detection of light and of infrared radiation. In Volume III specific optical components are considered, together with their manufacture, construction, and use. In Volume IV some of the principal types of optical instruments are described, particularly those relating to photography. In Volume V spectrographs and many other optical instruments are covered in detail.

There are, of course, numerous aspects of applied optics and optical engineering that will not have been covered in these volumes. However, it is hoped that the footnotes and occasional special bibliographies will lead the reader to other sources where he will be able to find the information he needs.

RUDOLF KINGSLAKE

Rochester, New York

Preface to Volume V

This volume completes the series of 50 chapters contributed by 51 authors. The Editor and Publishers express their thanks to these authors, many of whom are extremely busy men, but who nevertheless found time to submit carefully written papers on their particular subjects. Circumstances made it impossible for three of the originally announced authors to write their chapters, but other well-recognized authorities in their fields generously agreed to substitute at short notice. Because of the necessity of keeping the five volumes approximately equal in length, and because of a certain amount of last-minute reshuffling of chapters which became necessary to accommodate authors who needed longer time to complete their assignments, the volume titles do not in all cases accurately represent the contents.

It will be noted that this volume, in addition to its own index, also contains a general cumulative index covering the contents of all five volumes. It is hoped that readers will find this of value.

RUDOLF KINGSLAKE

Rochester, New York

Contents

CHAPTER 4

Spectrophotometers

Walter G. Driscoll

CHAPTER 5

Colorimeters

Harry K. Hammond III

CHAPTER 6

Astronomical Telescopes

A. B. Meinel

CHAPTER 7

Military Optical Instruments

Francis B. Patrick

CHAPTER 8

Surveying and Tracking Instruments

M. S. Dickson and D. Harkness

CHAPTER 9

Medical Optical Instruments

John H. Hett

CHAPTER 10

Opthalmic Instruments

Henry A. Knoll

CHAPTER 11

Motion Picture Equipment

Robert M. Corbin

Contents of Other Volumes

Volume I: Light: Its Generation and Modification

Volume II: The Detection of Light and Infrared Radiation

Volume III: Optical Components

Volume IV: Optical Instruments—Part I

CHAPTER 1

Dispersing Prisms

RUDOLF KINGSLAKE
Eastman Kodak Company, Rochester, New York

I. INTRODUCTION

Because of their great light efficiency, single-order spectrum, ruggedness, and ease of manufacture, prisms have long been favored as a dispersing medium in spectroscopes, spectrographs, and monochromators. Disadvantages of prisms are their nonlinear dispersion and the very limited wavelength range for which they are transparent.

The refractive index of a spectroscope prism is basically unimportant, except that if it is high, there will be a serious loss of light by surface reflection upon entering and leaving the prism; on the other hand, the dispersion of the prism material is of very great importance. The dispersive power of a material is defined by the index difference taken over a given wavelength range divided by the mean refractive index minus one. Thus, in the visible part of the spectrum, the dispersive power of a glass is defined by

$$\text{Dispersive power} = \omega = (n_F - n_C)/(n_D - 1)$$

where wavelength $C = 0.6563\ \mu$, $D = 0.5896\ \mu$, and $F = 0.4861\ \mu$. The dispersive powers of some common materials in the visible spectrum are given in Table I.

Spectroscopists, however, generally define dispersion by the ratio $dn/d\lambda$, which, of course, varies from point to point in the spectrum. A typical flint glass, e.g., has the measured indices given in Table II. To obtain the dispersion, we substitute the index values in a suitable formula connecting

TABLE I

DISPERSIVE POWERS IN THE VISIBLE REGION

Material	Refractive index			$\omega = \frac{n_F - n_C}{n_D - 1}$
	n_C	n_D	n_F	
Liquids:				
Water	1.3312	1.3330	1.3371	0.0180
Methyl salicylate	1.5304	1.5363	1.5528	0.0418
Bromnaphthalene	1.6499	1.6588	1.6824	0.0493
Ethyl cinnamate	1.5522	1.5598	1.5804	0.0505
Carbon disulfide	1.6201	1.6295	1.6544	0.0544
Methylene iodide	1.7275	1.7559	1.7750	0.0628
Solids:				
Fluorite	1.4325	1.4338	1.4370	0.0105
Crystal quartz (*o*)	1.5419	1.5442	1.5497	0.0143
Crystal quartz (*e*)	1.5509	1.5533	1.5590	0.0146
Fused quartz	1.4564	1.4585	1.4632	0.0148
Crown glass	1.5204	1.5230	1.5293	0.0170
Methylmethacrylate	1.4892	1.4917	1.4978	0.0174
Rock salt	1.5407	1.5443	1.5534	0.0234
Flint glass	1.6122	1.6170	1.6290	0.0273
Polystyrene	1.5848	1.5902	1.6039	0.0323
Extra-dense flint glass	1.7131	1.7200	1.7377	0.0341
Silver chloride	2.0526	2.0664	2.1030	0.0473

TABLE II

REFRACTIVE INDICES OF A TYPICAL FLINT GLASS 1.617/36.6

	Wavelength λ	Refractive index n	Calculated dispersion $dn/d\lambda$
M	1.0140 μ	1.60007	−0.0201
A'	0.7665	1.60684	−0.0387
C	0.6563	1.61218	−0.0605
D	0.5893	1.61700	−0.0846
F	0.4861	1.62904	−0.1610
g	0.4358	1.63887	−0.2383
h	0.4047	1.64740	−0.3150
m	0.3650	1.66280	−0.4750

refractive index with wavelength, which may then be differentiated to give the dispersion. One such formula is that proposed by Herzberger,[1] in which

$$n = a_1 n_M + a_2 n_C + a_3 n_F + a_4 n_m$$

where the a are particular functions of λ, and wavelengths M and m are 1.014 μ and 0.365 μ, respectively. On differentiating this, we get an expression for the dispersion,

$$\frac{dn}{d\lambda} = n_M \frac{da_1}{d\lambda} + n_C \frac{da_2}{d\lambda} + n_F \frac{da_3}{d\lambda} + n_m \frac{da_4}{d\lambda}$$

The computed values of $dn/d\lambda$ are also given in Table II for several wavelength regions. It will be noticed that in the near infrared (at 1.014 μ), the dispersion of this glass is very low, while it is more than 20 times as great in the near ultraviolet, at 0.365 μ.

II. THE GEOMETRY OF A PRISM

A. Refraction of a Ray at an Inclined Plane Surface

In Fig. 1, we let ϕ represent the tilt of the surface from its normal

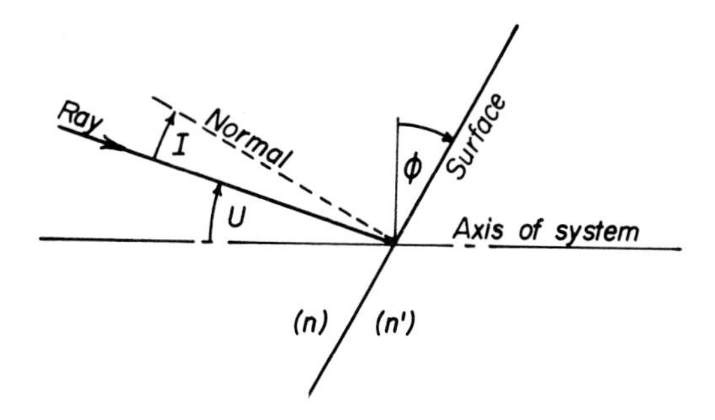

Fig. 1. A ray refracted at an inclined plane surface.

position perpendicular to the axis of the system, and we define the slope of the entering ray by the angle U. Both these angles will be considered positive if a clockwise rotation takes us from the axis to the ray or to the surface normal. The angle of incidence I is therefore given by $(\phi - U)$, and the ray-tracing equations are simply

$$I = \phi - U, \qquad \sin I' = (n/n') \sin I, \qquad U' = \phi - I'$$

---

[1] M. Herzberger, *Opt. Acta* **6**, 197–215 (1959).

B. Passage of a Ray of Light through a Prism

We shall suppose that the prism has an angle A and is made of material having a refractive index n (Fig. 2). A ray of monochromatic light will

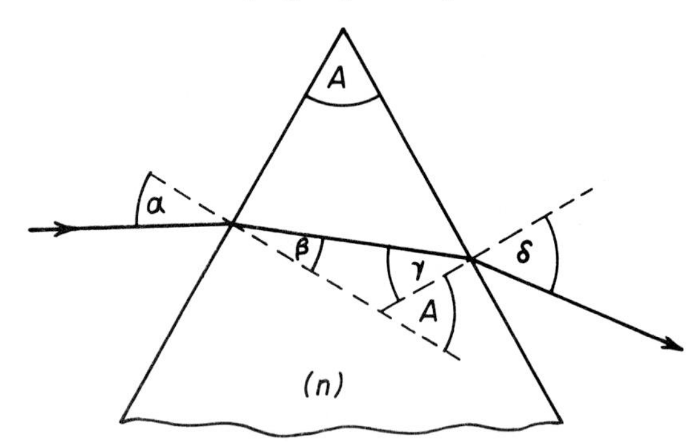

Fig. 2. Passage of a ray through a prism.

enter the first surface at an incident angle α, the angle of refraction being β. The ray then crosses the prism and strikes the second surface at an angle of incidence γ, finally leaving at an angle of refraction δ. These four angles are related by

$$\sin\beta = (\sin\alpha)/n, \qquad \gamma = A - \beta, \qquad \sin\delta = n\sin\gamma \tag{1}$$

Two useful expressions may be obtained which give δ in terms of α and A. Since $A = \beta + \gamma$, $\sin A = \sin\beta\cos\gamma + \cos\beta\sin\gamma$. By squaring this, replacing $\cos^2$ by $(1 - \sin^2)$, and recalling that $\cos A = \cos\beta\cos\gamma - \sin\beta \sin\gamma$, we find for the internal angles that

$$\sin^2 A = \sin^2\beta + \sin^2\gamma + 2\sin\beta\sin\gamma\cos A$$

Including the refraction at each surface, this equation becomes

$$n^2\sin^2 A = \sin^2\alpha + \sin^2\delta + 2\sin\alpha\sin\delta\cos A \tag{2}$$

Another simple relation is found by writing

$$\sin\delta = n\sin(A - \beta) = \sin A\,(n^2 - \sin^2\alpha)^{1/2} - \cos A\sin\alpha \tag{3}$$

The deviation of the emerging ray from its incident direction is given by

$$D = (\alpha - \beta) + (\delta - \gamma) = \alpha + \delta - A \tag{4}$$

C. Minimum Deviation

If we take a typical prism and plot a graph of the deviation D *vs* the angle of incidence α (Fig. 3), we find that at one particular value of α, the

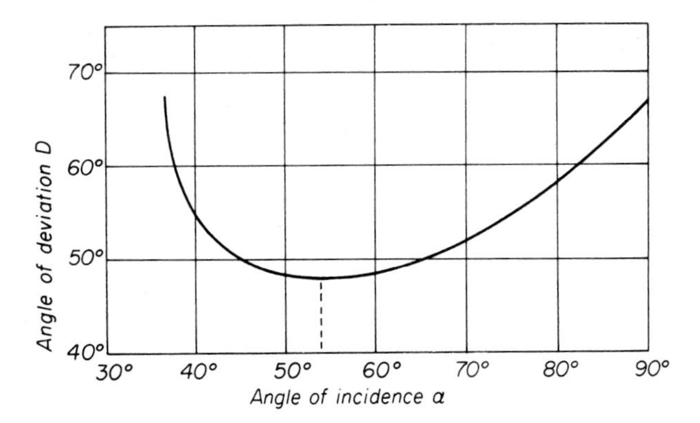

FIG. 3. Angle of deviation *vs* angle of incidence for a 60° prism of dense flint glass.

deviation reaches a minimum value. To determine the values of α and D at this minimum, we set $dD/d\alpha$ equal to zero in the usual way. By Eq. (4), when $dD/d\alpha = 0$, we see that $d\delta/d\alpha = -1$. Now we can evaluate $d\delta/d\alpha$ by

$$\frac{d\delta}{d\alpha} = \frac{d\delta}{d\gamma}\frac{d\gamma}{d\beta}\frac{d\beta}{d\alpha} = \frac{n \cos \gamma}{\cos \delta}(-1)\frac{\cos \alpha}{n \cos \beta} = -\frac{\cos \alpha \cos \gamma}{\cos \beta \cos \delta} \tag{5}$$

Hence, at the minimum deviation case when $d\delta/d\alpha = -1$, we have

$$\cos \alpha \cos \gamma = \cos \beta \cos \delta$$

By writing $\cos = (1 - \sin^2)^{1/2}$, this condition reduces to

$$\alpha = \delta \qquad \text{and} \qquad \beta = \gamma$$

The ray therefore passes symmetrically through the prism, and $\beta = \gamma = \frac{1}{2}A$. In this case $\alpha = \delta = \frac{1}{2}(D_{\min} + A)$, and

$$n = \frac{\sin \frac{1}{2}(D_{\min} + A)}{\sin \frac{1}{2}A} \tag{6}$$

This is the well-known spectrometer formula which is commonly employed when measuring the refractive index of a prism.

The effect of small errors in the measurement of the prism angle A and the angle of minimum deviation D can be found by differentiating Eq. (6):

$$\frac{\partial n}{\partial A} = \frac{\sin \frac{1}{2}D}{2 \sin^2 \frac{1}{2}A}$$
$$\frac{\partial n}{\partial D} = \frac{\cos \frac{1}{2}(A + D)}{2 \sin \frac{1}{2}A} \tag{7}$$

From these equations we see that if a 60° prism having a refractive index of 1.6 is used, an error of 1 sec of arc in the measurement of A and D leads to errors in the computed refractive index of 0.000004 and 0.000003 respectively. Thus, to obtain an accuracy of 1 in the fifth decimal place of refractive index, it is necessary to measure A and D to about 1 sec of arc. The precautions to be observed when using a precision spectrometer for this purpose have been discussed by Martin,[2] Tilton,[3] Guild,[4] and others.

D. Passage of a Converging Light Beam through a Prism

Because the surfaces of the prisms considered here are plane, all rays of a strictly parallel beam will be refracted identically, and hence the parallelism of a beam of monochromatic light will be maintained after passage through any number of prisms in succession. No aberrations such as spherical, chromatic, coma, or astigmatism therefore arise. There will, however, be some distortion and lateral color even though the beam is parallel.

The case is otherwise if a converging or diverging beam passes through a prism. In this case, the various rays in the beam will be differently refracted by the prism, and all types of aberrations will arise.

Suppose that an optical system forms a plane aberrationless image of some object, and we insert in the beam between the lens and its image a

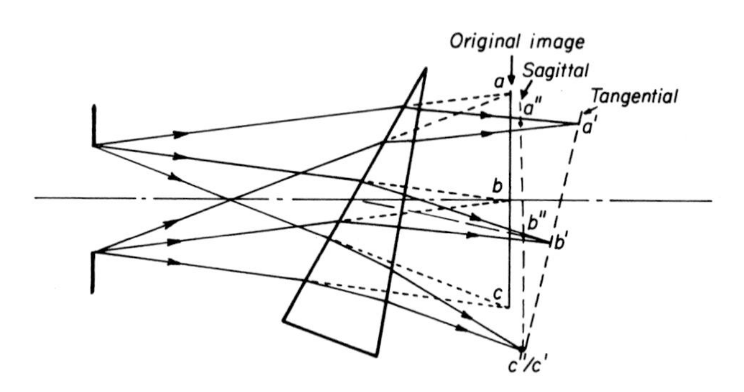

Fig. 4. Inserting a prism between a lens and its image.

---

[2] L. C. Martin, "Optical Measuring Instruments," pp 124–135. Blackie, London, 1924.

[3] L. W. Tilton, *Natl. Bur. Std. (U.S.) Res. Papers* Nos. 64, 262, 575, 776, and elsewhere.

[4] J. Guild, Spectroscopes and refractometers, *in* "Dictionary of Applied Physics" (Sir R. Glazebrook, ed.), Vol. IV, pp. 754–778. Macmillan, London, 1923.

20° glass prism, as shown diagrammatically in Fig. 4. Three typical converging beams are shown, proceeding from the lens toward the top (a), the middle (b), and the bottom (c) of the original image, respectively. After refraction by the prism, these beams come to a focus in an inclined, approximately plane surface at the points a', b', and c', respectively, and it should be noted that the image surface is nearly, but not quite, perpendicular to the emerging axial ray. However, there is now some negative coma in the image and a large amount of overcorrected astigmatism, the sagittal foci being at the points a'', b'', and c'', respectively. It should be noticed that the principal ray of the beam through c happens to pass through the prism at approximately minimum deviation, and very little astigmatism will arise under these conditions. In the remainder of the present discussion, only parallel light will be considered.

E. The Prism Diopter

In ophthalmic work, a prism which deviates a ray by 1 cm measured at a distance of 100 cm from the prism when the light is incident perpendicularly upon the first prism face is said to have a power of 1 prism diopter. The dioptric power of a prism used under these conditions is therefore 100 times the tangent of the angular deviation of the prism. This is indicated in Fig. 5, the distance x being numerically equal to the number of prism

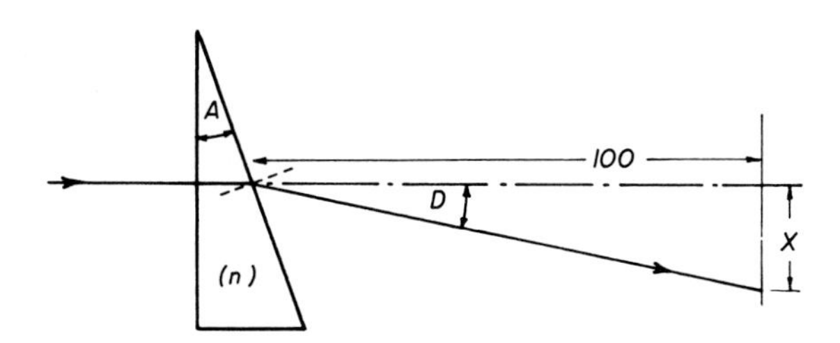

FIG. 5. A prism of x diopters power.

diopters. Since all the deviation occurs at the second surface, we see that the angular deviation $D = \delta - A$, where $\sin \delta = n \sin A$. Assuming $n = 1.523$ for ophthalmic purposes, we can construct a series of typical cases as given in Table III.

It will be seen from this table that the dioptric power of a prism is approximately equal to the prism angle in degrees, at least for prisms having an angle less than 20°.

TABLE III

RELATION BETWEEN PRISM DIOPTERS AND PRISM ANGLE

A (deg)	β (deg)	D (deg)	Prism diopters
1	1.523	0.523	0.913
2	3.047	1.047	1.828
5	7.628	2.628	4.590
10	15.335	5.335	9.338
20	31.392	11.392	20.149
30	49.597	19.597	35.602
40	78.227	38.227	78.768

F. PRISM MAGNIFICATION

If a prism is used in a position decidedly away from the minimum-deviation condition, the emerging parallel beam will be either broader or narrower than the entering parallel beam (Fig. 6). This is the essential

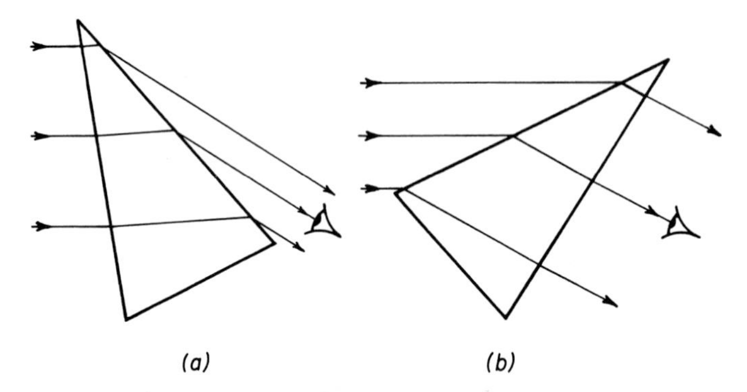

FIG. 6. (a) Prism magnification. (b) Prism demagnification.

property of a telescope, and if such a prism is held before the eye, objects seen through it will exhibit a magnification (Fig. 6a) or a demagnification (Fig. 6b). However, this occurs only in the direction perpendicular to the refracting edge of the prism, there being no change in the apparent size of objects in the meridian parallel to the refracting edge. The result is some degree of anamorphic stretching or compression, combined with the normal deviation of the prism.

The telescopic magnifying power (M.P.) of a prism under any given set of conditions can be readily computed by

$$\text{M.P.} = -\frac{d\delta}{d\alpha} = \frac{\cos\alpha\cos\gamma}{\cos\beta\cos\delta} \tag{8}$$

and, as shown above, this becomes unity when the deviation is a minimum. For a 30° prism of index 1.523, the relation between magnifying power and the angle of incidence α is shown graphically in Fig. 7.

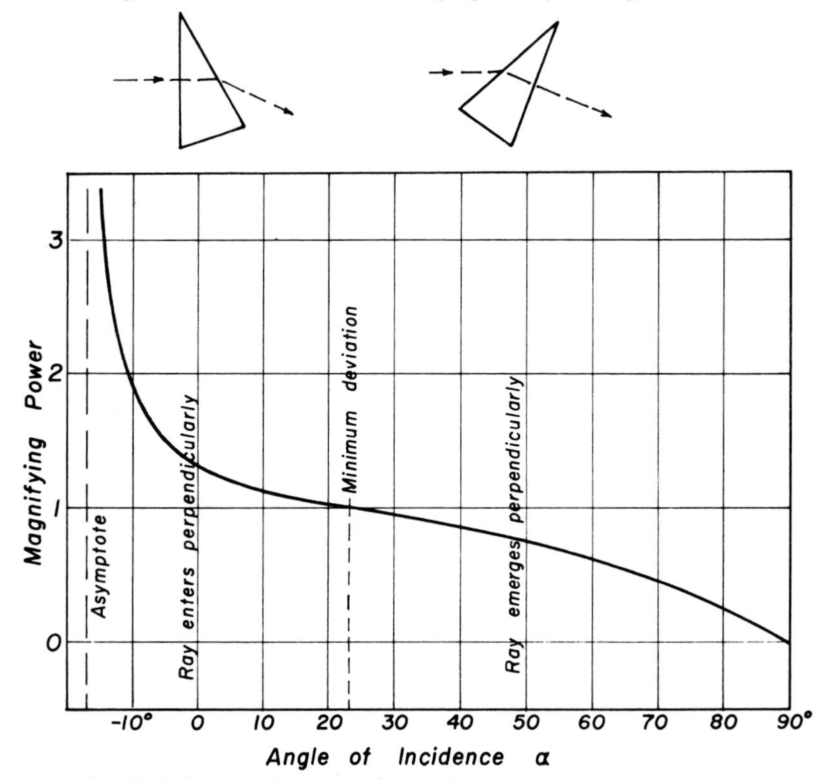

FIG. 7. Prism magnification *vs* angle of incidence. $A = 30°$, $n = 1.523$.

Early in the nineteenth century, Sir David Brewster[5] studied this situation and observed that if two similar prisms are used in succession in parallel light, the angle of incidence being the same at each, then the deviations will cancel out, but the anamorphic compression will be doubled. In this way, it is possible to produce an anamorphic effect with a straight-through beam (Fig. 8). Many recent inventors have applied this idea to

[5] Sir D. Brewster, "A Treatise on Optics," 1st American Ed., p. 302. Carey, Lea, and Blanchard, Philadelphia, 1833.

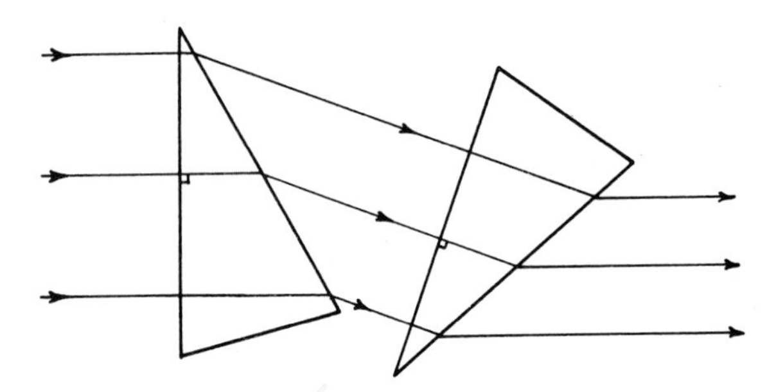

FIG. 8. Parallel light incident normally on two oppositely oriented similar prisms in succession.

motion-picture cameras and projectors, to compress a long, narrow picture into the standard film format and then stretch it back to its original shape on the screen.

Brewster also proposed the use of two such anamorphic prism pairs in succession in perpendicular meridians, so that an undistorted telescopic magnification of variable power could be obtained by using prisms only and no lenses. He called this device a "teinoscope." Because these prisms give good definition only when used in strictly parallel light, it is necessary to use a collimator lens in front of the prisms when the object is not at infinity.

III. THE DISPERSION OF A PRISM

In spectroscopic work, the dispersion of a prism is of prime importance. By dispersion is meant the change in the emergent angle corresponding to a given change in the wavelength,

$$\text{Dispersion} = \frac{d\delta}{d\lambda} = \frac{d\delta}{dn}\frac{dn}{d\lambda} \tag{9}$$

On differentiating Eq. (3) with respect to n, regarding the prism angle A and the entering angle α as constants, we find

$$\begin{aligned}\cos\delta\,(d\delta/dn) &= \tfrac{1}{2}\sin A\,(n^2 - \sin^2\alpha)^{-1/2}2n \\ &= (\sin A)/(\cos\beta).\end{aligned}$$

Hence

$$d\delta/dn = \sin A/(\cos\beta\cos\delta)$$

and

$$\text{Dispersion} = \frac{d\delta}{d\lambda} = \frac{\sin A}{\cos\beta\cos\delta}\frac{dn}{d\lambda} \tag{10}$$

For a 60° prism in minimum deviation, $\beta = \frac{1}{2}A = 30°$, and Eq. (10) is simplified to

$$\text{Dispersion} = \frac{d\delta}{d\lambda} = \frac{dn}{d\lambda} \sec \delta \tag{11}$$

The value of $dn/d\lambda$, of course, depends upon the type of glass and the wavelength region (see Table I); sec δ can be found if the refractive index is known:

n	1.50	1.55	1.60	1.65	1.70	1.75	1.80
δ (deg)	48.59	50.81	53.13	55.59	58.21	61.04	64.16
sec δ	1.511	1.582	1.667	1.769	1.898	2.065	2.294

It is of interest to note that the dispersion of a prism is nearly linear with wave number ($1/\lambda$), whereas the dispersion of a grating is linear with wavelength.

IV. THE RESOLVING POWER OF A PRISM SPECTROSCOPE

Assuming an infinitely narrow slit and no aberrations, so that the whole optical system is diffraction-limited, we may regard the resolving power as the closest resolvable difference in wavelength, $d\lambda$, at a wavelength λ. It is customary to express the resolving power by

$$\text{Resolving power} = \lambda/d\lambda \tag{12}$$

In Fig. 9, we assume that the limiting aperture of the system is the rim of the second lens, L_2, of focal length f_2, and that parallel light passes through the prism. The first lens and the prism itself are assumed to be larger than necessary, so that they do not limit the aperture in any way.

By diffraction theory, the least resolvable separation of two adjacent line images in the focal plane of the lens is about $\lambda f_2/w$. As our spectroscope

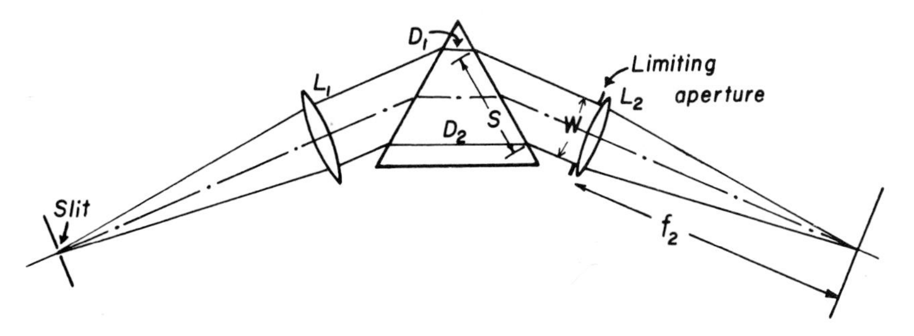

FIG. 9. Resolving power of a prism spectrograph.

is just able to resolve two wavelengths separated by $d\lambda$, it is clear that $\lambda f_2/w$ must be equal to $f_2\, d\delta$, where $d\delta$ is the angular separation of the emerging rays in wavelengths λ and $\lambda + d\lambda$, respectively. Therefore

$$\frac{\lambda f_2}{w} = f_2 \frac{d\delta}{dn} \frac{dn}{d\lambda} d\lambda$$

or

$$\frac{\lambda}{d\lambda} = w \frac{d\delta}{dn} \frac{dn}{d\lambda}$$

Hence

$$\begin{aligned}\text{Resolving power} &= w \times (\text{prism dispersion}) \\ &= \frac{w \sin A}{\cos \beta \cos \delta} \frac{dn}{d\lambda} \\ &= \frac{s \sin A}{\cos \beta} \frac{dn}{d\lambda} \qquad (13)\end{aligned}$$

where s is the slant width of the beam of light emerging from the prism and entering the lens L_2. The focal length of this lens has cancelled out.

If the prism is in minimum deviation, $\beta = \frac{1}{2}A$, and the resolving power is $2s \sin \frac{1}{2}A\ (dn/d\lambda)$. Now $2s \sin \frac{1}{2}A$ is the difference between the longest and the shortest paths inside the prism, namely, $D_2 - D_1$ in Fig. 9. Hence in this case,

$$\text{Resolving power} = \lambda/d\lambda = (D_2 - D_1)(dn/d\lambda) \qquad (14)$$

To resolve the two sodium lines at 5890 and 5896 Å, a resolving power of $5893/6 = 980$ is required, and assuming that the prism is made of the glass described in Table II, then $dn/d\lambda = -0.0846$ in micron units, and we find that the smallest prism of this glass capable of resolving the sodium lines is one having a base length of 12 mm ($\frac{1}{2}$ in.).

V. CURVATURE OF SPECTRUM LINES

Because light originating at the two ends of the slit passes through the prism in a direction which is slightly inclined to the principal or meridian plane, this light will suffer slightly more deviation than the light from the midpoint of the slit, which lies in that plane. Hence the slit images in various wavelengths, the spectrum lines, will be curved with their ends displaced toward the blue end of the spectrum.

As it is sometimes important to know the radius of curvature ρ of the spectral lines, e.g., when constructing the curved exit slit of a prism monochromator, we need a formula for ρ in terms of the other data of the instrument.

A ray passing obliquely through a prism is really a skew ray, and we can project its path into the principal plane of the prism, and also into a plane which is parallel to the refracting edge and to the internal ray inside the prism. This plane contains the "ground face" of the prism (Fig. 10).

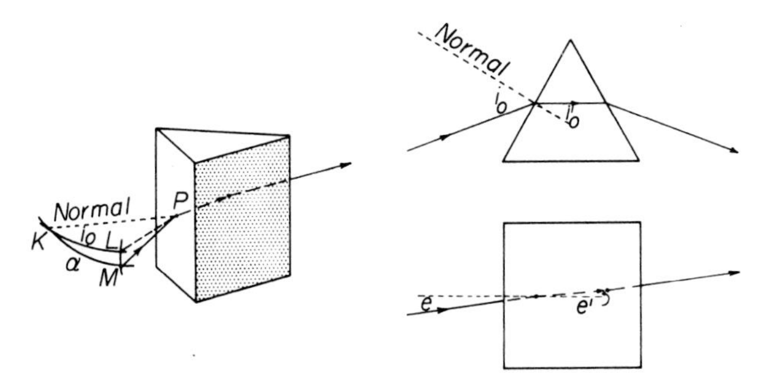

FIG. 10. A skew ray through a prism.

The angle i_0 is the projection of the true incidence angle α into the principal plane, and i_0' is the projection of the angle of refraction β.

In the spherical triangle KLM shown in Fig. 10, the angle KLM is a right angle; α is the true angle of incidence in the plane of incidence KPM (containing the incident ray, the refracted ray, and the normal at the point of incidence); KL is the projection of α into the principal plane, namely, i_0 ; and LM is the projection of α into the plane containing ground face, namely, the small angle e.

By the well-known equation for a right-angled spherical triangle,

$$\sin a \cos B = \cos b \sin c$$

where A is a right angle, we have for the incident skew ray in air,

$$\sin \alpha \cos MKL = \cos e \sin i_0$$

and for the refracted ray inside the prism

$$\frac{\sin \alpha}{n} \cos MKL = \cos e' \sin i_0'$$

On eliminating ($\sin \alpha \cos MKL$), we find

$$\sin i_0 = \sin i_0' \frac{n \cos e'}{\cos e} \tag{15}$$

Thus we can follow the path of an oblique ray through the prism by considering only the projection of the ray into the principal plane, provided we replace the true refractive index n by the quantity ($n \cos e'/\cos e$).

For the case of a spectrograph having the usual short slit, the angle e will be very small, and we may write $e' = e/n$. We may also replace the cosines by the first term of their expansions, namely

$$\cos e = 1 - \tfrac{1}{2}e^2 \qquad \text{and} \qquad \cos e' = 1 - \tfrac{1}{2}e'^2$$

Hence, approximately,

$$\begin{aligned}(\cos e')/(\cos e) &= (1 - \tfrac{1}{2}e'^2)(1 + \tfrac{1}{2}e^2) = 1 + \tfrac{1}{2}e^2 - \tfrac{1}{2}e'^2 \\ &= 1 + \tfrac{1}{2}e^2(n^2 - 1)/n^2.\end{aligned}$$

Applying the approximate spherometer formula to the curved slit image in the focal plane of the lens L_2, we may write $\rho = y^2/2x$, where

$$y = f_2 e \qquad \text{and} \qquad x = f_2\, \Delta\delta$$

Here, $\Delta\delta$ is the difference between the axial angle δ and its projection onto the principal plane. Thus

$$\rho = f_2{}^2 e^2/(2 f_2\, \Delta\delta) \tag{16}$$

and $\Delta\delta = (d\delta/dn)\, \Delta n$, where $\Delta n =$ "oblique n" — true n,

$$\begin{aligned}\Delta\delta &= \frac{\sin A}{\cos\beta\cos\delta}\, n\left(\frac{\cos e'}{\cos e} - 1\right) \\ &= \frac{\sin A}{\cos\beta\cos\delta}\, n\left(\frac{1}{2}\, e^2\, \frac{n^2 - 1}{n^2}\right) \\ &= \frac{(\sin A)\, e^2(n^2 - 1)}{2n\cos\beta\cos\delta}\end{aligned}$$

Hence, by Eq. (16),

$$\begin{aligned}\rho &= \frac{f_2 e^2}{2\, \Delta\delta} \\ &= \frac{1}{2} f_2 e^2 \left[\frac{2n\cos\beta\cos\delta}{e^2(n^2 - 1)\sin A}\right] \\ &= \frac{f_2 n\cos\beta\cos\delta}{(n^2 - 1)\sin A}\end{aligned} \tag{17}$$

This general result may be simplified for a prism in minimum deviation, for which $\delta = \alpha$ and $\beta = \tfrac{1}{2}A$. Then

$$\rho = \frac{f_2 n^2}{2(n^2 - 1)} \cot\text{an}\, \alpha \tag{18}$$

$$= \frac{f_2 n}{n^2 - 1}\, \frac{(1 - n^2 \sin^2 \tfrac{1}{2}A)^{1/2}}{2 \sin \tfrac{1}{2}A}$$

and if $A = 60°$

$$\rho = \frac{f_2 n}{n^2 - 1}\left(1 - \frac{1}{4} n^2\right)^{1/2} \tag{19}$$

As an example, using a dense flint 60° prism in minimum deviation and a telescope lens of 10 in. focal length, the radius of curvature ρ of the spectrum lines is related to the refractive index n of the glass as follows

n	1.70	1.71	1.72	1.73	1.74
ρ(in.)	4.738	4.608	4.481	4.356	4.231

The radius of curvature is therefore noticeably shorter in the blue end of the spectrum than in the red end.

VI. ACHROMATIC AND DIRECT-VISION PRISMS

By combining two prisms made from types of glass having different dispersive powers, it is possible to construct a prism combination which deviates light without chromatic dispersion. By analogy with lenses having similar properties, this combination is called an achromatic prism. Conversely, we can develop a combination which has a significant residual of color dispersion with no deviation for the mid ray; this is the prism system used in direct-vision spectroscopes, and it is analogous to a doublet lens having zero power but a considerable residual of chromatic aberration.

Typical combinations of these two prism types are shown in Fig. 11. The glasses are C-1 (1.523/58.6) and EDF-3 (1.720/29.3). The direct-vision prism (Fig. 11a) deviates the F line upward by 0.30° and the C line downward by 0.11°; the achromatic prism (Fig. 11b) deviates both the C and F lines downward by 12.045°.

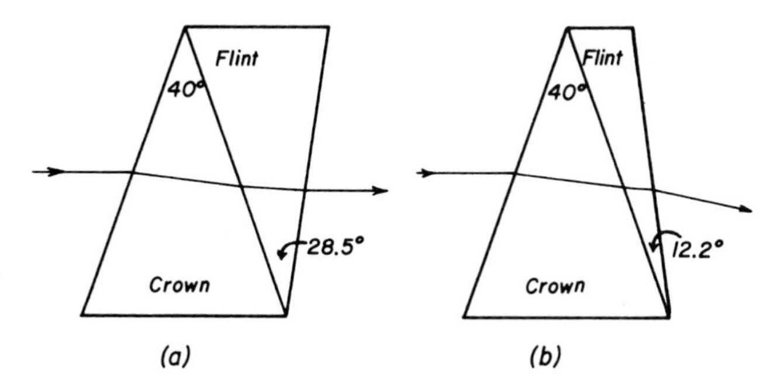

FIG. 11. (a) A direct-vision prism. (b) An achromatic prism.

In 1881, Wernicke[6] pointed out that certain liquids are known which have very high dispersive powers with only a moderate refractive index. By selecting a glass having the same index as the liquid for the *D* spectral line, a prism can be constructed with perpendicular end faces (Fig. 12),

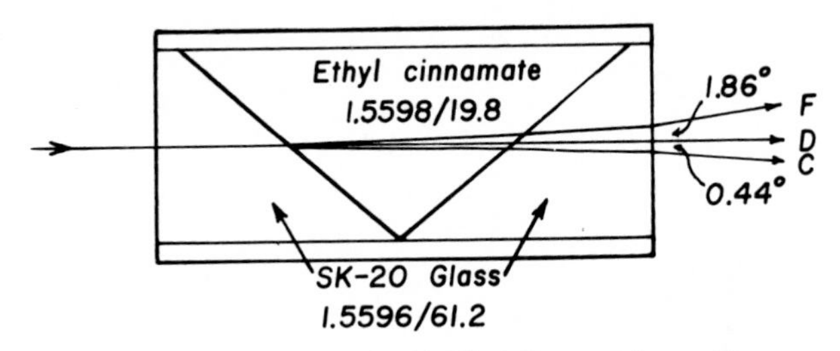

FIG. 12. A Wernicke liquid-filled direct-vision prism.

which does not deviate the *D* ray but which produces considerable dispersion for other colors, and thus constitutes a convenient direct-vision prism for projection and other spectroscopic purposes. The refractive indices of some suitable liquids are given in Table I.

[6] W. Wernicke, *Z. Instrumentenk.* **1**, 353–357 (1881).

CHAPTER 2

Diffraction Gratings

DAVID RICHARDSON*
Bausch and Lomb, Inc., Rochester, New York

I. INTRODUCTION

A. Preliminary Remarks

A diffraction grating is a plane or curved optical surface with many straight, parallel, and equally spaced grooves which cover the optical aperture of the instrument in which it is used. Gratings operate by the mutual interference of the light beams diffracted by the grooves. For the

---

* Deceased.

grating to operate efficiently, it is necessary for the incident radiation to have a wavelength less than the groove spacing.

When light is incident on the ruled surface of a diffraction grating, it is reflected at angles that depend only on the spacing of the grooves of the grating, the angle of incidence, and the wavelength of the light. Light thus separated into its components according to wavelength forms a spectrum that can be observed, photographed, or scanned photoelectrically or thermoelectrically. Gratings are now used as the dispersive elements in most spectroscopes, spectrometers, and spectrographs.

The principle of the diffraction grating was discovered in the late eighteenth century. The dual requirements for fine ruling (or wire lattice spacing practiced by some of the pioneers) as well as a large total number of lines were formidable, and the history of the grating can almost be told by plotting these numbers against time. In 1785, Rittenhouse ruled 53 grooves in $\frac{1}{2}$ in.; in 1823, Fraunhofer ruled 4000 grooves in $\frac{1}{2}$ in.; in 1846, Norbert ran up 6000 grooves in 1 in.; L. M. Rutherfurd, a New York lawyer by profession, but an amateur astronomer at heart, ruled 35,000 grooves on a speculum mirror 2 in. wide in 1870. By this time, the advantages of a reflection grating were appreciated. An excellent summary of the development of gratings has been given by Harrison.[1]

Glass prisms were used for most optical spectroscopy until H. A. Rowland at Johns Hopkins University in 1882 built an engine capable of ruling 6-in. gratings of the quality needed to produce a very pure spectrum. He also invented the concave grating, which made it possible to build spectrographs covering wide wavelength ranges without the use of either prisms or lenses. An account of the contributions to the art of making gratings made by the Johns Hopkins University has been given by Strong.[2]

B. Choice of Grating

1. *Plane vs Concave Gratings*

The first decision to be made by the prospective user of a grating is whether it shall be on a plane or a spherical concave blank. If he is ordering a grating for a particular type of instrument, the choice is an easy one, but if he is considering the design of a new instrument, the comments that follow may be helpful to him.

In general, it can be said that plane gratings are used in scanning-type instruments, in which a simple rotation of the grating allows a succession of wavelengths to emerge from the exit slit of the instrument. These include

[1] G. R. Harrison, *J. Opt. Soc. Am.* **39**, 413–426 (1949).

[2] J. D. Strong, *J. Opt. Soc. Am.* **50**, 1148–1152 (1960).

most spectrometers, spectrophotometers, and monochromators. The advantage of the plane grating over concave gratings for this type of use is in the fact that no focus adjustment is needed as the spectrum is scanned. Plane gratings are used for almost all of the infrared and for most of the visible and ultraviolet grating instruments. With the advent of high-reflectance coatings[3-6] for the 1200–2000 Å region, the use of plane gratings has increased rapidly in this vacuum-ultraviolet spectral band.

Another general advantage of plane gratings over concave gratings is the fact that plane grating mountings are stigmatic,[7] while most concave grating mountings are not. Stigmatic mountings are usually preferred because of greater efficiency in the use of the available light. Other advantages of plane over concave gratings include lower price for a given type and size, availability in larger sizes, and ability to cover completely the usual square or round aperture of the optical system.

Concave gratings,[7,8] on the other hand, find their principal use in photographic and direct-reading instruments, in which a wide region of the spectrum is in focus. In the vacuum ultraviolet, they are used even in scanning instruments, primarily to avoid the need for additional mirrors or lenses to focus the spectrum. Plane gratings always require additional focusing optics.

2. *Transmission Gratings*

Once it has been decided that a plane grating can be used, the possible advantages of using a transmission grating should be considered. This type of grating can be used in both the visible and ultraviolet regions of the spectrum.

Transmission gratings have found their most important use for photographing the spectra of stars, meteors, and reentrant rockets. A suitable camera with a transmission grating over the lens becomes a spectrograph for distant luminous objects. The incident light is parallel and the object itself acts as both the entrance slit and the collimator of the system.

Whenever it is desired to have separate collimating and camera lenses, it is frequently desirable to use transmission gratings. They offer advantages in speed and in freedom from scattering. If desired, they can be formed on the hypotenuse face of a prism[9] for a straight-through system.

---

[3] G. Hass, W. R. Hunter, and R. Tousey, *J. Opt. Soc. Am.* **46**, 1009–1012 (1956).

[4] J. A. R. Samson, *J. Opt. Soc. Am.* **52**, 525–528 (1962).

[5] P. G. Wilkinson and D. W. Angel, *J. Opt. Soc. Am.* **52**, 1120–1122 (1962).

[6] W. R. Hunter, *Opt. Acta* **9**, 255–268 (1962).

[7] H. G. Beutler, *J. Opt. Soc. Am.* **35**, 311–350 (1945).

[8] T. Namioka, *J. Opt. Soc. Am.* **49**, 446–465, 951–961 (1959).

[9] N. A. Finkelstein, C. H. Brumley, and R. J. Meltzer, *J. Opt. Soc. Am.* **43**, 335 (1953) (abstract only).

Until very recently, transmission gratings have been limited to wavelengths longer than 3000 Å by the fact that they were made on optical glass and in a resin that absorbed all shorter wavelengths. By forming the replica in a resin that transmits well down to 2000 Å on a blank of clear fused quartz, it has been possible to make experimental transmission gratings that work throughout the ultraviolet as well as the visible region of the spectrum.

3. *Ruled Area*

In commercial practice, the ruled area of a grating is given in millimeters, with the figure for the groove length first and the width of the ruled area second. For example, a grating with ruled area 65×76 mm has grooves 65 mm long and a 76-mm width of ruling. It is customary to rule a somewhat larger area than the specifications stipulate in order to make centering less critical.

The ruled area of a grating should be large enough to intercept all the incident light even when the grating is turned to its extreme angular position. Thus, for a square incident beam with side h, or a circular beam of diameter h, the ruled area should be h millimeters high and w millimeters wide, where

$$w = h/(\cos \alpha_{\max})$$

and $\alpha_{\max}$ is the maximum angle that the incident beam makes with the normal to the grating. Any smaller area will decrease the useful light in the spectrum and increase that going into the zero order.

II. THEORY

A. Grating Equation

If the light is incident at angle α relative to the surface normal of a reflecting grating ruled with spacing d, the path difference for the light incident on any two adjacent grooves is $d \sin \alpha$ (Fig. 1). When this light is diffracted from the rulings at some other angle β, the path difference for the light is further increased by the amount $d \sin \beta$. Thus, for these two rays, the path difference is the algebraic sum $d(\sin \alpha \pm \sin \beta)$. The plus sign is used when both rays are on the same side of the normal to the surface of the grating, while the minus sign applies when they are on opposite sides.

The reflected light of wavelength λ will be in phase over the entire wavefront when the path difference for rays incident on adjacent grooves is

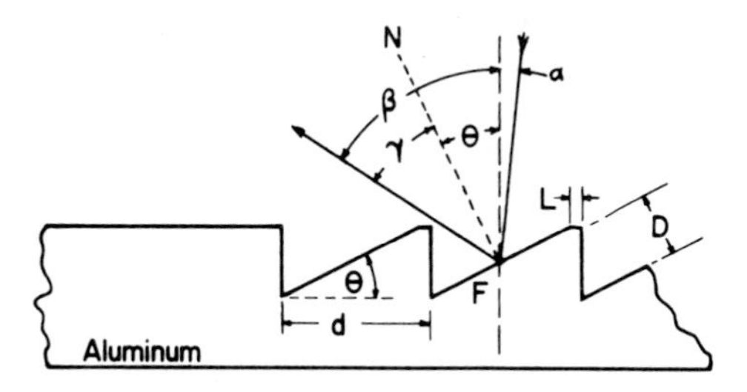

FIG. 1. Blazed groove profile. The thickness of the aluminum is generally equal to the blaze wavelength, and the depth of the groove about half of the thickness. The dimensions shown are approximately those for a grating ruled with 1200 grooves per millimeter, blazed at 7500 Å. D: depth of ruling; d: groove spacing; L: unruled land; α: angle of incidence; β: angle of diffraction, γ: angle of reflection; θ: blaze angle. The line N is normal to the groove face F.

an integral multiple of the wavelength. For light of a given wavelength incident at a particular angle, the light reflected from all the grooves will be in phase only at certain angles. The number of wavelengths of path difference for adjacent grooves is called the order of interference, m. From the foregoing, we arrive at the grating equation:

$$m\lambda = d(\sin \alpha \pm \sin \beta) \tag{1}$$

Using this equation, it is possible to calculate the angles at which light of a given wavelength will appear in various orders when the light is incident at angle α to the normal to the grating. When m is zero, $\alpha = \beta$, and the grating acts as a mirror. All wavelengths present in the incident light will be superposed in this zero order.

For each other order, the value of β is a function of the wavelength. Thus a grating forms a spectrum, the various wavelengths present in the incident light being diffracted at different angles.

B. DISPERSION

The angular dispersion of two different wavelengths of light is given by the expression:

$$\partial\beta/\partial\lambda = m/(d \cos \beta) \tag{2}$$

This is obtained by differentiating the grating equation (1) assuming that α, the angle of incidence, is fixed.

It might seem possible to obtain extremely high dispersion in the first order by using a grating with a large number of grooves per millimeter. There are two factors limiting the number of grooves per millimeter that can be used for a given wavelength. First is the fact that neither α nor β can be greater than 90°. The light cannot penetrate a reflectance grating. Second, the groove aspect as "seen" by the light cannot be substantially less than the wavelength of the light. When the groove dimensions are small compared to the wavelength of the light, the grating acts as a mirror, reflecting the light rather than dispersing it. This property is valuable when it is desired to use finely-spaced gratings as rejection filters for work in the infrared.

From the foregoing, it is evident that for first-order work, the spacing should not be substantially less than the wavelength of the light being used if one is to avoid serious loss of light into the zero order. From this, it follows that a grating should not be blazed (Section II, D) much above 30° for first-order use. Of course, one can increase both dispersion and resolving power by working in higher orders at greater angles. If a value of β is selected, the ratio m/d is constant. Thus the dispersion really depends only on β if it is acceptable to work in a high order ($m > 1$).[10]

There is also an upper limit to the practical number of grooves per millimeter for transmission gratings. This arises from the fact that as the angle of diffraction approaches the face angle of the groove, more and more of the light is lost by internal reflection. Thus the maximum possible blaze angle for a transmission grating using a resin with index of refraction of 1.566 is about 38° at normal incidence. From this, it follows that 900 grooves/mm is the practical upper limit for transmission gratings in visible light.

C. Resolving Power

The resolving power of a grating is a measure of its ability to separate adjacent spectrum lines. It is expressed as $\lambda/\Delta\lambda$, where $\lambda + \Delta\lambda$ is the wavelength of a spectrum line that can just be distinguished from a line of wavelength λ. To obtain an expression for the theoretical resolving power of a given grating, we may use simple Fraunhofer diffraction theory to determine the resolution of two adjacent narrow spectral lines of equal

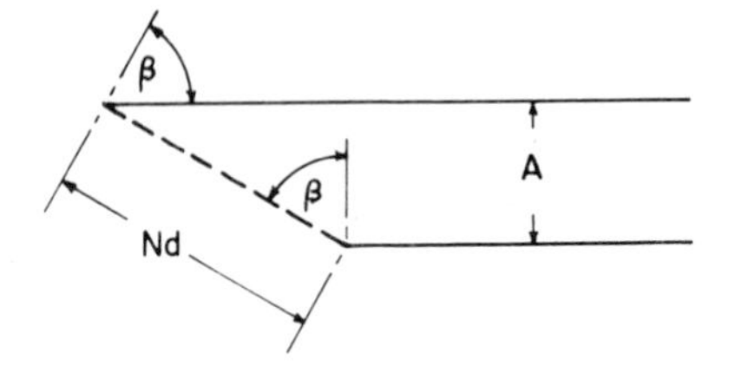

Fig. 2. The projected aperture of a grating.

[10] A. E. Douglas and G. Herzberg, *J. Opt. Soc. Am.* **47**, 625–628 (1957).

intensity. By Fig. 2, a grating having N grooves at a spacing d has a projected width in the direction of the image equal to $A = Nd \cos \beta$. The angular resolution of such an aperture is $\lambda/A = \lambda(Nd \cos \beta)$ radians. However, by Eq. (2), the angular dispersion of two spectral lines separated in wavelength by $\Delta\lambda$ is $m\, \Delta\lambda/(d \cos \beta)$ radians. At the limit of resolution, we may equate these two angles, giving the resolving power of the grating as

$$\lambda/\Delta\lambda = mN \tag{3}$$

or the simple product of the order number of the spectrum and the number of grooves in the grating.

Combining this with Eq. (1), we see that

$$\frac{\lambda}{\Delta\lambda} = \frac{Nd(\sin \alpha \pm \sin \beta)}{\lambda} \tag{4}$$

From this equation, it is evident that large gratings used at high angles are needed to achieve high resolving power. The number of grooves per millimeter simply determines the order to be used for a given wavelength.

The actual attainment of high resolving power[11,12] with a grating depends upon the optical quality of the grating surface, the uniformity of the spacing of the grooves, and the associated optics, if any. Any departure from true flatness of a plane grating, or sphericity of a concave grating, results in a loss of resolving power. The grating spacing must be constant with a tolerance of one hundredth of the wavelength at which near-theoretical performance is desired.

From its definition, it would seem that the theoretical resolving power of a grating could be increased indefinitely by increasing the total number of grooves in a given width. However, combining the definition of resolving power with the grating equation,

$$m\lambda = W(\sin \alpha \pm \sin \beta)/N$$

where W is the width of the grating, Michelson[13] showed early in this century, that the maximum resolving power was limited by the width of the grating and the wavelength. Since $(\sin \alpha + \sin \beta)$ can have a maximum value of 2, the maximum resolving power at any wavelength turns out to be equal to $2W/\lambda$. At 5000 Å, the maximum resolving power of a 6-in. grating is around 600,000 regardless of the diffraction order or the number of grooves.

---

[11] D. H. Rank, *J. Opt. Soc. Am.* **42**, 279–280 (1952).

[12] D. H. Rank, J. N. Shearer, and J. M. Bennett, *J. Opt. Soc. Am.* **45**, 762–766 (1955).

[13] A. A. Michelson, *Astrophys. J.* **8**, 37 (1898).

D. Blaze

The distribution of light into the various orders of a grating depends upon the microscopic shape[14–17] of the grooves of the grating. By ruling the groove with flat faces and controlling the face angles of the grooves, it is possible to concentrate the spectral energy in any desired angular region. This is called "blazing" a grating, and the groove-face angle θ is called the blaze angle (Fig. 1). The wavelength of the light for which the angle of reflection from the groove face and the angle of diffraction from the grating are the same is the blaze wavelength. This value is usually specified as the first-order wavelength, although the grating is also blazed for the second order of one-half that wavelength and the third order of one-third of that wavelength, etc.

It is customary to calculate and specify the blaze angle and blaze wavelength assuming that the angle of incidence is the same as the angle of diffraction. For this case, called the Littrow operation, the grating equation reduces to

$$m\lambda = 2d \sin \beta$$

and when β is equal to the blaze angle θ, then λ_B, the first-order blaze wavelength, is

$$\lambda_B = 2d \sin \theta$$

Here, d is the grating spacing in the same units as the wavelength. For other angular relationships, where α is the angle of incidence and β is the angle of diffraction, we use the full grating equation $m\lambda_B = d(\sin \alpha \pm \sin \beta)$, and then the blaze angle θ is

$$\theta = (\alpha \pm \beta)/2 \tag{5}$$

As before, the plus sign is used when both rays are on the same side of the grating normal, while the minus signs apply when they are on opposite sides of the normal.

If the blazed wavelength λ_B for the condition $\alpha = \beta$ is known, and the blazed wavelength λ_B' for other combinations of α and β is wanted, the following equation applies:

$$\lambda_B' = \lambda_B \cos[(\alpha' - \beta')/2] = \lambda_B \cos \tfrac{1}{2}\delta$$

---

[14] H. D. Babcock, *J. Opt. Soc. Am.* **34**, 1–5 (1944).
[15] R. F. Stamm and J. J. Whalen, *J. Opt. Soc. Am.* **36**, 2–12 (1946).
[16] G. K. T. Conn, *Proc. Cambridge Phil. Soc.* **43**, 240 (1947).
[17] J. J. Staunton, U.S. Patent 3,045,532 (1962); reviewed in *Appl. Opt.* **3**, 1346 (1964).

Here, δ is the angle between the incident and blazed diffracted rays. For example, a grating that is blazed for 7500 Å when $\alpha = \beta$ will be blazed for 6459 Å when the difference between α and β is 60°. For 90° between the rays, the blazed wavelength becomes 5303 Å.

Reflectance gratings can be blazed for very short wavelengths, even down to 10 Å, when the gratings are used at grazing incidence. This means that the incident ray is at an angle greater than 80° from the grating normal, and often at 88° or 89°. Under these conditions, the same equations can be used to find the blaze. As an example, if we consider a 1200 grooves/mm grating blazed at 1200 Å for $\alpha = \beta$, the blaze angle θ is 4°. From this, we find that when used with $\alpha = 88°$ or $\frac{1}{2}\delta = 84°$, the blazed wavelength will be 126 Å.

In the case of a transmission grating, the grooves are formed in a transparent resin that is strongly adherent to the surface of the glass grating blank. The blaze angle is calculated so that the light of the blazed wavelength will be refracted at the constructive interference angle for that wavelength. In order to calculate the blaze angle for a transmission grating, it is necessary to know the index of refraction n of the resin used to make the grating. For the resin used in Bausch and Lomb "CP" transmission gratings, e.g., $n = 1.566$ for 5461 Å. This figure can be used for calculatng the blaze throughout the visible spectrum.

In the simplest case, the light is incident normal to the unruled plane surface of the transmission grating and leaves the grating at the angle of diffraction β from the ruled surface. For this condition,

$$\sin \beta = m\lambda/d$$

since $\sin \alpha = 0$. Knowing β for the blazed wavelength λ, and the grating constant d, the blaze angle θ is given by

$$\tan \theta = \sin \beta/(n - \cos \beta)$$

For other angles of incidence, and for the reverse direction of the light through the grating, the situation is complicated by the fact that the light is refracted on passing through the plane back surface of the grating and α is not zero. The index of the BSC-2 glass commonly used as grating blanks is 1.519 at 5461 Å.

It is practical to make transmission gratings on the hypotenuse face of a right-angle prism[9] in cases where it is desired to have some particular wavelength go straight through the grating. In this case, the blaze angle is made equal to the prism angle, and this angle θ for a grating constant d is given by

$$\sin \theta = m\lambda/d(n - 1)$$

This is exactly analogous to Michelson's transmission echelon.[13] The grating is oriented so that the light of the desired wavelength passes normally both through the face of the prism and through the set of parallel faces of the grating grooves. The dispersion is not linear for this type of grating, because the prism dispersion is added to that of the grating. There are some occasions when the use of a striaght-through grating leads to a more convenient spectroscopic arrangement.

A blazed grating will frequently throw more than 80% of the light incident on it into the order lying at the blaze angle. In the first order, the useful range is usually from two-thirds of the blazed wavelength to twice this wavelength. For higher orders, it is from $\frac{2}{3}$ to $1\frac{1}{2}$ times the blazed wavelength.

E. Rowland Ghosts

Spurious spectrum lines seen in grating spectra originating in periodic errors in the spacing of the grooves are called Rowland ghosts.[18–20] These lines are usually symmetrically located with respect to the parent line at a spectral distance from it depending upon the period of the error and with an intensity depending upon the amplitude of the error curve.

If this curve is not simply sinusoidal, there will be a number of ghosts on each side of the parent line, representing harmonics of the principal frequency of the error curve. Rowland ghosts are associated with errors in the lead of the precision screw used in the ruling engine or with its bearings. As a consequence, their location depends upon the number of grooves ruled for each complete turn of the screw. For example, if the ruling engine has a lead of 2 mm, and a ruling is made at 1200 grooves/mm, the ghosts in the first order are expected to lie at $\pm(1/2400)\lambda$, with additional ghosts at integral multiples of this value.

The intensities of Rowland ghosts for a given amplitude of the error curve usually increase as the square of the order in which the parent line is observed. Since they are also proportional to the square of the amplitude of the error, the formula for ghost intensity can be expressed as

$$I_G/I_P = \pi^2 m^2 e^2/d^2 \tag{6}$$

where I_G/I_P is the ratio of the intensity of the ghost line to that of its parent, m is the spectral order, e is the amplitude of the sinusoidal ruling error, and d is the separation between grooves, or the grating constant.

[18] H. A. Rowland, *Phil. Mag.* **V, 35**, 397–419 (1893).

[19] A. A. Michelson, *Astrophys. J.* **18**, 278 (1903).

[20] J. A. Anderson, *J. Opt. Soc. Am.* **6**, 434–442 (1922).

Using this equation, it can be shown that the intensity of the first-order ghost for a periodic error with an amplitude of 0.01μ is 0.14% of the parent line for a 1200 grooves/mm grating. This is an indication of the importance of making every effort to minimize periodic errors of ruling.

F. Lyman Ghosts

When ghost lines are observed at large distances from their parent lines, they are called Lyman ghosts.[20, 21] These are the result of compounded periodic errors in the spacing of the grating grooves. They differ from Rowland ghosts in that each period consists of very few grooves. Lyman ghosts can be said to be in fractional-order positions. Thus, if every other groove is misplaced, ghosts are seen in the half-order positions. The number of grooves per period determines the fractional-order position of Lyman ghosts. It is usually possible to find the origin of the error in the ruling engine once its periodicity is determined.

G. Satellites

Satellites are misplaced spectrum lines usually occurring very close to the parent line. Individual gratings vary greatly in the number and intensity of satellites which they produce. They may be so numerous as to be called "grass," or they may be absent, as in a "clean" grating. In contradistinction to Rowland ghosts, which usually arise from errors extending over large areas of the grating, each satellite usually originates from a small number of misplaced grooves in a localized part of the grating. If there are one or two troublesome satellites, it is often possible to eliminate them by masking appropriate small areas of the grating. The Foucault knife-edge test can be used to locate the areas that must be masked.

H. Appearance

A perfect grating is a thing of beauty with its many brilliant colors and its bright mirror surface outside the ruled area. Unfortunately, it is not always possible to produce gratings completely free of obvious beauty defects of one sort or another. Blemishes which detract from the appearance of the grating often have no adverse effect on its optical performance.

Blemishes inside the ruled area are usually caused either by small droplets of metal, which cause lifting of the diamond, or streaks along the

---

[21] C. Runge, *J. Opt. Soc. Am.* **6**, 429–434 (1922).

grooves due to adhesion of aluminum to the tip of the diamond. The target pattern seen on many concave gratings originates from truly submicroscopic defects in the curved edge of the ruling diamond.

III. GRATING RULING

Grating ruling is without doubt the most precise continuous mechanical operation carried out in practice. The difficulties involved do not lie in the ruling of great numbers of closely-spaced grooves, but rather in the uniformity with which they are spaced and formed. It is in the degree of uniformity measured in small fractions of an optical fringe, as well as in techniques for controlling groove shape, that modern precision gratings differ from past efforts. These give performance levels close to the theoretical limits.

In a well-made grating, the tolerances for periodically-repeated errors typically run at about 1/100 wave, while progressively accumulating errors should not exceed 1/10 wave, (500 Å or 2 μin. for green light, or half this for $\lambda = 2537$). Such figures are at least 100 times less than what one associates with the finest machine tools, and provide the major problems involved in the design, construction, and operation of a ruling engine. Most of these engines[1, 22–24] operate with an indexing carriage on which the grating blank rests, while a reciprocating carriage moves a suitably-shaped diamond tool across with just enough pressure to burnish the required groove into the metal coating. No cutting is involved. On some machines, the role of indexing and diamond reciprocation is reversed.[25] Most ruling engines use screws as the basic indexing means, but require correction mechanisms to compensate for the apparent natural limit of accuracy of even the finest screws, which is around 10–20 μin. If the engine has sufficiently stable bearings, carriageways that allow perfect repetitive motion without stick-slip, and in general behaves in a repeatable fashion, correction by the use of a suitable cam is an excellent solution. Interferometric fringe counting is now the preferred method for determining the residual screw errors (Fig. 3).[26, 27]

---

[22] R. W. Wood, *Nature* **140**, 723 (1937).

[23] R. W. Wood, *J. Opt. Soc. Am.* **34**, 509–516 (1944).

[24] H. D. Babcock and H. W. Babcock, *J. Opt. Soc. Am.* **41**, 776–786 (1951).

[25] J. D. Strong, *J. Opt. Soc. Am.* **41**, 3–15 (1951).

[26] G. R. Harrison, and J. E. Archer *J. Opt. Soc. Am.* **41**, 495–503 (1951).

[27] N. A. Finkelstein, C. H. Brumley, and R. J. Meltzer, *J. Opt. Soc. Am.* **42**, 121–126 (1952).

FIG. 3. A typical ruling engine equipped with interferometric control. (Courtesy Bausch and Lomb, Inc.)

In recent years, several ruling engines have been constructed that use a built-in interferometer system[28–33] to sense screw errors and provide appropriate feedback at regularly-spaced short intervals of time, or to store the accumulated error during the ruling stroke and use the information only during the indexing stroke. Error-sensing is usually in units of about 0.1 μin. Engines of this type can produce to their limits of capability only when provided with a proper environment.

Vibration and, especially, variations in temperature are the primary sources of disturbance. Isolation against external vibrations in the important region of 5–100 cps can be achieved very effectively by mounting the engines

[28] G. R. Harrison and G. W. Stroke, *J. Opt. Soc. Am.* **45**, 112–121 (1955).

[29] G. R. Harrison, N. Sturgis, S. C. Baker, and G. W. Stroke, *J. Opt. Soc. Am.* **47**, 15–22 (1957).

[30] G. R. Harrison, *Proc. Am. Phil. Soc.* **102**, 483 (1958).

[31] H. W. Babcock, *Appl. Opt.* **1**, 415–420 (1962).

[32] G. W. Stroke, *J. Opt. Soc. Am.* **51**, 1321–1339 (1961).

[33] R. J. Farrell and G. W. Stroke, *Appl. Opt.* **3**, 1251–1262 (1964).

on concrete blocks weighing several tons and supporting these blocks above their center of gravity on standard vibration-isolation springs. Temperature problems are best handled by a high-grade air conditioning system using cascade control to provide stability, with well-insulated walls, floor, and ceiling. Average temperature in the ruling room must remain constant within 0.01°F for long periods of time, and, in addition, the engine itself is usually given an aluminum housing to damp out minor temperature variations. During the setup for the ruling of a new grating (Fig. 4), the engine temperature may rise because of the operator's body heat. A prerun of 18 or more hours is usually allowed at the start of each ruling to permit the engine to arrive at its operating temperature before the diamond begins ruling.

Original gratings are most commonly ruled in a coating of aluminum evaporated on the optically polished surface of a plane or concave glass blank. The aluminum coating must be thicker than the depth of the grooves to be ruled. Some gratings are ruled in gold rather than aluminum.

For gratings blazed for the extreme ultraviolet, the coating thickness may be as little as 1000 Å. At the present time, the thickest coating of

FIG. 4. Adjusting the diamond on a ruling engine. (Courtesy Bausch and Lomb, Inc.)

aluminum that can be made to adhere properly to the blank limits the burnishing-type ruling of gratings for the infrared to those blazed for 45 μ.

Although aluminizing itself might seem to be one of the easiest steps in the complex production of a diffraction grating, even this is an art hounded with provoking, interacting, and critical variables. Among these are the pressures in the evaporator, purity of the aluminum, depth and speed of its deposition, cleaning treatment of the base glass, geometry of boat or filament, and presence and strength of an electric field. All influence the "ruleability" of the aluminum, and only a test ruling can tell whether the surface of the blazed face is proper for the production of a grating of acceptable quality.

The diamond tools are shaped to form grating grooves by burnishing the aluminum coating rather than cutting it. This avoids troublesome metal chips, and produces gratings with grooves having optical surfaces matching the high-quality blank. The use of mineral oil on the surface of the blank during ruling helps to produce gratings with smooth grooves, and prevents sticking of the aluminum to the tool during ruling.

In addition, the diamond tools are very carefully shaped so as to form smooth grooves with the specified face angles. In the past, diamond wear during ruling frequently resulted in undesirable changes in the groove contours as the ruling progressed. This problem has been overcome through the use of X-ray diffraction methods to orient the diamond in the tool so as to take advantage of the superhard crystal planes present in each diamond. The availability of a large number of differently shaped diamonds makes it possible to alter groove contour by changing diamonds. Once the correct diamond has been selected, hours of adjustment are needed to achieve the proper loading and orientation of the tool.

The proper orientation and loading of the diamond prior to ruling requires that the operator observe the shape of a series of test grooves. An interference microscope which produces fringes with light reflected from the grooves is often used for this purpose. Electron microscope methods[34] are used in cases where the groove spacing is smaller than the wavelength of visible light (Fig. 5).

Modern grating-ruling engines are designed to be able to rule grooves of any spacing, as gratings are used with a very wide range of wavelengths. It has been found possible to rule regularly-spaced burnished grooves as coarse as 20 per mm and as fine as 10,800 per mm. This range produces gratings usable from 75 μ in the infrared to 10 Å X rays.

---

[34] W. A. Anderson, G. L. Griffin, C. F. Mooney, and R. S. Wiley, *Appl. Opt.* **4**, 999–1003 (1965).

FIG. 5. An electron micrograph of the terminal groove of a grating blazed for the extreme ultraviolet. The black line is a natural asbestos fiber deliberately laid on the grating and shadowed obliquely with platinum. The lower edge of the white shadow is an image of the groove profile. (Courtesy Bausch and Lomb, Inc.)

For even coarser rulings, designed for far-infrared work, it is possible to rule useful gratings in aluminum or brass metal blanks. In this case, it is necessary to use a special engine with a diamond tool designed to cut and remove the metal from the grooves instead of burnishing. With care, optical surfaces can be produced which serve excellently in the long-wavelength infrared.

IV. REPLICATION

A grating-ruling engine normally rules from 50 to 75 gratings a year. It would require many engines to rule the thousands of gratings needed each year if original gratings were used. It is evident that the output of one good engine can be multiplied many times if a replica process is available that can make copies that are fully the equal of the master gratings from which they are made.

1. *Reflectance Gratings*

Wood[22, 23] succeeded in making useful echelette-type replicas by forming collodion films on master gratings ruled in copper plates. The films were stripped from the copper gratings and cemented to flat glass plates.

A similar method has been used to make replicas from masters ruled as a fine screw or helix on a cylindrical rod. The use of these gratings for infrared work has been described by Cole.[35]

Modern high-quality replica gratings are made by methods similar to those described by White and Frazer.[36] The basic steps in replica making are illustrated in Fig. 6. After the master grating (C) has been ruled in an evaporated aluminum coating (B) on the glass master blank (A), it is over-coated with a second coating of evaporated aluminum (D). Next, the glass blank (E) for the replica is cemented to the replica coating. After curing, the replica grating (F) is separated from the master grating. It has all the exact details of the master reproduced on its aluminum coating. Such replicas frequently show higher efficiency than the master grating from which they are made. This type of replica process has made gratings of many types and sizes available in unlimited quantities throughout the world.

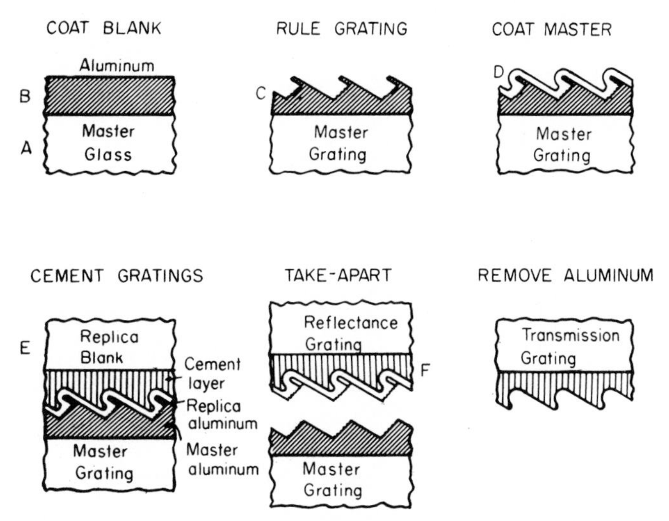

FIG. 6. The replication process for gratings. Here the long face is the one blazed. Replication removes the overhanging burr which is seen to intercept a portion of the incoming light to the original. The result is an increase of efficiency often of the order of 100%.

[35] A. R. H. Cole, *J. Opt. Soc. Am.* **44**, 741–743 (1954).

[36] J. V. White and W. A. Frazer, U.S. Patent 2,464,738 (1949).

2. *Transmission Gratings*

Transmission gratings are made in the same way, except that the glass blank is optically flat on its back surface and the aluminum is chemically removed from the ruled surface after separation. The grating grooves are formed in the transparent cement layer. Gratings made in this way are of much higher quality than those made by stripping transparent films and cementing them to a glass plate. To make transmission gratings for the ultraviolet region of the spectrum, ultraviolet-transparent cement must be used on a blank of clear fused silica.

3. *Special Coatings*

Although aluminum has a very high reflectance for ultraviolet, visible, and infrared light, there are better materials for wavelengths shorter than 1500 Å. Canfield *et al.*[37] have described the use of a fast-fired coat of pure aluminum followed immediately in the same vacuum by a thin coating of magnesium fluoride to enhance the vacuum-ultraviolet reflectance of both gratings and mirrors, particularly in the 1200-Å region of the spectrum. For shorter wavelengths and for grazing-incidence gratings, both platinum and gold coatings have been shown to be of value.

V. GRATING TESTING

A. Appearance Conditions

Gratings intended to be identical twins and ruled on the same engine can usually be distinguished. In practice, optical tests of a grating are used to see how the ruling engine and other technical parts of the system are changing, and to provide the necessary guide for alterations to be made before the next ruling. The periodicity of the grating exposes its spatial-frequency performance clearly, in contrast to mirrors, prisms, and other forms of optical element.

The result of working near the achievable limit of precision is that the efficiency, resolving power, and spectral purity produced by a grating depart somewhat from the nominal values predicted from the specified geometry. Fortunately, an exceptional grating, once ruled, can be duplicated with fidelity by replication procedures.

The replicating process always involves minute material deficiencies which add to the blemishes always present locally. Fortunately, spectrum

---

[37] L. R. Canfield, G. Hass, and J. E. Waylonis, *Appl. Opt.* **5**, 45–50 (1966).

slit images are formed only where the grating surface itself is out of focus, so surface blemishes are observed only in terms of their effect on the source light. Whether to discard a given grating because of its limitations, or to retain it for its virtues, is a decision that must be made regarding every good grating.

The quality of a grating is controlled by three aspects of its optical performance:

1. Completeness of the optical interference reinforcement at the spectrum line images of the entrance slit, measured by spectral resolving power.
2. Energy delivery by the grating to the diffracted spectrum lines, measured by diffraction efficiency relative to the specular reflection by a polished mirror coated with the same material.
3. Spectral purity or homogeneity of the light in a spectrum line, measured by scattered-light filtering methods.

B. Resolving Power and Line Profile Examination

A real grating used with a spectrum line source and with a finite slit will depart in detail from the ideal case, and there will be "wings" extending well away from the parent line and "grass" that is similar in magnitude to the secondary maxima of the ideal grating.

The groove-to-groove variations in a grating are important in spectrum details far removed from the spectrum line. Direct visual examination of a grating oriented for viewing between specular reflection and the first-order visible spectrum provides a good viewpoint for observing the far wings and "grass" associated with the visible lines of a bright mercury source.

The other spectral artifacts that occur close to the main line position

TABLE I

Spectral Artifacts; Origin and Effect on Performance

Artifact	Origin	Effect
Broadening	Progressive change of spacing	Low resolution
Satellite	Abrupt change of spacing	Displaced or extra line
Ghost	Periodic change of spacing	Spurious lines

are listed in Table I. These artifacts can be examined with natural mercury-lamp light; an isotope lamp is desirable for high-resolving-power gratings. The blue line of mercury is often used for quantitative resolving-power tests because the photoelectric and photographic detection means employed by many commercially-available systems are most efficient in this spectral region.

Two yellow lines of the three that can be seen, those at 5790.65 and 5789.65 Å, are used for resolving-power tests with low-resolving-power gratings, and particularly for ghost intensity evaluation; they have a strength ratio near 40/1 when an isotope mercury lamp is operated at a 2450 MHz frequency and at a temperature near 80°C. The strength of ghosts from a given grating is likely to rise as the square of the order number. Thus ghost-strength measurements can be evaluated from a comparison of the intensity of the 5790.65 ghost to the 5789.65 line.

Figure 7 gives the theoretical strength[38] and wavelengths of the isotope and hyperfine structure of the natural green line of mercury. The resolving power ascribed when the various line neighbors appear resolved is also listed under Fig. 7.

Another factor contributing to low resolving power is the optical figure of the surface. Spherical-power error in a plane grating surface does not broaden the spectrum lines, but it does extend their height and cause the point-to-point image identity to be lost. More complex irregularity of the surface figure precludes good optical imagery.

Another type of examination that can be used to test the resolving power of a grating is to create optical interference between a diffracted wavefront and a good reference wavefront.[39]

Despite the larger information content of the interferogram relative to direct spectrum examination, a single interferogram does not distinguish blank-figure errors from ruling errors, and it does not note full-fringe displacement in adjacent parts of the aperture. The interferogram must be carefully interpreted. The ability to locate defective regions from the interferogram is fully matched by the direct examination of a grating while admitting only the ghost or satellite or wing light in a spectrum to the eye by using a slit to block the parent spectral line. The interferogram method is especially sensitive to grating-spacing errors when the grating is used at a large diffraction angle; therefore it has been extensively used in laboratories whose main interest is echelles.[40]

---

[38] H. Schüler and T. Schmidt, *Z. Physik* **98**, 239–251 (1935).

[39] G. W. Stroke, *J. Opt. Soc. Am.* **45**, 30–35 (1955).

[40] G. R. Harrison and G. W. Stroke, *J. Opt. Soc. Am.* **50**, 1153–1158 (1960).

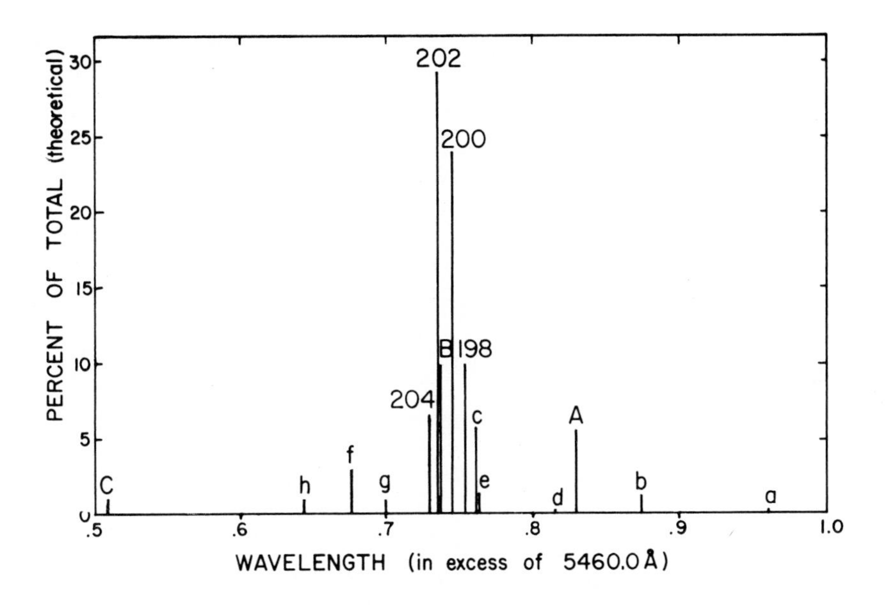

Component	Wavelength (Å)	Resolving power	
		Interval	Thousands
198	5460.753	199*C*–201*h*	40.5
199*A*	5460.831	201*h*–201*f*	165.0
199*B*	5460.737	201*f*–201*g*	260.0
199*C*	5460.509	201*g*–204	182.0
200	5460.745	204–202	910.0
201*a*	5460.961	202–199*B*	1820.0
201*b*	5460.875	199*B*–200	685.0
201*c*	5460.762	200–198	685.0
201*d*	5460.816	198–201*c*	606.0
201*e*	5460.763	201*c*–201*e*	5461.0
201*f*	5460.677	201*e*–201*d*	103.0
201*g*	5460.698	201*d*–199*A*	364.0
201*h*	5460.644	199*A*–201*b*	124.0
202	5460.734	201*b*–201*a*	63.5
204	5460.728	—	—

FIG. 7. Structure of mercury green line used for resolving-power tests.

C. Geometric and Optical Efficiency

Grating efficiency is the overriding problem of most spectroscopic applications. It is properly measured as the amount of monochromatic light diffracted in the various orders relative to the specular reflection from a polished blank coated with the same material.

Efficiency measurements provide the traditional control for groove depth, for blaze angle, and for groove-face flatness (see Table II).

TABLE II

Information from Efficiency Measurements

Aspect	Efficiency Measurement
Groove depth	Amount of light in zero order
Groove slope angle	Wavelength of maximum intensity
Groove face flatness	Half-width of efficiency *vs* wavelength curve

In many instruments, the ultimate of efficiency is required for some narrow wavelength band; in others, usable efficiency is required for a broad wavelength band. The efficiency at numerous wavelengths is therefore measured in order to have a record of the efficiency *vs* wavelength for the grating under test. The proportion of the light in each order depends primarily on the shape of the grating grooves. Besides observing the shape of the grooves using the interference microscope, it is possible to determine the effective shape by measuring the efficiency of the grating at a series of angles using light of suitable wavelengths in an appropriate series of grating orders.

To measure the efficiency of a grating for light of a particular wavelength in a given order, it is necessary to measure the incident and diffracted light and to specify the angular relations between these two rays of light. The intensity of the incident monochromatic light is measured by substituting a plain aluminized mirror for the plane grating and turning it to the zero-order angular position. As a result, the efficiency is recorded and reported as a percentage of the energy that is reflected by a good aluminized mirror.

The blaze angle determined by finding the angle for which efficiency measurements are maximum for the higher orders is approximately the same as that determined by using the first order.

There is no simple method for evaluating the efficiency of concave gratings. It is possible to intercompare the efficiency of a series of concave gratings with identical specifications. Over a period of years of use, it is to be expected that the efficiency of a grating used in the ultraviolet region of the spectrum may decrease somewhat. If this loss in efficiency becomes appreciable, it is frequently possible to recover the lost efficiency by recoating the grating with a thin, fast-fired coating of aluminum.

D. Spectral Purity and Stray Light

The composite of misplaced spectral energy is called stray or scattered light. Near-scatter is due to the satellites and ghosts described previously. Far-scatter is due to the secondary spectrum inherent in gratings and to the entire gamut of groove-position defects. One source of random spacing error in gratings with closely-spaced grooves occurs because interferometric position control of the grating carriage was not exact.

Measurement methods used to determine far-scatter light intensities typically use filters to block the proper spectral region while transmitting other regions. The ratio of radiometric readings made with and without the filter measures the stray energy from outlying spectral regions relative to some reference line strength. Such methods are arbitrary, influenced by the grating efficiency, and depend on the test spectrograph, on the source excitation, and on the spectral sensitivity of the detection system. The definition and measurement of far-scatter is therefore a continuing field of research.

VI. GRATINGS FOR SPECIAL PURPOSES

A. Vacuum Ultraviolet

In general, concave gratings are used for instruments designed for use at wavelengths shorter than 2000 Å. This is done to reduce the number of reflecting surfaces, because uncoated aluminum surfaces decrease in reflectivity upon oxidation.[3–5]

In recent years, this situation has changed through the development of special coatings that enhance and stabilize the inherent high reflectance of freshly-deposited aluminum. This is accomplished by overcoating the aluminum with a carefully controlled coat of magnesium fluoride before it comes into contact with air.[37] For the wavelength region above 1100 Å, it is now practical to use plane gratings, as the losses due to the additional reflections are no longer excessive.

For the regions from 500 to 1100 Å, however, it is still best to use concave gratings. These can be coated with a noble metal to enhance their reflectance in this region. For all gratings used in the 1100–2000 Å region, the special magnesium fluoride coating is highly recommended.

Tripartite concave gratings[41] are ruled with three areas, each with a different blaze angle. These blaze angles are calculated to maximize the efficiency for each area for the specified wavelength. Tripartite gratings have higher efficiency at the blazed wavelength than similar concave gratings ruled with the same blaze angle over the entire area. The resolving power is not as high, however, because the three sections are not ruled in phase with each other. It is seldom desirable to change a blaze angle that is large compared to the maximum slope of the curved blank.

For soft X rays and the far-ultraviolet wavelengths shorter than 500 Å, use is made of concave gratings working at very high angles of incidence.[20] These grazing incidence gratings can be used down to wavelengths as short as 10 Å by using grazing angles of 1° or 2° ($\alpha = 89°$ or 88°). To work at these angles, it is important that the grating be rectangular, so that the edge of the blank will not interfere with the incident light. Gratings can be blazed for grazing incidence work if the grazing angle and desired wavelength region are known. To improve reflectivity, it has been recommended[42] that such gratings be coated with gold to enhance their performance.

B. Far Infrared

Very coarse rulings are needed for work in the far infrared because the wavelengths are so long (50 μ or more). Wood[43] gave the name echelette (little ladder) to coarse gratings blazed for use in the infrared. It is not practical to rule gratings with spacings greater than 50 μ by the usual process of burnishing thick coats of aluminum on glass. Both because of the difficulty of producing such thick adherent coatings of aluminum and because of the heavy loading on the diamond tool it is impractical to rule gratings coarser than 20 grooves/mm.

For this purpose, use is made of a special ruling engine designed to cut grooves in plano aluminum or magnesium alloy blanks for ruling far-infrared gratings. These gratings are commercially available in any size up to 13 in. $\times$ 14 in. and with a variety of groove spacings and blaze angles. It is usually possible to resolve the orders with these gratings when they are illuminated with visible light.

Filter Gratings

It is frequently necessary to use gratings as reflectance filters when working in the far infrared in order to remove unwanted second and higher

---

[41] R. Tousey, *Appl. Opt.* **1**, 685 (1962).

[42] D. O. Landon, *Appl. Opt.* **2**, 450 (1963).

[43] R. W. Wood, "Physical Optics," 3rd ed., p. 265. Dover, New York, 1967; see also R. D. Hatcher and J. H. Rohrbaugh, *J. Opt. Soc. Am.* **46**, 104–110 (1956).

orders from the light incident on the far-infrared grating. For this purpose, small plane gratings are used which are blazed for the wavelength of the unwanted radiation. The grating acts as a mirror, reflecting the light wanted into the instrument and diffracting the shorter wavelengths out of the beam. These are then known as "filter gratings."

C. Echelles

The echelle grating was suggested by Harrison[44] in 1949 (also see Harrison and co-workers[45, 46] and Finkelstein[47]) as an intermediate step between the virtually unmakeable reflective echelon[48] and an ordinary grating. Whereas the steps of an echelon are several millimeters high, and those of an ordinary grating are measured in angstroms, the steps of an echelle may have a height D of about 5.6 μ and a width t of about 13 μ. (Fig. 8). Echelles are usually ruled with spacings such as 300, 75, or 30

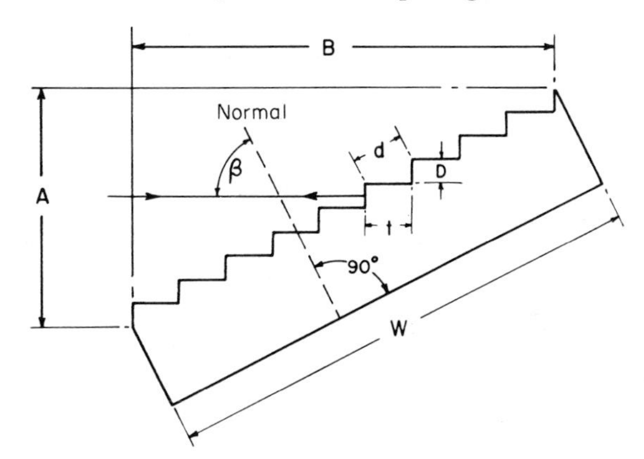

Fig. 8. Diagram for making echelle calculations.

grooves/millimeter and blazed at 63°26′. The 254-mm-wide echelles frequently exhibit resolving powers of one million or more for ultraviolet wavelengths. Correspondingly high resolving powers can be obtained at wavelengths extending to 20 μ in the infrared. Efforts are being made to rule larger and coarser (20 lines/mm) echelles so as to achieve even higher resolving powers and the ability to work at even longer wavelengths.

---

[44] G. R. Harrison, *J. Opt. Soc. Am.* **39**, 522–528 (1949).

[45] G. R. Harrison and C. L. Bausch, Echelle spectroscopy, *in* "Optical Instruments" (*Proc. London, Conf. 1950*), p. 77. Chapman and Hall, London, 1951.

[46] G. R. Harrison, S. P. Davis, and H. J. Robertson, *J. Opt. Soc. Am.* **43**, 853–861 (1953).

[47] N. A. Finkelstein, *J. Opt. Soc. Am.* **43**, 90–96 (1953).

[48] L. R. Griffin, *Proc. Phys. Soc.* (*London*) **62**, 93 (1948).

The "free spectral range" is the range of wavelengths from λ to $(\lambda + \delta\lambda)$ such that the mth order of wavelength $(\lambda + \delta\lambda)$ coincides with the $(m + 1)$th order of wavelength λ. Since an echelle operates in the Littrow mode with $\alpha = \beta$, the grating equation becomes $m\lambda = 2d \sin \beta = 2t$. We therefore determine the free spectral range by writing $m(\lambda + \delta\lambda) = (m + 1)\lambda$, from which $\delta\lambda = \lambda/m$. However, since m is given by the grating equation as $2t/\lambda$, we find that the free spectral range

$$F_\lambda = \delta\lambda = \lambda^2/2t \tag{7a}$$

Expressed in terms of wave numbers, or reciprocal wavelengths, this becomes

$$F_\sigma = \delta(1/\lambda) = 1/2t \tag{7b}$$

If a collimator lens of focal length F is used to project the echelle spectrum upon a photographic plate, the linear dispersion on the plate is given by

$$F\frac{\partial\beta}{\partial\lambda} = F\left(\frac{m}{d \cos \beta}\right) = \frac{Fm}{D} = \frac{F}{D}\frac{2t}{\lambda} \tag{8}$$

Therefore, the useful length of spectrum between two consecutive orders is equal to the product of the linear dispersion and the free spectral range, or

$$\left(\frac{F}{D}\frac{2t}{\lambda}\right)\left(\frac{\lambda^2}{2t}\right) = \frac{F\lambda}{D} \tag{9}$$

The resolving power of the echelle is found from Eq. (3), to be

$$\lambda/\Delta\lambda = mN = 2Nt/\lambda = 2B/\lambda \tag{10}$$

Since echelles are always used in high orders, it is necessary to provide cross dispersion in order to minimize overlapping orders. As there is always some order for any given wavelength that is found near the blaze, it can be said that echelles are blazed for all wavelengths. Echelles make it possible to design very compact spectrographs having both high dispersion and resolving power as well as high efficiency over a very wide range of wavelengths.

It is extremely difficult to rule satisfactory echelles, because the tolerances are in terms of the ultraviolet wavelengths, while the grooves are ruled in the thick coatings commonly used for infrared gratings. It is evident that periodic errors must be kept to a very low amplitude if Rowland ghosts are to held to acceptable levels. This requirement makes it necessary to rule echelles under interferometric control on blanks that are coated with special care to achieve a flatness of $\frac{1}{4}$ fringe.

The maximum efficiency for echelles is measured for wavelengths having an order that falls on the blaze. These usually measure in the range 50–60%. The wavelengths for which the two strongest orders are equal are said to fall in the double-order position. For these wavelengths, the maximum efficiency is between 30 and 35%. To find the strongest order for a given wavelength, we calculate the order falling nearest the blaze. As an example, we can determine that for an echelle with 300 grooves/mm and blazed at 63°26′, the 10th order of 5461 Å falls at 55°00′, while the 11th order is at 64°18′.

D. Laminar Gratings

Reflection gratings consisting of grooves with a rectangular cross section, both the bottoms and tops of the grooves being flat and parallel to the blank surface, are called laminar gratings.[49] Recent work by Sayce and Franks[50] has invigorated interest in this type of grating. Such gratings are blazed for zero order, but interference of the light from the groove bottoms with that from the tops can affect the distribution significantly. Ideally, such a grating would produce the same flux distribution when the grating is rotated 180° about its perpendicular.

E. Beam Dividers and Interferometers

One interesting application of diffraction gratings relates to their use as elements in an interferometric system. The diffraction grating may be used in an interferometer as a beam divider and recombiner[51] or as a direct replacement for any mirror in the system.[52] This interesting application is derived from a consideration of the grating as an "order disperser" rather than as a wavelength disperser.

If a plane monochromatic wave is incident on a diffraction grating, the wave will be split into several plane diffracted waves corresponding to the several orders of interference present (Fig. 9). The angles at which the

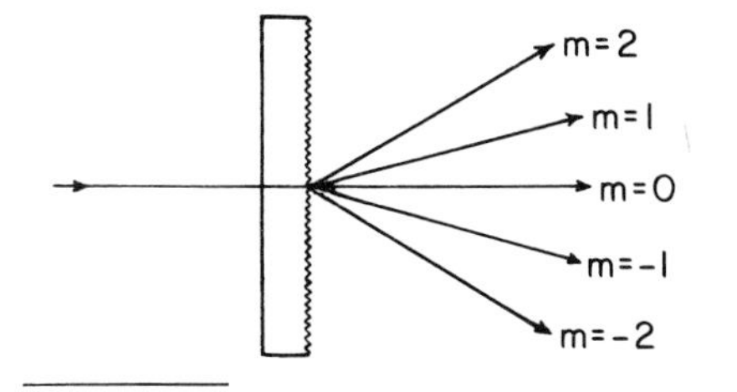

Fig. 9. Dispersion of various orders by a diffraction grating.

[49] C. H. Cartwright, *J. Opt. Soc. Am.* **21**, 785–791 (1931).
[50] L. A. Sayce and A. Franks, *Proc. Roy. Soc.* (*London*) **A282**, 353 (1964).
[51] R. Kraushaar, *J. Opt. Soc. Am.* **40**, 480–481 (1950).
[52] B. P. Ramsay, *J. Opt. Soc. Am.* **24**, 253–258 (1934).

different waves leave the grating can be predicted from the grating equation.

For normal incidence, the angle of diffraction β can be expressed for the monochromatic case, as

$$\beta_m = \sin^{-1}(mK)$$

where $K = \text{const} = \lambda a^{-1}$, $m = 0, \pm 1, \pm 2, \pm 3, \ldots$, and $|m| \leq K^{-1}$.

All of the diffracted waves are coherent, and therefore any two orders can be combined to produce interference effects. The relative intensity of the two chosen orders depends on the groove profile or "blaze" characteristics of the grating.

Several grating interferometers have been designed and successfully reduced to practice. Most of them have been used to analyze localized variations of optical path in a cross section of air[51, 53, 54] (e.g., aerodynamic flow and heat flow). Another grating interferometer, due to Chisholm [55] has been used to investigate the flatness of large surfaces. This interferometer shown in Fig. 10, is of particular interest in that it uses a grating

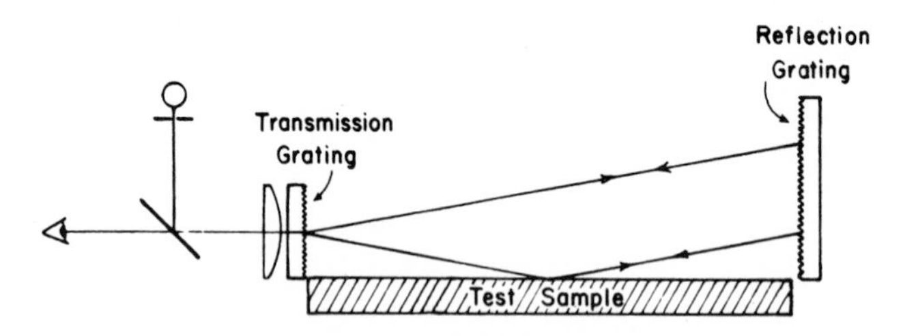

FIG. 10. Grating interferometer for checking the flatness of large surfaces.

both as a primary beam divider and in the "arms" of an interferometer. It also illustrates the utility of both reflection- and transmission-type diffraction gratings.

When gratings are used in an interferometer, they exhibit the following principal advantages:

1. They make the interferometer into a pseudomonochromator; hence the interferometer can often be used without external filtering.
2. The interferometer becomes remarkably insensitive to misalignment in comparison with an equivalent mirror system. Weinberg and Wood[53] have shown this for a grating-equivalent Mach–Zehnder system.

---

[53] F. J. Weinberg and N. B. Wood, *J. Sci. Instr.* **36**, 227–230 (1959).

[54] J. R. Sterret and J. R. Erwin, Tech. Note 2827. Nat. Advisory Comm. for Aeron., Washington, D.C., 1952.

[55] J. J. Chisholm, U.S. Patent 3,090,279 (1963).

3. All conventional beam dividers involve the transmission of at least one of the beams (after splitting) through a dispersive medium. The grating beam divider, on the other hand, permits the design of an interferometer in which the split beams encounter only surface reflections. Hence, compensating plates are not required and homogeneity (refractive index) of the elements is not a problem.

4. Improvements in the geometrical arrangement or layout of the interferometer are commonly allowed by the use of diffraction gratings.

F. Other Nonspectrometric Uses of Gratings

The uniformity with which diffraction grating lines are spaced makes them a natural choice for reference scales in linear measuring.[56] The grooves can be observed individually with suitable microscopes, and settings can be made visually. Alternatively, a single transmitting slit can serve as the reference index, and photoelectric observation can take the place of the eye. The light intensity will vary sinusoidally if only the zero and first diffraction orders are observed by the imaging optical system. Two or more measuring slots set to look at different parts of the sine wave are frequently used to obtain a direction-sensitive signal and to help subdivide the grating groove separation accurately.

There are obvious advantages to using a multiple slit for observing the moving grating. Such a multiple slit, with the same effective spacing as the grating, not only results in much higher light intensities, but also serves to average out any minor irregularities due to ruling errors, damaged areas, or particles of dust. Such "crossed" gratings give rise to so-called moiré fringes,[57] the number of fringes depending on the angle between the rulings. The literature on moiré fringes is increasing rapidly, a suggestion of widespread application in the future.

Similar to moiré fringes is the Ronchi[58] test of optical wavefronts. In this test, an image of a grating, deformed to some extent by the optical system that creates the converging wavefront, is formed on the same or on a similar grating. Deformations of the wavefront appear as deformed moiré fringes. One of the most relevant uses of this is the crossed-ruling procedure for observing periodic error. After a grating area has been ruled,

---

[56] D. Richardson and R. M. Stark, *J. Opt. Soc. Am.* **47**, 1–5 (1957).

[57] J. Guild, "The Interference Systems of Crossed Gratings; Theory of moiré Fringes." Oxford Univ. Press (Clarendon), London and New York, 1956.

[58] G. Toraldo di Francia, Geometrical and interferential aspects of the Ronchi test, *in* Optical Image Evaluation, *Natl. Bur. Std. Circ.* 526, pp. 161–170. U.S. Dept. of Commerce, Washington, D.C. 1954.

the blank is rotated through a small angle in its own plane and translated half the error period, and the same area is ruled again. The error period of the engine system appears visually as a sine-wave pattern in the ruling.

Ruled grids consisting of two lightly-ruled sets of orthogonal grooves have been used as electron microscope specimens.[59] The magnifications and the residual astigmatism or distortion of the electron optics can then be accurately measured.

Circular and spiral diffraction gratings have been studied[60] because they may prove useful for the alignment of mechanical and optical systems.

VII. GRATINGS GENERATED WITHOUT A RULING ENGINE

Gratings occur naturally in some minerals.[61] Debye and Sears[62] observed the grating behavior of a homogeneous medium traversed by compression waves. The molecular structure of crystals provides the three-dimensional grating used for electron diffraction[63] and short-wavelength X-ray diffraction.[64] Longer-wavelength (soft) X rays are often diffracted by pseudocrystals formed as a multilayer structure,[65] but grazing-incidence gratings are being used with increasing frequency for this spectral region.

Gratings often appear in photographs of interference fringes, easily produced today with laser beams.

Each of these phenomena is periodic, and none is blazed to concentrate the diffracted light. The burnished grooves of ruled gratings can concentrate much of the light in one spectrum order; for this reason primarily, ruling engines are a continuing field of development.

Since this chapter was written, a lengthy and very complete discussion of the theory of gratings, their methods of manufacture, and their applications has been published by Stroke.[66] The reader is referred to this article for further information on this subject.

---

[59] C. F. Oster and D. C. Skillman, Determination and control of electron microscope magnification. *Intern. Congr. Electron Microscopy, 5th*, paper EE-3. Academic Press, New York, 1962.

[60] J. Dyson, *Proc. Roy. Soc.* (*London*) **A248**, 93 (1958).

[61] F. T. Jones, *Lapidary J.* **20**, 37–40 (April 1966).

[62] P. Debye and F. W. Sears, *Proc. Natl. Acad. Sci. U.S.* **18**, 409 (1932).

[63] C. Davisson and L. Germer, *Proc. Natl. Acad. Sci. U.S.* **14**, 619 (1928).

[64] W. L. Bragg, *Proc. Cambridge Phil. Soc.* **17**, 43 (1912).

[65] K. B. Blodgett, *J. Am. Chem. Soc.* **57**, 1007 (1935).

[66] G. W. Stroke, Diffraction Gratings (in English), *in* "Handbuch der Physik," Vol. XXIX, Optische Instrumente (S. Flügge, ed.), pp. 426–735. Springer, New York, 1967.

CHAPTER 3

Spectrographs and Monochromators

ROBERT J. MELTZER

Dynamic Optics, Inc., Fairport, New York

I. INTRODUCTION

There are, generally speaking, two classes of dispersing instruments: spectrographs and monochromators. The spectrograph acts to disperse the incident light into one- or two-dimensional spatial arrays. At the output of a spectrograph, there is a correspondence between spatial position and wavelength. The monochromator acts to bring each wavelength of the incident light sequentially to the same output position. The spectrograph thus acts to allow one to record all or many parts of the spectrum simultaneously, while the monochromator isolates one wavelength at a time for examination. The spectrograph thus obviously lends itself to spectral recording by such multiple detectors as the photographic plate, the eye, television, or several photomultipliers, and the monochromator to such single detectors as photomultipliers, thermocouples, and photoconductive devices.

The most common use of spectrographs is in spectrochemical analysis. Either photomultipliers or photographic film is used. In general, the design of the spectrograph is roughly the same for both photomultiplier and for photographic film, even though the detectors are substantially different, and each detector could, in principle, profit from a different system design. The profit would, however, be small enough in a practical sense that a radical change in the design of instruments which both now use is not likely to be worthwhile.

Monochromators are most commonly used in analyses where light absorption or reflectance is the most informative characteristic. The principal reason for using sequential rather than simultaneous recording is that the light sources for absorption or reflectance (or, for that matter, fluorescence) are far more time-stable than are those for emission.

A. The Prism as Disperser

Most of the properties of prisms as dispersers have already been described in Chapters 1 and 2. The appropriate equations for dispersion, resolution, line curvature, and magnification are given in Chapter 1.

In quantitative spectroscopy, it is often useful to know the polarization properties as well. The polarizing property of a dispersing prism arises from the Fresnel reflections at the prism surfaces and can be calculated from the equations given in Volume 1, Chapter 9 of this series.

The materials used for dispersing prisms depend on the spectral region for which the dispersion is desired. The prism must not only be transparent in the spectral region of interest, but must also have a high dispersion in that region. For example, even though quartz is transparent in the red and near visible infrared, it is very nearly useless as a disperser because of its low dispersion.

In the vacuum ultraviolet, the optical material which transmits to shortest wavelengths is lithium fluoride, useful as a prism down to 1200 Å. Calcium fluoride is useful to 1400 Å. Quartz is excellent down to 1850 Å, the present-day fused quartz having higher transmission even though its dispersion is lower. In the visible, crystalline materials are generally inferior to the flint glasses.

Used in the infrared are the crystalline materials (with long-wavelength limits given in parentheses) calcium fluoride (9 μ), sodium chloride (16 μ), potassium bromide (25 μ), and cesium bromide (40 μ). A substantial amount of information on all these materials can be found in the publications of the Harshaw Chemical Company.[1] Prism materials are sometimes difficult

[1] Harshaw Chemical Company, Catalog of Crystalline Optical Materials, 1967.

to obtain in pieces large enough to fill larger apertures. In such instances, recourse can be had to Fresnel-type multiple-prism arrays. Liquid prisms have found use particularly in the visible region. Clearly their, value is, that they can be made substantially without size limit, and that some liquids have very high dispersion (ethyl cinnamate is a particular favorite in this respect). The difficulty with such prisms is that it is extraordinarily difficult to prevent convection currents within the prism from interfering with the theoretically-achievable high performance. Extraordinary temperature uniformity and stability must be achieved to realize the potential of a large liquid prism.

B. The Diffraction Grating as Disperser

1. *Resolution and Dispersion*

The resolution and dispersion of diffraction gratings have been discussed in Chapter 2.

2. *Curvature of Lines*

Just as the spectral lines produced by a prism are curved, so are those produced by a grating. As the direction of incident light departs from a plane perpendicular to the grating grooves, the apparent groove spacing decreases. The light from the ends of a slit will thus be deviated more than the light from the center of the slit (just as with a more finely-spaced grating). The apparent decrease in spacing is proportional to the cosine of the angle between the plane of incidence and the plane perpendicular to the grooves.

3. *Magnification of a Grating*

Just as a prism will magnify if not used at minimum deviation, so a grating will magnify if not used with the angle of diffraction equal to the angle of incidence. The angular magnification is given by $d\beta/d\alpha = -(\cos\alpha)/(\cos\beta)$, where α is the angle of incidence and β is the angle of diffraction.

4. *The Energy Distribution from a Diffraction Grating*

A number of papers, from Rowland[2] to the present, contain equations which give the energy distribution from a grating. These equations have generally proven less than useful over an extended wavelength range,

[2] H. A. Rowland, *Astronomy Astrophys.* **12**, 129 (1893). See also R. F. Stamm and J. J. Whalen, *J. Opt. Soc. Am.* **36**, 2–12 (1946); G. K. T. Conn, *Proc. Cambridge Phil. Soc.* **43**, 240 (1947).

chiefly because variations in groove shape too small to be adequately measured have a profound effect on the energy distribution. Close to the blaze, an adequate description of the energy distribution is given by the single-slit diffraction pattern as calculated from the subtended groove widths.

5. *Polarization Effects*

When the groove width of a diffraction grating becomes commensurate with the wavelength of the light falling on it, the light diffracted by the grating becomes noticeably polarized, the direction of polarization of the diffracted light then being perpendicular to the groove direction. The intensity of the undiffracted polarization is then added to the intensity of the zero order.

6. *Grating Anomalies*

The spectrum from a grating may show a sharp loss in energy at wavelength λ in order m_0 if in some other order m_A there is a solution to the equation $m_A \lambda / a = (\sin x) \pm 1.0$, where a is the grating space. These effects are called grating anomalies, and were first investigated by Wood[3] and by Rayleigh[4] and, most recently, by Palmer[5]. The light in the region of an anomaly is most commonly polarized.

7. *Stray Light from Gratings*

Light will frequently appear in grating spectra where no light should appear. Supposing the grating to be illuminated by monochromatic light, this stray light will appear in two different ways, as either well-defined lines in the "wrong" place, or as a general scatter of light throughout the spectrum, and, indeed, out of the plane of the spectrum.

These two kinds of scatter arise from two different causes. The appearance of distinct lines is always the result of some error in grating ruling. A periodic error, a random error, or an abrupt error will cause the formation of spurious spectra close to the parent line. Periodic errors, however, can cause spurious spectra at some distance from the parent as well as nearby.

The general kind of scattering can be caused by dirt or corrosion, by a general roughness of the groove, or by material piled up along the groove by the diamond during the ruling. Groove roughness is an uncommon difficulty today (except in the far ultraviolet), but appears often in older glass or speculum gratings.

---

[3] R. W. Wood, *Phil. Mag.* **4**, 396 (1902).

[4] Lord Rayleigh, *Proc. Roy. Soc.* (*London*) **A79**, 399 (1907).

[5] C. H. Palmer, *J. Opt. Soc. Am.* **42**, 269–276 (1952); **46**, 50–53 (1956).

II. PRISM SPECTROGRAPHS

Reference to the construction of prism spectrographs can be found in many places.[6, 7]

1. *Direct-Vision Spectrographs*

The simplest type of spectrograph can be made small enough to hold in the hand, and uses a nondeviating Amici prism. This prism is made of several opposed prisms alternately of flint and crown glass. The deviation of one prism is removed by the deviation of the next, but sufficient dispersion remains to be useful. The spectrograph then consists of a slit, a collimating lens, and the Amici prism (Fig. 1). There is no telescope, the

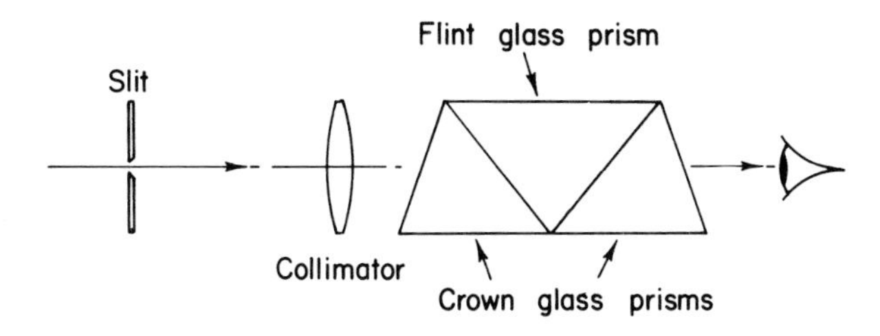

FIG. 1. Amici prism in a direct-vision spectroscope.

eye itself imaging the spectrum on the retina. Sometimes a wavelength scale is projected into the field of view by a side tube and reflecting prism.

2. "*Medium Quartz*"-*Type Spectrographs*

One of the commonest of all spectrograph types is shown in Fig. 2. It

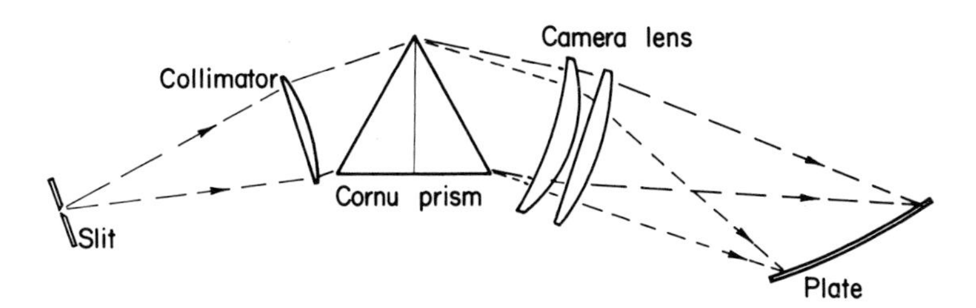

FIG. 2. Medium quartz spectrograph.

---

[6] R. A. Sawyer, "Experimental Spectroscopy." Dover, New York, 1963.

[7] G. R. Harrison, R. C. Lord, and J. R. Loofbourow, "Practical Spectroscopy." Prentice-Hall, Englewood Cliffs, New Jersey, 1948.

is called the "medium quartz spectrograph" irrespective of manufacturer. It consists of a slit, a collimator lens of about 600 mm focal length and 50 mm aperture, a prism, a camera lens, and the photographic plate. All of the optics are commonly of crystal quartz, and the instruments have a spectral range, covered in one exposure, from 2100 Å to 8000 Å. The quartz prism has a high dispersion in the ultraviolet, and the quartz lens has a short focal length in the same spectral region. The longer focal length in the visible tends to compensate for the lower dispersion, but does not compensate sufficiently to prevent a highly nonlinear spectrum. Since crystal quartz is doubly refracting, the optical axis of a prism made from crystal quartz should coincide with the optical axis of the crystal or else the spectrum lines will appear doubled. Even though this coincidence is obtained, some doubling will remain because of the rotary dispersion of crystal quartz. This difficulty can be removed by making half the prism of left-hand rotating quartz and half of right-hand rotating quartz. A prism made in this way is known as a Cornu prism.

3. *Littrow Arrangement*

An arrangement which allows increased dispersion for the same size prism, and eliminates the effects of optical activity, is the Littrow spectrograph shown in Fig. 3. Focal lengths of the collimator-camera lens are of the order of 1700 mm and the spectrum from 1850–8500 Å is commonly photographed in ten overlapping steps. The successive regions are obtained by focusing the lens for the region desired. A cam simultaneously rotates the prism properly. The photographic plate is also rotated for each region.

An inherent difficulty with the Littrow arrangement is that reflection and scattering of light from the front face of the collimator may fog the photographic film objectionably. The false reflections can be eliminated by tipping the lens, but general scatter is more difficult to remove, and requires careful baffling and attention to cleanliness to prevent it from becoming objectionable.

A further simplification of the autocollimation principle used in the Littrow spectrograph was made by Féry.[8] His spectrograph combines in a single optical element the lens, prism, and mirror of the Littrow spectrograph, and is shown in Fig. 4. The great advantage of the system is in the far ultraviolet, where the elimination of material and of air–glass interfaces increases the system's light efficiency. Much efficiency is lost, however, by the enormous astigmatism, since there is no focusing whatever in a direction perpendicular to the optical plane.

---

[8] C. H. Féry, *Astrophys. J.* **34**, 79 (1911).

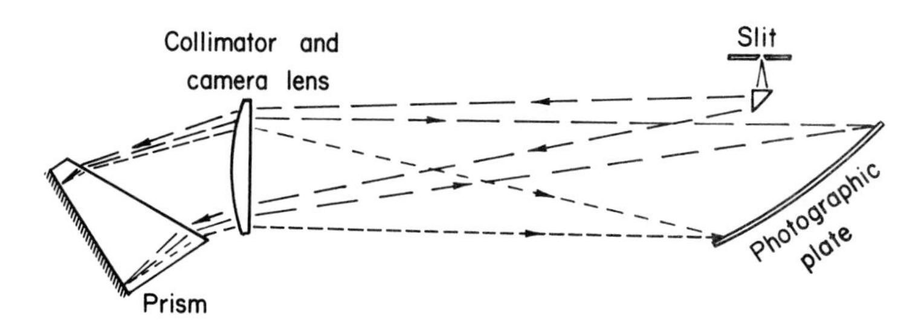

FIG. 3. Littrow spectrograph.

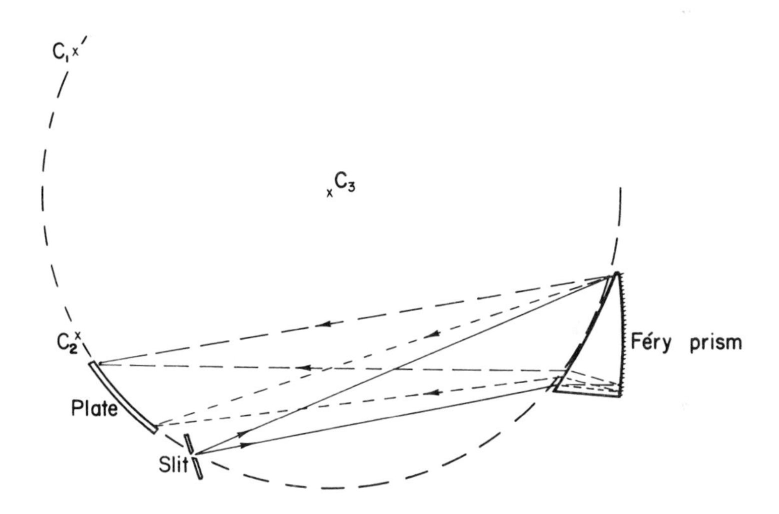

FIG. 4. Féry spectrograph. Here C_1 is the center of curvature of the front prism face, and C_2 that of the rear reflecting face. The two centers, the slit, and the photographic plate all lie on a circle with center at C_3.

4. *Gaertner Arrangement*

To eliminate the difficulty with scattered light found in the Littrow arrangement, but to achieve the compactness of a folded system, the Gaertner spectrograph was designed. It is illustrated in Fig. 5. It exchanges the material economy of a single lens and a 30° prism for the freedom from scatter of a once-through optical system. The folding is obtained with a single mirror as shown. The arrangement also eliminates the right-angle prism immediately behind the slit in the conventional Littrow arrangement.

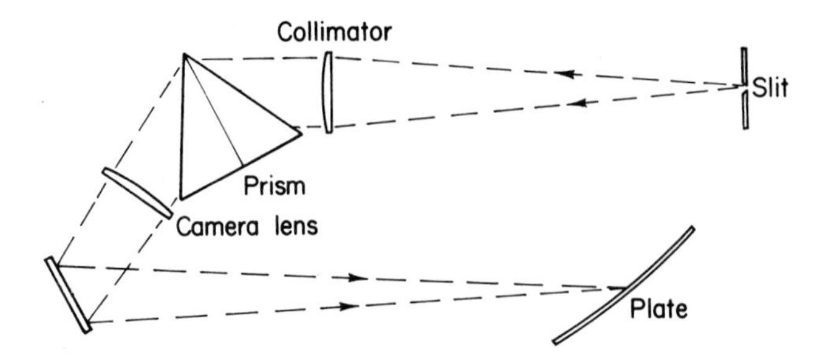

FIG. 5. Gaertner spectrograph.

5. *The ISP-22 Spectrograph*

A system originated in the Russian ISP-22 spectrograph is shown in Fig. 6. In this instrument, the collimator is a concave mirror, the prism is of the conventional Cornu kind, and the camera lens is of multielement quartz construction (not shown here). In a later version, the ISP-28, an off-axis parabolic mirror is substituted for the spherical mirror of the original.

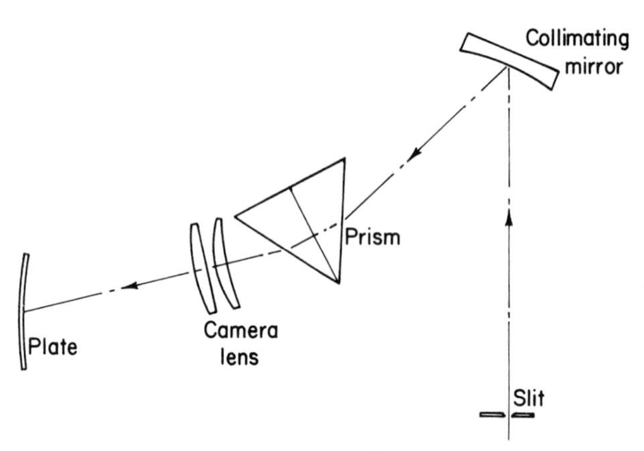

FIG. 6. ISP-22 spectrograph.

III. PLANE GRATING SPECTROGRAPHS

1. *Lens Arrangements*

The simplest of all plane-grating spectrographs is an arrangement using a simple collimator lens and a conventional camera lens. Such an arrangement is shown in Fig. 7. In the visible, where such lenses are available, it is the most convenient.

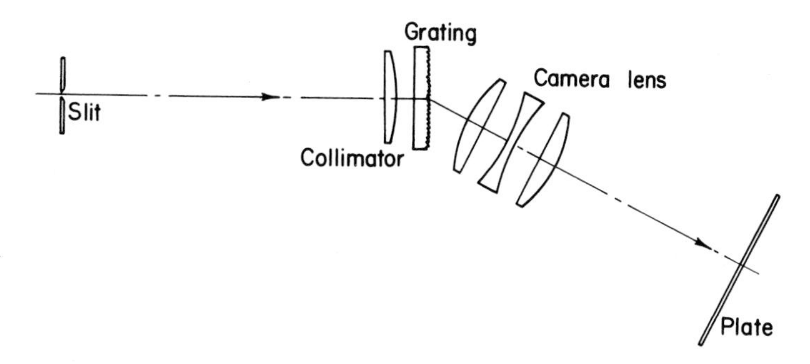

FIG. 7. Plane grating with conventional camera.

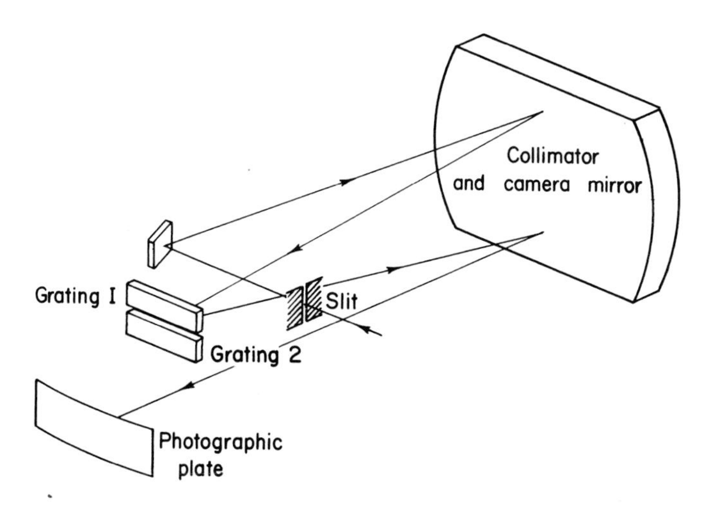

FIG. 8. Dual-grating spectrograph. The gratings cover different spectral regions, the two spectra being separated on the plate by a slight upward tilt of one grating relative to the other.

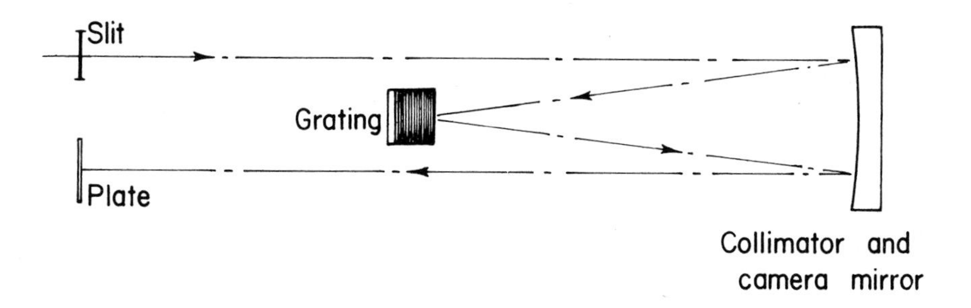

FIG. 9. Ebert–Fastie spectrograph.

2. *Ebert Arrangement*

The Ebert–Fastie monochromator will be discussed in Section VID1. However, several adaptations of the arrangement have been made for spectrographic purposes, one of which is shown in Fig. 8. This over-and-under arrangement with two gratings is used in the Bausch and Lomb dual-grating spectrograph. One disadvantage of this arrangement is the very large mirror required, equal to the width of the grating plus the width of the photographic plate. To reduce this size, the arrangement of Fig. 9 has been used in which the grating is much closer to the mirror. The size of the mirror is now reduced to the width of the grating plus half the width of the photographic plate. In exchange for this gain, the field angle of the concave mirror is increased to some degree, worsening the image quality. The curvature of the focal plane depends on the position of the grating. The curvature of the arrangment in Fig. 9 is far greater than the curvature of the other arrangements. A side-by-side arrangement has been used by Applied Research Laboratories. In this instance, the two mirrors are of different focal length, with some gain in image quality. The advantage of all these arrangements is their compactness and stigmatic image formation.

3. *Echelle Arrangements*

The echelle has been discussed in Chapter 2. To make effective use of the echelle, some method of separating the overlapping spectral orders of this coarse, high-angled grating must be used. This separation is accomplished by crossing the dispersion of the echelle with that of a relatively low-dispersion grating or prism spectrograph. Two methods of using a prism spectrograph have been used, an internal arrangement (Fig. 10) and an external arrangement (Fig. 11). An internal arrangement with a Wadsworth spectrograph (see Section IV,C,2) has also been used, (Fig. 12).

IV. CONCAVE GRATING SPECTROGRAPHS

Rowland showed that a diffraction grating ruled with grooves equally spaced on the chord of a spherical mirror would combine dispersion and focusing into one element.

A. TANGENTIAL FOCAL CURVE

For maximum resolution, both the object and image formed by a concave grating must lie on a circle whose diameter is the principal radius of the spherical surface on which the grating is ruled. This circle of diameter R, called the Rowland circle, is illustrated in Fig. 13. As shown, the

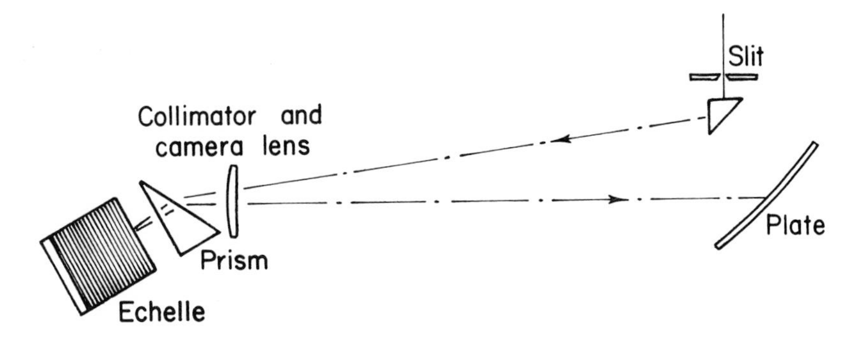

FIG. 10. Echelle Littrow spectrograph.

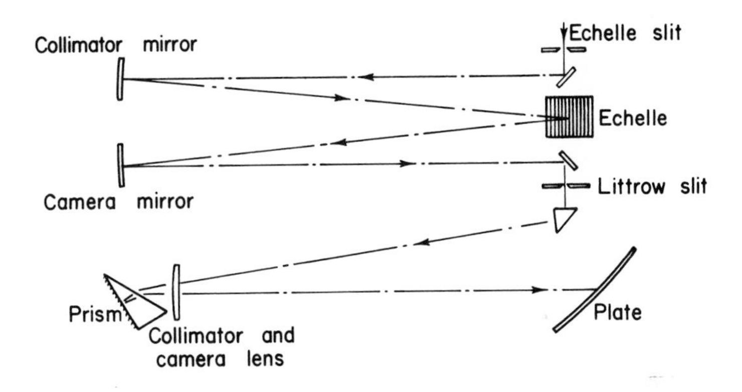

FIG. 11. Echelle; external arrangement.

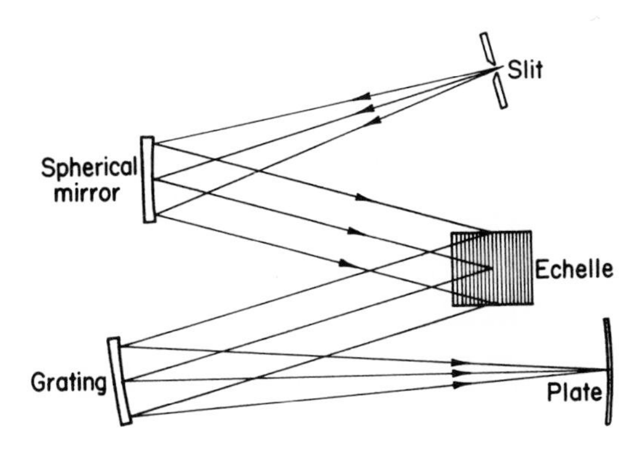

FIG. 12. Wadsworth echelle mounting.

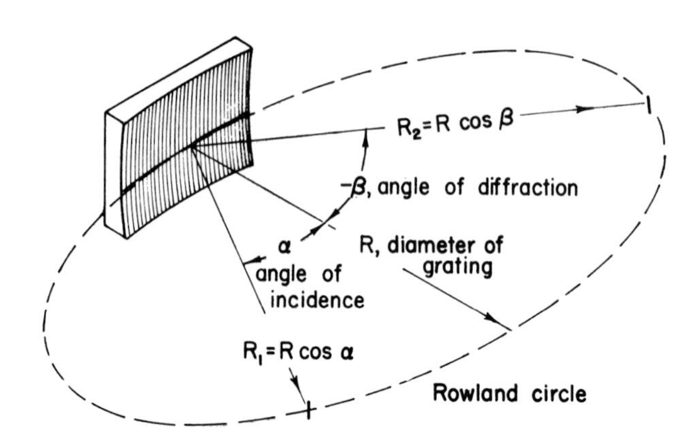

FIG. 13. Focal curve of the concave grating.

object distance is $R_1 = R \cos \alpha$, where α is the angle of incidence. The image distance is $R_2 = R \cos \beta$, where β is the angle of diffraction. The angle of diffraction is considered to be negative when it is on the opposite side of the grating normal from the incident light.

If a concave grating is used off the Rowland circle, the focal conditions in the tangential plane (the plane of the Rowland circle) are given by

$$(\cos \alpha)\{[(\cos \alpha)/R_1] - (1/R)\} + (\cos \beta)\{[(\cos \beta)/R_2] - (1/R)\} = 0$$

However, off the Rowland circle, aberration problems become quite severe, small departures giving rise to a serious loss in resolution. Under circumstances in which relatively low resolution will serve, there are some advantages of using non-Rowland circle mountings. One non-Rowland circle mounting which does give maximum resolution is the Wadsworth mounting, in which the concave grating is illuminated with parallel light. The aberrations are then as low as or lower than in the Rowland circle mounting. The Wadsworth mounting is described in Section IV,B5.

The tangential magnification of a Rowland circle mounting is $-(R_2/\cos \beta)/(R_1/\cos \alpha)$ because the focal plane is not perpendicular to R_2. Since $R_1 = R \cos \alpha$ and $R_2 = R \cos \beta$, the magnification is -1. For this reason, the dispersion of a Rowland-circle concave grating mount is the same for as any other grating, namely $dl/d\lambda = mR/(a \cos \beta)$, and so long as aberration losses do not overwhelm the diffraction losses, the theoretical resolution is also the same.

B. MOUNTINGS

A number of mountings have been designed to take advantage of the focusing properties of the concave grating. The range of incidence and diffraction angles covered by some of these mountings are shown in Fig 14 (after Beutler).[9]

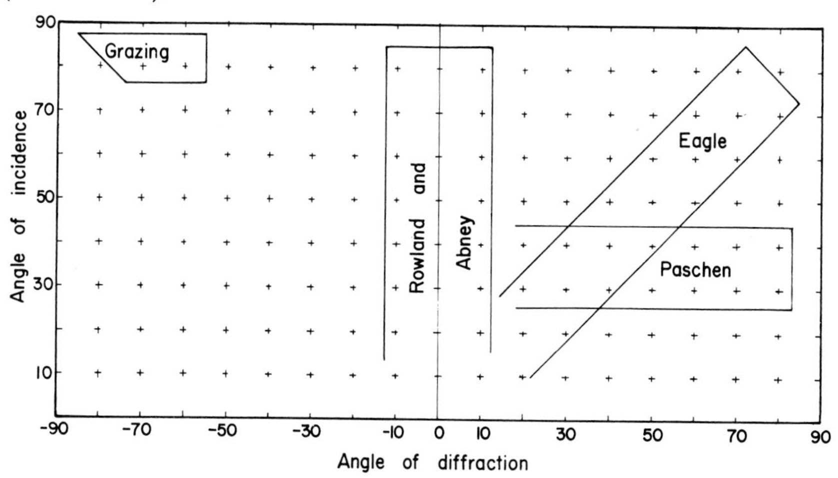

FIG. 14. Rowland-circle concave-grating mounts. Ranges of the mountings.

1. *Rowland Mounting*

In the Rowland mounting, the slit is located at the intersection of two perpendicular rails (Fig. 15). The grating and photographic plate are

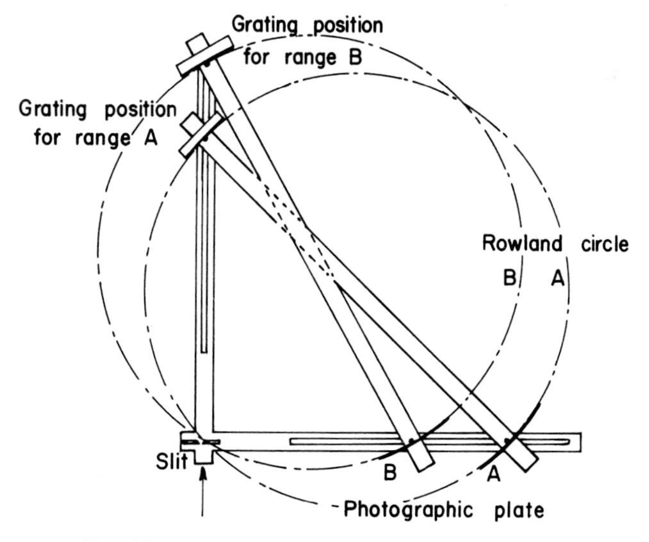

FIG. 15. Rowland mounting for a concave grating.

[9] H. G. Beutler, *J. Opt. Soc. Am.* **35**, 311–350 (1945).

mounted on a common rail separated by the distance R. A carriage under vertex of the grating slides along one rail and the carriage under the vertex of the photographic plate slides along the other rail, the whole assembly pivoting as necessary. It is obvious that this mounting always satisfies the Rowland equations, and, indeed, can easily be arranged to automatically and continuously change range. The principal disadvantage is that it is wasteful of space and cumbersome for long radii.

2. *Abney Mounting*

In the Rowland mounting, the center of the plate is always on the grating normal. The dispersion over the restricted range of a single plate is therefore almost linear. Another such mounting is the Abney mounting[10] (Fig. 16), in which slit and source are moved instead of plate and grating. This requirement can be a disadvantage if the source is cumbersome.

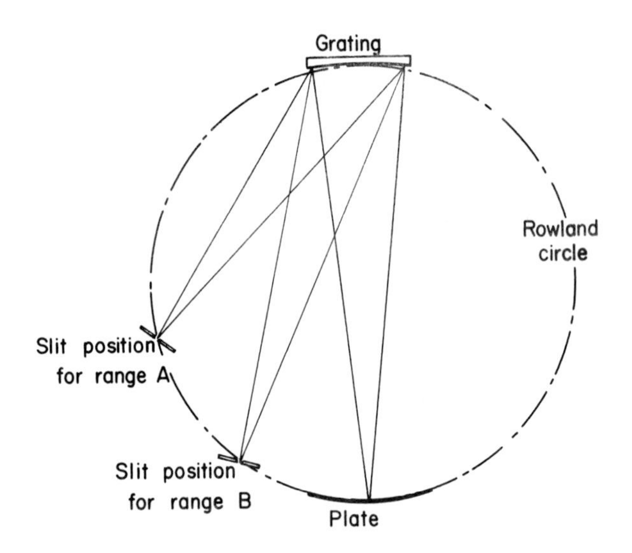

FIG. 16. Abney mounting for a concave grating.

3. *Paschen Mounting*

In the Paschen mounting, the slit is fixed, as is the grating, commonly on concrete pillars. The positions of the two determine a Rowland circle around which is a set of rails on which plates may be located at any position. The mounting takes a large amount of space, but permits a wide range of spectrum and many orders to be photographed at once.

---

[10] Sir Wm. Abney, *Phil. Trans.* **177**, 457 (1886).

4. *Eagle Mounting*

The most commonly used concave grating mounting is that of Eagle.[11] In this mounting, the angle of diffraction is kept almost equal to the angle of incidence. The mounting is thus analogous to the Littrow mounting for a prism, and so gives a very compact arrangement. To change the range, neither the slit nor the source need be moved, but the plate is rotated about an axis collinear with the slit, and the grating is both rotated and translated (Fig. 17). The positions for each range are commonly calibrated with either scales or counters, but a linkage can be used as in Fig. 18.

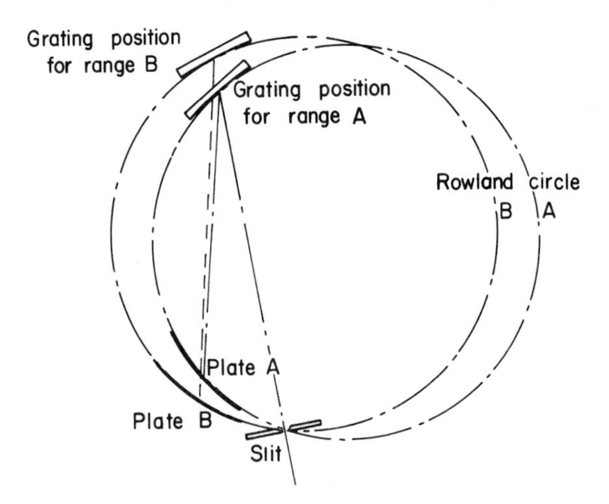

FIG. 17. Eagle mounting for a concave grating.

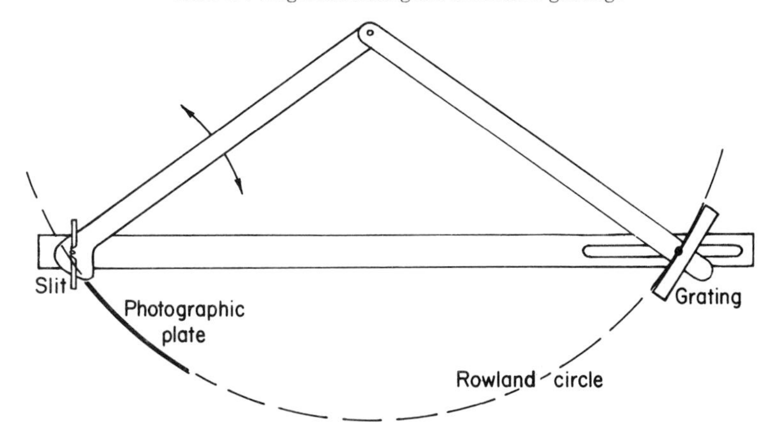

FIG. 18. Eagle mounting. Mechanism for positioning plate and grating.

---

[11] A. Eagle, *Astrophys. J.* **31**, 120 (1910).

5. *Wadsworth Mounting*

The Wadsworth mounting, also called the stigmatic mounting, uses a grating illuminated with parallel light.[12] The diffracted light is always used close to the grating normal, as illustrated in Fig. 19. The light is usually collimated by an off-axis spherical mirror. Wadsworth collimators have also been made using off-axis parabolas and lens systems.

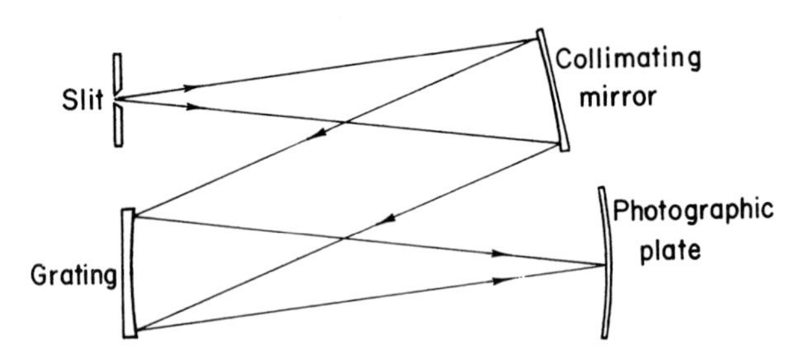

FIG. 19. Wadsworth mounting for a concave grating.

A number of commercial instruments, mostly of fixed range, have been produced using the Wadsworth arrangement. As is apparent from Fig. 19, changing the angle of incidence involves moving the collimating mirror, the slit, and the source, or moving an auxiliary plane mirror. In addition, the focal curve, which is parabolic, changes shape as the angle of incidence is changed. In the Wadsworth mounting, the tangential focal curve is given by $R_2 = R(\cos^2 \beta)/(1 + \cos \alpha)$. Because of this focal curve, the dispersion equation is somewhat different from that given in Section IVA. The Wadsworth dispersion is $dl/d\lambda = mR/a[\cos \alpha + 1]$.

Because the grating is in parallel light, a number of auxiliary arrangements are possible. One of these[13] uses a plane echelle the dispersion of which is crossed with that of a concave grating, as in Fig. 12. Another arrangement, suggested by Hulthen and Lind,[14] puts a Fabry–Perot etalon in the parallel beam.

6. *Grazing Incidence Mounting*

At very short wavelengths, the reflectivity of all materials is very low and the grating efficiency in any of the mountings so far described is low in consequence. However, the reflectivity increases markedly at very low angles of incidence. For this reason, the grazing-incidence mounting

[12] F. L. O. Wadsworth, *Astrophys. J.* **2**, 370 (1895); **3**, 54 (1896).

[13] G. R. Harrison, J. E. Archer, and J. Camus, *J. Opt. Soc. Am.* **42**, 706–712 (1952).

[14] E. Hulthen and E. Lind, *Arkiv Fysik* **2**, 253–270 (1950).

(Fig. 20) is in common use for wavelengths shorter than 500 Å. The dispersion of such a mounting is very nonlinear, but the instrument, even for long-radius gratings, is quite compact. The aberrations of the concave grating can become quite severe in this arrangement. For the limitations imposed thereby, see Beutler[9].

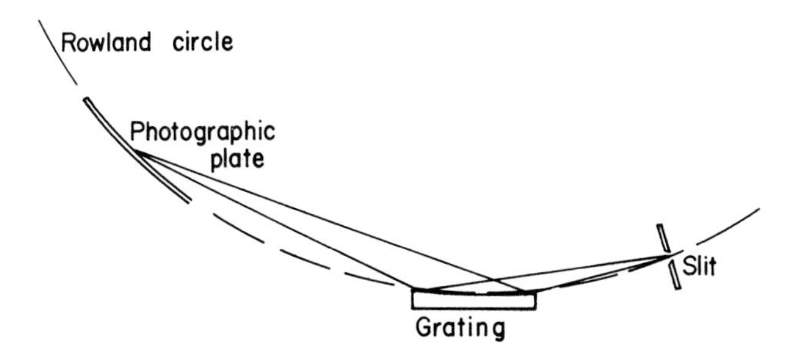

FIG. 20. Grazing-incidence mounting for a concave grating.

C. Astigmatism of the Concave Grating

The most serious image defect of the concave grating is astigmatism. The focal equation in the sagittal direction is not the same as that for the tangential direction. The sagittal focal equation is

$$(1/R_1) - [(\cos\alpha)/R] + (1/R_2) - (\cos\beta)/R = 0$$

in which the symbols have the same meaning as before except they are now applied in a plane perpendicular to the Rowland circle. The equation may be written in a form which is sometimes more useful (see Fig. 21);

$$[1/(S + R\cos\alpha)] + [1/(F + R\cos\beta)] = (\cos\alpha + \cos\beta)/R$$

A point on the slit, for which $S = 0$, will not be imaged on the Rowland circle in the sagittal direction, and so will appear as a line. The length of this line depends on the length of the grating grooves, and is given in units of groove lengths by Beutler.[9] The expression for the length of line into which a point is converted is

$$Z' = L(\sin^2\beta + \sin\alpha\tan\alpha\cos\beta)$$

where L is the groove length.

Because each point on the entrance slit of a spectrograph is transformed into a line, the entrance slit must be precisely parallel to the grating or the spectrum line in the image plane will be broadened and resolution will be lost. An advantage of astigmatism is that small imperfections in the slit are not seen in the spectrum line. Spectra photographed with an astigmatic spectrograph have a clean appearance seldom seen in stigmatic spectrographs of equal resolution.

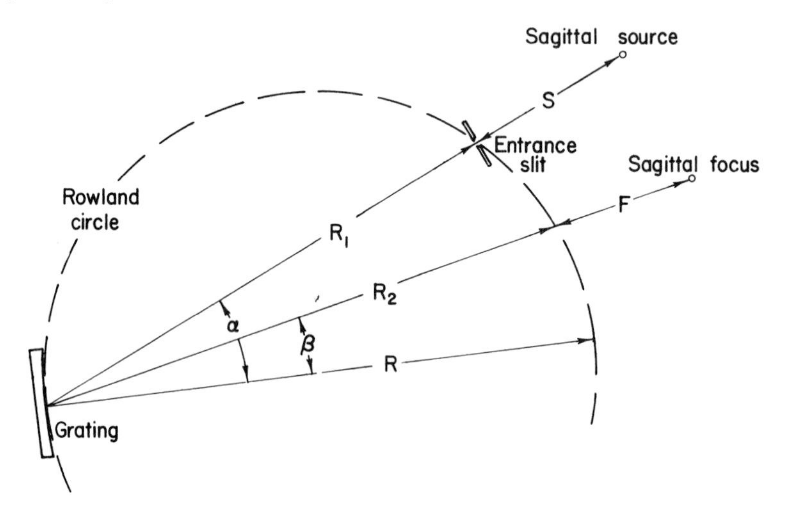

FIG. 21. Concave grating; sagittal focal conditions.

1. *Energy Effects of Astigmatism*

A further consequence of the astigmatism is that the spectrum lines are nonuniform and, in some instances, their intensity is much reduced. The situation is illustrated in Fig. 22. The values of $Z'n$, $Z'u$, and $I_{Z'u}$ are

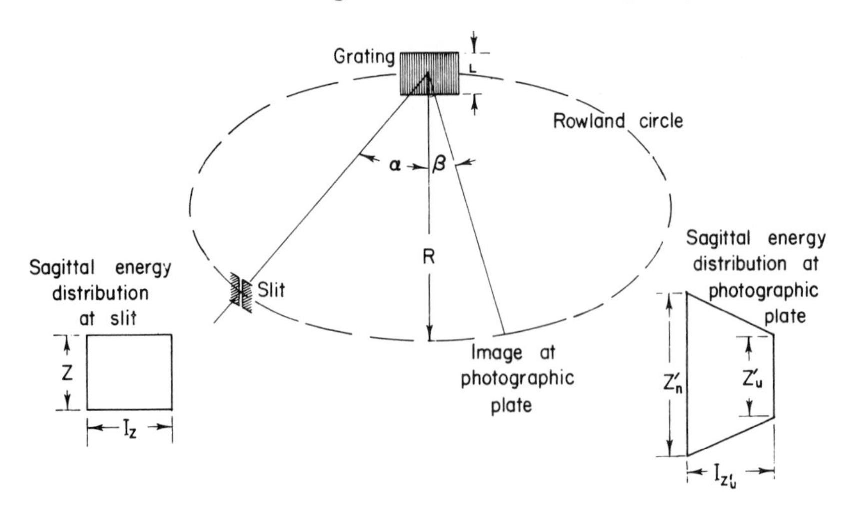

FIG. 22. Astigmatism of a concave grating; sagittal energy distribution.

given by

$$Z'n = L(\sin^2\beta + \sin x \tan x \cos \beta) + Z(\cos \beta/\cos x)$$
$$Z'u = L(\sin^2\beta + \sin x \tan x \cos \beta) - Z(\cos \beta/\cos x)$$
$$I_{Z'u} = I_Z Z'n/2Z$$

2. *Compensation of Astigmatism*

The chief disadvantage of astigmatic mountings, apart from energy loss, is that Hartmann slides, step wedges, logarithmic spirals, characteristics of the source, or the like cannot be conveniently placed at the slit, or cannot be focused on the slit, because they will not appear in focus at the plate.

A number of schemes have been devised to prevent, minimize, or correct the astigmatism of the concave grating. Most commonly, the astigmatism is compensated using the Wadsworth mounting. In that mounting, the residual astigmatism is given by $Z' = L \sin^2\beta$. Other astigmatism compensations are illustrated in Figs. 23 and 24. Still other schemes using toroidally-deformed gratings and using a compensating second grating also employed as predisperser have been used.

D. Blaze of a Concave Grating

If the central groove of a concave diffraction grating is ruled at such angles that it is well blazed for some deviation angle, then the grooves at the edges of the grating aperture will not be exactly blazed for the same deviation, a simple consequence of the divergence of the incident beam and the convergence of the diffracted beam, as shown in Fig. 25. In some circumstances the effect is advantageous, as it spreads the blaze over a wider wavelength range. The calculated comparison among three grating mounts is shown in Fig. 26.

Occasionally, particularly in gratings for the vacuum ultraviolet, the angular aperture of the grating will exceed that commensurate with the desired sharpness of blaze. Is is then not possible to rule a grating with the same blaze angle over the entire width, because there will not be clearance for the diamond, and because a single blaze angle will not produce the maximum efficiency in the spectral range where energy is at a premium. This combination of circumstances has given rise to the multipartite grating, in which the grating is ruled in a number of bands (often three, but up to as many as 12) each of which has a slightly different blaze angle. The resolution of such a grating is only the resolution of a single band, but the efficiency is very high.

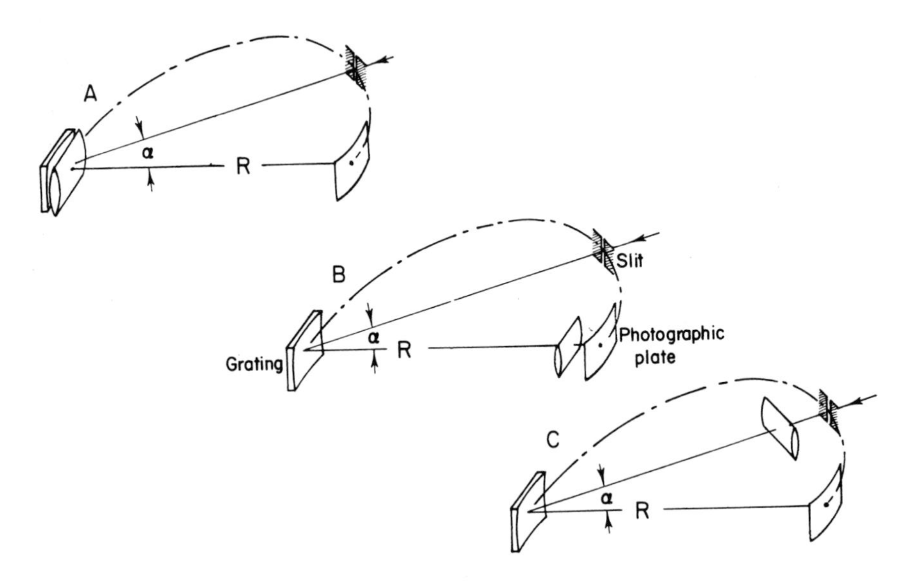

FIG. 23. Use of cylindrical lenses to compensate the astigmatism of a concave grating. (A) Lens in front of grating. (B) lens in front of plate. (C) Lens in front of slit.

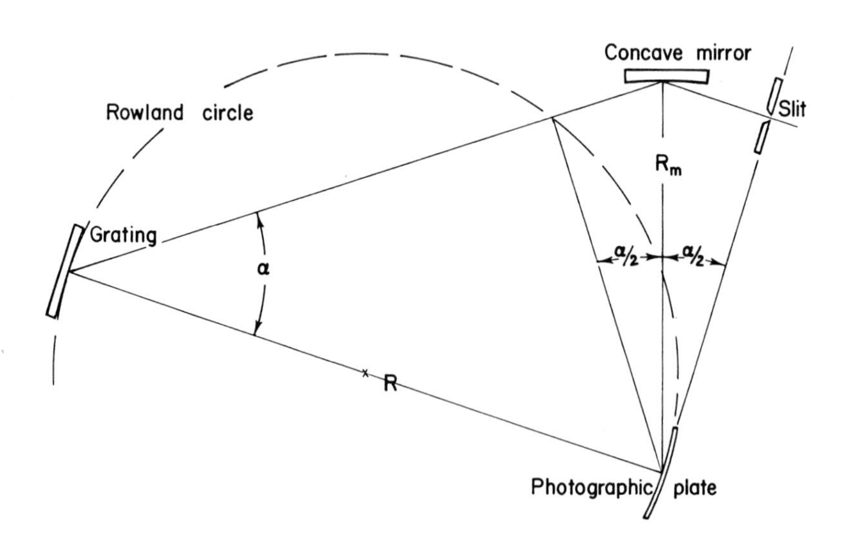

FIG. 24. Use of a concave mirror to compensate the astigmatism of a concave grating.

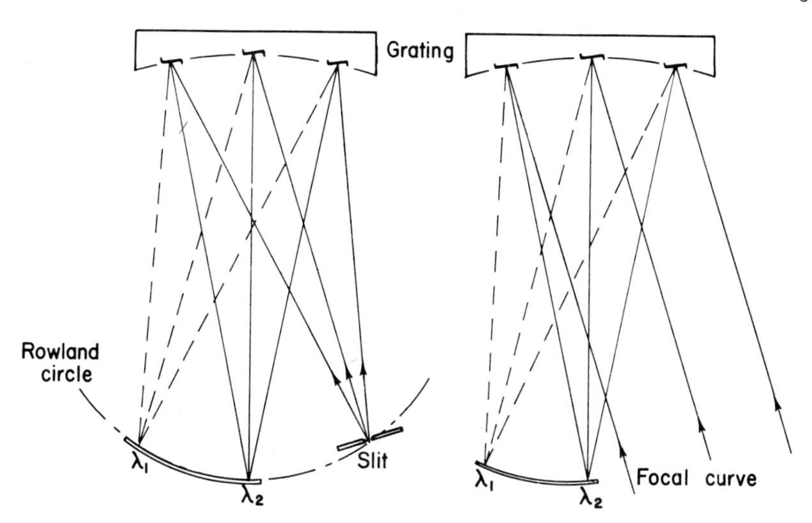

FIG. 25. Blaze variation across a concave grating.

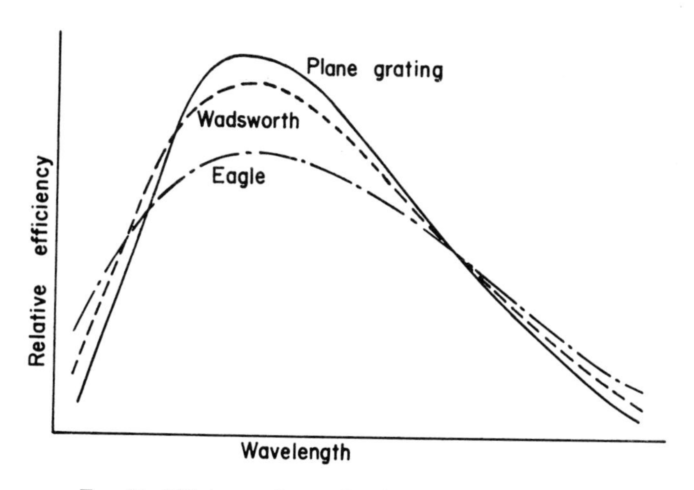

FIG. 26. Efficiency of a grating for three different mounts.

V. PROBLEMS ASSOCIATED WITH THE SPECTROGRAPH

A. UNIFORM ILLUMINATION OF THE ENTRANCE SLIT

The spectral line image formed by a spectrograph is an image of the slit. In consequence, the spectral line image will usually show up any defects of the slit. Dirt is the greatest offender, but nicks in the slit jaws will sometimes also occur. Similar difficulties can arise from a tapered slit.

One of the great advantages claimed for concave-grating astigmatic spectrographs is that such defects of the slit do not appear on the photographic plate. Stigmatic instruments, however, will produce clean lines only with care. Such clean lines are usually only of interest in photographic spectroscopy, and even there, esthetics aside, are of consequence only when the lines are to be densitometered for quantitative spectrochemical analysis.

Such mechanical inadequacies are not the greatest source of difficulty with photographic spectrographs. The major problem arises when attempting to form uniform spectral lines from the nonuniform sources commonly used for spectrochemical analysis. To produce uniform lines in this circumstance requires that the source be focused on to the aperture stop of the spectrograph. To focus the source at the aperture stop, a field lens at the slit is used. The field lens acts to make the image uniform, as shown in Fig. 27. Here, it can be seen in somewhat exaggerated form that if the

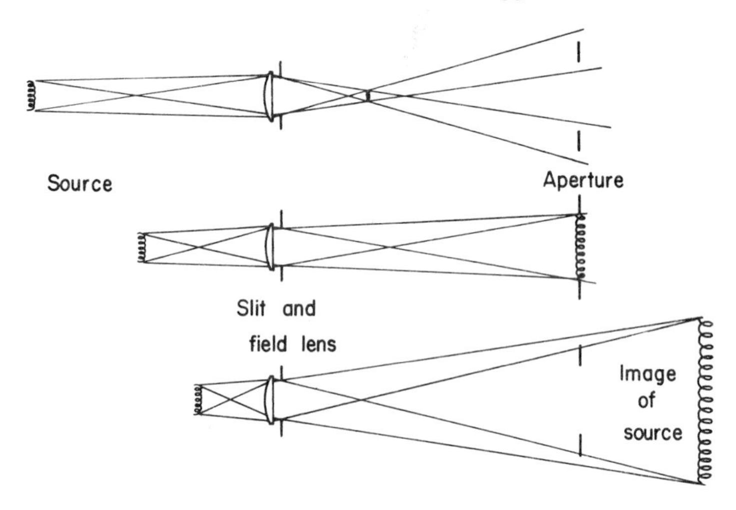

FIG. 27. Spectrograph illumination. Effect of focusing the source.

image of the source is focused either in front of the aperture or in back of the aperture stop, no light from the ends of the slit will pass the aperture stop. If no light from the ends of the list can pass the aperture stop, obviously, then, the ends of the line will not appear on the plate. More exact focusing may be required than is commonly supposed, and occasionally even chromatic aberration of the field lens can be detected. Chromatic aberration can be particularly evident when two overlapping orders of a grating spectrograph are photographed simultaneously. The usual cure is to use a field lens of long enough focus so that the aperture stop is much underfilled. The focusing requirement is then much less stringent. Similarly, a shortened entrance slit much minimizes the problem, as does a source with appreciable longitudinal depth. However, when long slits are needed (as with step attenuators) or when weak sources are used (needing maximum aperture), then only care will produce uniform lines.

B. Critical Slit Width

As the entrance slit to a spectrograph is narrowed, the spectral lines become finer and the apparent resolution increases. Below some particular slit width, resolution no longer increases, and the only consequence of further narrowing of the slit is a decrease in the available energy. The slit width below which no increase in resolution can be observed is called the critical slit width, and is given by $S = K(f/w)\lambda$. The quantity (f/w) is the f-number of the spectrograph in the same direction in which the slit width is measured, and K is a number between 1 and 2 which depends on the coherence of the source. The value of K is 2 for a coherent source and 1 for a totally incoherent source.

Generally, the coherence is not known, and so the critical slit width is most easily determined by observing the diffraction pattern of the source at the aperture. The critical slit width has been reached when the first diffraction minima are at the edges of the aperture.

For quantitative analysis, the slits are commonly four to ten times the critical slit width, A slit width wider than critical will both increase the total light available and will also give a line image with a relatively flat-topped energy distribution for ease of densitometering.

To fill the aperture stop, spherical condenser lenses are used to form an image of the source on the slit. If the source is uniform, no difficulty arises, but if the source is not uniform, the energy distribution of the source will appear on the slit, and, in stigmatic spectrography, on the plate. This kind of imaging can be useful if the characteristics of sources are under investigation, but otherwise is troublesome. A cylindrical condenser with its axis parallel to the slit can be used to form an image of the source on the slit in only one dimension, while the field lens forms an image of the source on the aperture stop in the other dimension.

C. Stray Light

Stray light in spectrographs arises from one of several causes: (1) leakage of light through the housing of the spectrograph; (2) reflection of light from the wall of optical mounts of spectrographs; (3) reflection of light from the optics themselves; and (4) scattering or fluorescence within or from the surfaces of the optics. Stray light is one of the most common reasons why spectrographs (or monochromators, for that matter) go out of quantitative calibration. The effects of day or night operation in a leaky spectrograph can easily be imagined. The easiest way to discover the source of leak in a leaky spectrograph is to look into the spectrograph while the outside is "painted" with a bright diffuse source.

Reflection or scattering from the walls or optical mounts of the spectrograph can only be minimized, never prevented entirely. Some of the sources of this kind of stray light are: (1) Reflection from an overfilled collimator; (2) reflection of light of a spectral region not intended to be recorded onto the plate; and (3) reflection from the edges of baffles intended to remove stray light. This can sometimes be unexpectedly intense because the light may be reflected quite strongly at the grazing incidence at which it will strike such surfaces. Sharpening these surfaces will prevent this kind of reflection. Once again, the best way to discover such sources is by looking into the spectrograph while an intense source is projected into the spectrograph through a wide slot.

There are three common ways in which light reflected from the optics of a spectrograph will cause difficulty. The problems with Littrow arrangements have already been discussed. Only a few degrees of tilt of the lens are allowed or else the resulting astigmatism may cause difficulty. Alternately, the light can be blocked by a small opaque strip across the center of the lens. Occasionally, such a strip will cause photometric difficulties when the source is imaged on the lens aperture, but this is rare enough that this solution for scattered-light removal is usually useful. When multielement lenses are used in spectrograph cameras, it will occasionally happen that light multiply reflected from the surfaces of the lenses will be close to being in focus at the normal focal plane. This "almost-focused" light can be very troublesome, and has led to redesign of camera lenses in which it has occurred. A third source of scattered light by reflection arises in prism systems in which the short-wavelength light is reflected off the base of the prism and then reflected in the same direction as the longer-wavelength light. To prevent this reflection from occurring, the bases of such prisms are commonly made diffuse or are serrated.

To prevent scattered light arising from optical inclusions or fluorescence is often beyond the power of the user. Happily, it is usually not a serious problem. Far more serious is the effect of dirt, fingerprints, and the slow accumulation of grime that takes place with the passage of time. A clean environment and a tightly-closed instrument are the best preventives. In general, the effects of dirt are more severe on reflecting surfaces than on transmitting surfaces. In designing such instruments, it is worth taking some pains to avoid horizontal surfaces on which dirt can settle. Another useful trick is using a tacky paint which never dries inside the instrument to provide a trap for such dirt as might enter.

D. Vibration

The effects of vibration in a spectrograph are easy to imagine. All spectrographs are very considerable optical levers with sensitive receivers on the output end. Slits of 20 μ are common, as are lever arms of 3 m are common. Vibrational motions of 1 arc sec are thus easy to detect. The most common symptom of vibration in a photographic spectrograph is a line doubling, in which every line on the plate is doubled. Occasionally, one will see a line tripling. Such tripling only means that the vibration stopped or started at some time during the exposure. Vibrations in photoelectric instruments are easier to pinpoint. Conventional oscilloscope techniques are entirely diagnostic. Rigid construction, conventional isolation, and, where possible, choice of location are the preventives.

E. Temperature and Barometer Effects

In these days of air conditioned laboratories, the effects of temperature are rarely seen in conventional instruments. Outside of such controlled environments, and in unconventional systems, the effects can be considerable.

In prism instruments, the refractive index of the prism can be a more or less sensitive function of the temperature, depending on the material. Whether this index shift is important depends on the design of the rest of the instrument. In monochromators, such a change will clearly affect the calibration of angles vs wavelength. In spectrographs, a change in index will have the effect of changing the angle of refraction, and hence the position on the plate. If the plate is perpendicular to the optical axis, as in achromatic spectrographs, no harm whatever would arise (except for such change of focus as the temperature change may cause). However, in monochromatic instruments, such as the Littrow spectrograph, a change in angle of refraction will change the longitudinal distance at which a given wavelength is incident on the plate. In consequence, an exceptionally large defocusing can occur in such instruments with change in temperature. Nonisotropic materials like crystal quartz will expand differently in different directions, and so can affect the angle of a prism made of such a material. In grating instruments, the grating spacing itself will change with change in temperature, this change being calculable from the expansion coefficient of the grating substrate. Since the optical characteristics of the substrate material are not important in reflection gratings, such gratings are commonly made of fused silica in critical applications. In connection with index or dimensional changes, it must be kept in mind that most optical materials are notoriously poor thermal conductors.

Another common cause of wavelength shift with temperature in both grating and prism instruments is the mechanical geometry by which the disperser is positioned. The change in dimension of the mechanical parts with temperature can frequently cause rotation of the prism or grating. The mechanical effects are not always easy to analyze unless a simple construction is used.

Even in those spectrographs in which wavelength shifts *per se* are not important, some consideration must be given to the problem of wavelength shift during long exposures and the consequent loss of resolution which will result.

The most important effect of barometer change in prism instruments will be the change in refractive index of the air. The effect can be troublesome with respect to long exposures, or less so with respect to wavelength calibration. More significant is the effect of barometric pressure on the apparent wavelength which is diffracted by a grating. In grating instruments, the effect can be so troublesome for long exposures, or for the exact wavelength calibration required of direct-reading spectrographs, that a variety of compensations have been devised to cope with the problem. Most common is the use of an aneroid to rotate a glass plate behind the entrance slit of a spectrograph. The apparent position of the entrance slit is thus changed to maintain the output wavelengths in constant position.

VI. MONOCHROMATORS

Monochromators are customarily used to isolate one wavelength band out of a continuous source. They are used in this sense in all manner of absorption and reflection spectrophotometers. The most interesting characteristics of a monochromator are its wavelength bandpass, its efficiency, and its geometry.

A. Band Pass of a Monochromator

When a continuous source is incident on the entrance slit of a monochromator, an infinite series of overlapping monochromatic images of the entrance slit are found at the exit-slit focal plane. In most instances, the slit widths are wide enough that diffraction effects may be neglected and the series of overlapping images can be treated as geometric images of the entrance slit. If the wavelength interval passed by the exit slit is small enough that within the interval the dispersion is constant, then the calculation of the bandpass is simple.

The width of the monochromator image of the entrance slit will depend on the magnification of the monochromator, i.e., on the ratio of the focal lengths of telescope to collimator (in many common monochromators, the ratio is unity, but in general, the ratio may be anything whatever.) The imaged slit width S' will be $(f_2/f_1)S$, where f_2 is the telescope focal length, f_1 the collimator focal length, and S the entrance slit width. If the angular dispersion of the monochromator is $d\beta/d\lambda$, then the spatial interval occupied by a wavelength interval $\Delta\lambda$ is $f_2 d\beta/d\lambda$, i.e., $S' = \Delta\lambda f_2(d\beta/d\lambda)$. This can be written in terms of the entrance slit as $\Delta\lambda = S/[f_1(d\beta/d\lambda)]$. Thus within the image of the entrance slit there will be other partial images of the slit corresponding to a wavelength range of $2\,\Delta\lambda$. It is easy to see that the range is $2\,\Delta\lambda$, for if we consider that λm is the wavelength at the middle of the slit image, then there will also be images $\lambda m + \Delta\lambda$ and $\lambda m - \Delta\lambda$, which are just sufficiently displaced so as not to overlap the image of λm. For wavelengths intermediate between λm and $\lambda m \pm \Delta\lambda$, there will be an intermediate amount of overlap.

If we now have an exit slit of width S_2 at the focal plane of the telescope, we can calculate the distribution of energy passing through the slit. Suppose as a start that $S_2 = S'$. In this instance, the exit slit will pass a range of wavelengths from $\lambda m - \Delta\lambda$ to $\lambda m + \Delta\lambda$. If the energy at the telescope focal plane is independent of wavelength, then the intensity distribution curve passing through the exit slit will be triangular with a maximum energy at λ_m and reaching 0 at $\lambda_m \pm \Delta\lambda$. The bandwidth at half-intensity is $\Delta\lambda$ and the total bandwidth is $2\,\Delta\lambda$. We now consider the situation in which the exit slit is different from the image of the entrance slit, i.e., $S_2 > S'$, or $S_2 < S'$. The exit slit will cover a wavelength band $\Delta'\lambda$ and the entrance slit will cover a wavelength band $\Delta\lambda$ as before. Under the same conditions as before, the intensity distribution will then be trapezoidal, the width of the top of the trapezoid being $|\Delta'\lambda - \Delta\lambda|$. The band pass at half-intensity is either $\Delta'\lambda$ or $\Delta\lambda$, whichever is the greater (see Fig. 28). For a given bandpass,

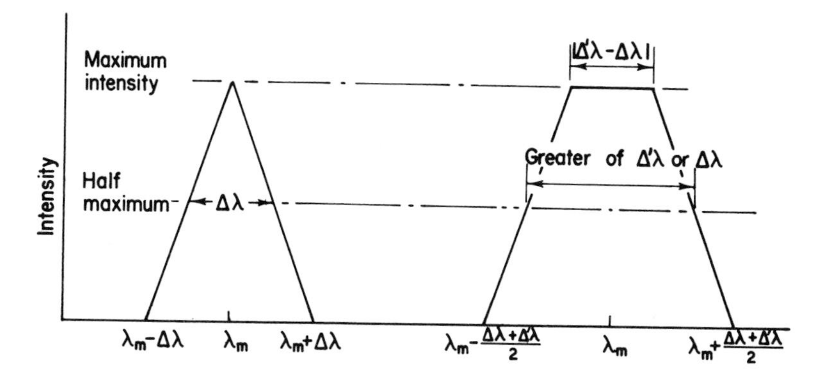

FIG. 28. Monochromator bandpass.

the maximum radiant power occurs when $S' = S_2$. However, there are many times when a broad, flat-topped energy distribution is desired. Then S_2 may be made larger or smaller than S'.

B. Efficiency of a Monochromator

The resolving power of a monochromator cannot be increased without limit. The conventional limitation is that imposed by diffraction. This limitation has been discussed elsewhere. In most monochromators, the limit is generally imposed by either geometric aberration of the image forming system, or by there being insufficient energy to achieve the desired signal-to-noise ratio. This failure to supply enough energy in the desired bandpass is a common difficulty in the infrared.

By "efficiency of a mcnochromator" we refer not only to the monochromatic transmittance of the monochromator, but also to the relative ability of a given monochromator to deliver a large flux of monochromatic radiation.

For a given spectral bandpass, the ability of the monochromator to deliver large flux is certainly linearly proportional to the monochromatic transmittance of the system. Thus a high transmission of the refractive elements, a high reflectance of the reflecting elements, and a high grating efficiency are all-important. Equally important is the ability of the system to achieve the desired bandpass with wide slits, high slits, and a large numerical aperture. When all of these other design parameters are taken into account, the efficiency of a monochromator is given by

$$E = (TlA/f)\, d\beta/d\lambda$$

where E is the efficiency, T is the monochromatic transmittance, l is the slit height, A is the area of the limiting aperture, f is the focal length, and $d\beta/d\lambda$ is the angular dispersion of the system. This equation is particularly useful for comparing the relative efficiency of different monochromator systems. In particular, it is easy to understand why grating monochromators are preferred over prism monochromators: The angular dispersion of a grating is so much larger than the angular dispersion of a prism that the energy for a given bandpass is far greater.

To use the equation to calculate the actual energy transmitted, one need only multiply the efficiency by the monochromatic light B_λ, in watts/cm^2/steradian. The equation also points out that for highly efficient systems, there is ultimately no substitute for size.

Any system which by ingenuity allows one to use a larger slit than would be conventionally calculated will clearly given a increase in energy. The first of such systems was described by Golay.[15] A more recent system using slit apertures shaped as families of hyperbolas is described by Girard.[16] Ultimately, the energy available with such slit systems approaches that available from interferometer systems.

C. Prism Monochromator Designs

There are a great number of designs for prism monochromators. Most desirable in any monochromator design is that both the entrance and exit slit position do not change, and that the directions in which the light enters and leaves the monochromator are also constant. Other arrangements have been employed, but they are generally not so useful.

1. *Littrow Arrangement*

The Littrow monochromator is shown in Fig. 29. The collimator shown is an off-axis parabola. The wavelength is changed by rotating the Littrow mirror. This is an arrangement commonly used in infrared spectrophotometers.

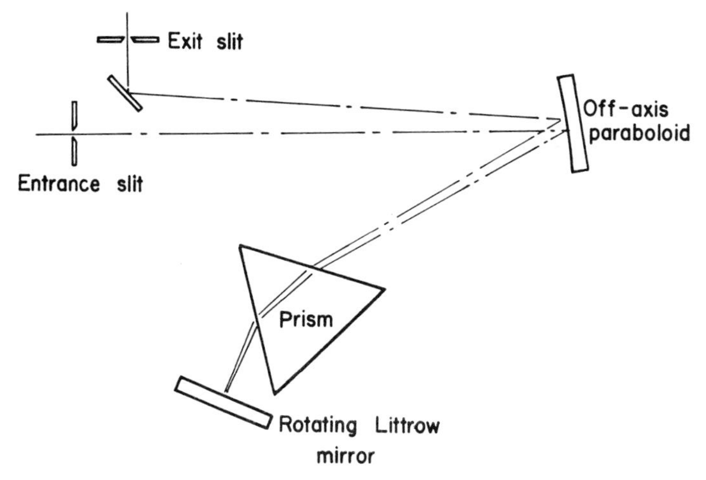

FIG. 29. Littrow monochromator.

2. *Wadsworth Arrangement*

The Wadsworth system combines one mirror with a prism on a turntable by which the combination is rotated together. If the axis of rotation is at the intersection of the angle-bisecting plane of the prism and the

---

[15] M. J. E. Golay, *J. Opt. Soc. Am.* **39**, 437–444 (1949); **41**, 468–472 (1951).
[16] A. Girard, *Appl. Opt.* **2**, 79–87((1963).

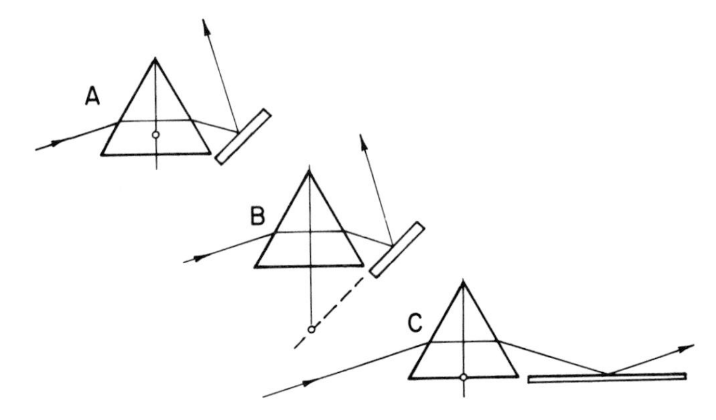

FIG. 30. Wadsworth monochromator arrangements. (A) Wavelength at minimum deviation is bent 90°, but the beam shifts laterally. (B) Wavelength at minimum deviation is perpendicular to entering ray, and the beam does not move. (C) Wavelength at minimum deviation is parallel to incident ray at a constant lateral displacement. (The axis of rotation in each case is marked by O).

extended surface of the mirror, then the combination introduces no lateral shift of the beam throughout the rotation. Some varieties of the system are illustrated in Fig. 30.

3. *Pellin Broca*

This prism (Fig. 31) is similar to the Wadsworth arrangement except that the 60° prism has been divided into two 30° prisms placed one on each side of the mirror. As the prism is rotated, the spectrum moves laterally across the slit, and at any position, the ray which emerges from the second slit is that which suffers the same refraction at entry and emergence as the minimum-deviation ray does in the complete 60° prism. This ray is incident

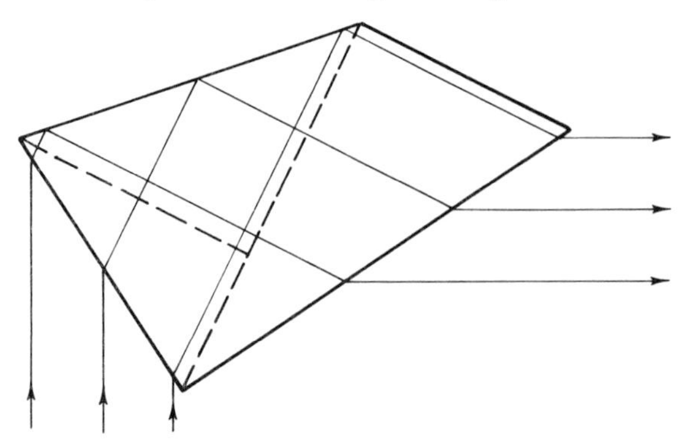

FIG. 31. The Pellin Broca prism.

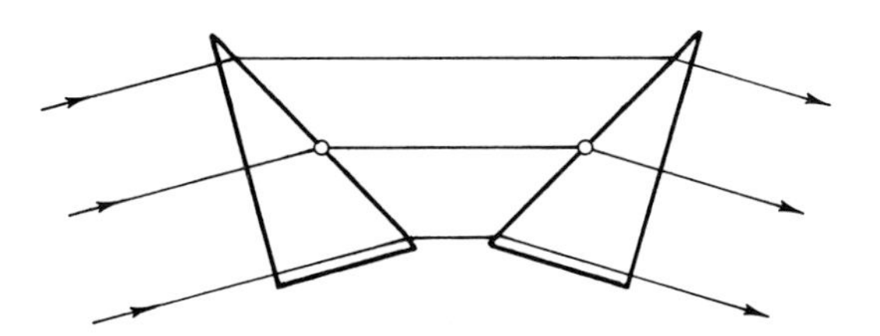

FIG. 32. The Young–Thollon prism arrangement.

at 45° on the intermediate mirror. For this reason, the Pellin Broca prism is often referred to as a constant-deviation prism, but there is no actual constancy of the angular deviation.

4. *Young–Thollon*

This compact arrangement is shown in Fig. 32. Once again the 60° prism has been divided into two, the halves being mounted symmetrically as shown. Light of some chosen intermediate wavelength enters and leaves the prisms perpendicular to their outer surfaces, and this wavelength emerges from the exit slit. To change the wavelength, both prisms are rotated in opposite senses about the points *O*. If crystal quartz is used, one prism is made from right-hand quartz and the other from left-hand quartz.

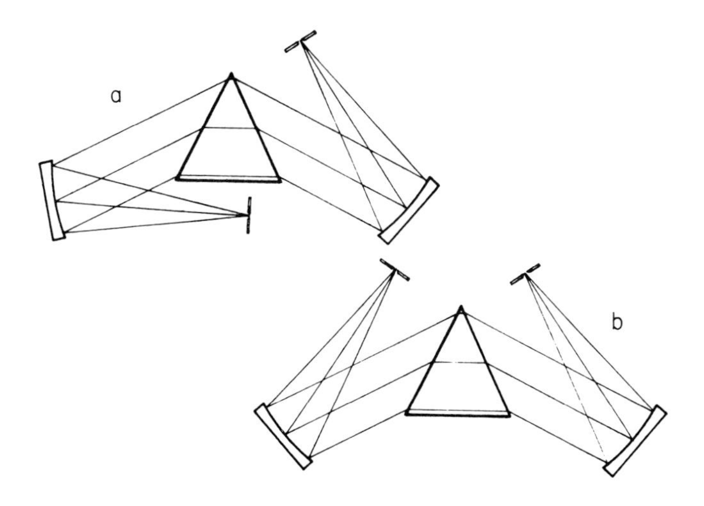

FIG. 33. The Czerny–Turner arrangement.

5. *Czerny–Turner*

In the infrared, where refracting optics are not useful, the collimator must be a concave mirror. Czerny and Turner[17] showed that in the arrangement in Fig. 33a, the coma of one mirror compensates that of the other, whereas in Fig. 33b, the two comas add rather than subtract.

D. GRATING MONOCHROMATORS

Because of the generally higher angular dispersion of gratings as compared with prisms, grating monochromators are far more generally used than prism monochromators, because the higher dispersion gives higher efficiency.

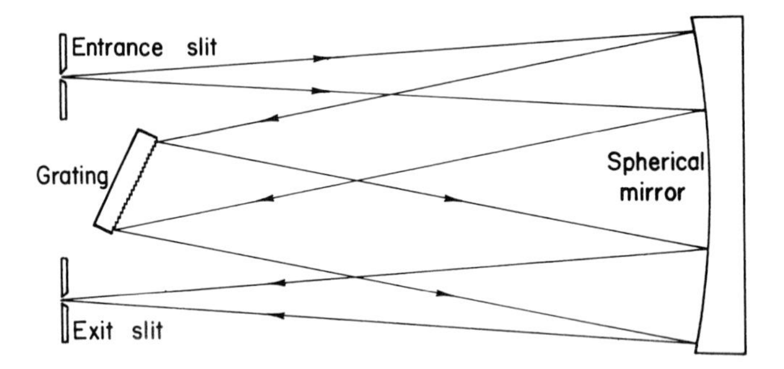

FIG. 34. Ebert–Fastie monochromator.

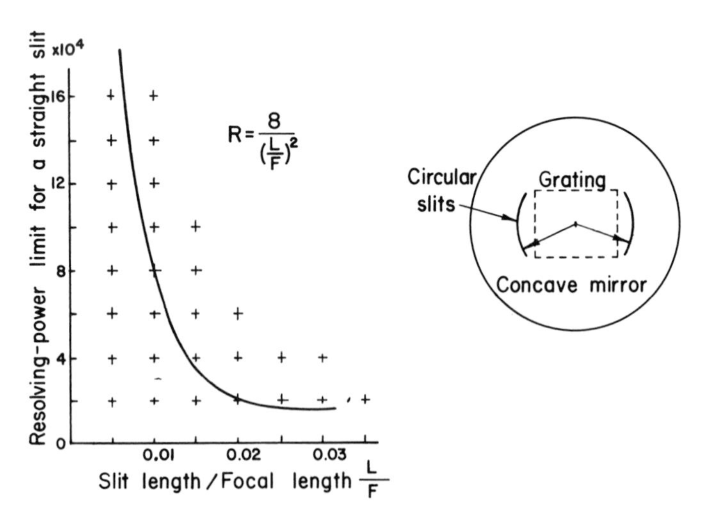

FIG. 35. Slit design in the Ebert–Fastie monochromator.

[17] M. Czerny, A. F. Turner, *Z. Physik* **61**, 792 (1930).

1. *Ebert–Fastie Monochromator*

In 1952, Fastie[18] showed the advantages of a grating monochromator arrangement first discussed by Ebert for prisms. This arrangement is shown in Fig. 34. Fastie showed that in the traditional arrangement, using straight entrance and exit slits, there is a distinct resolving-power limit imposed by the curvature of the spectral lines (see Section I,B,2). The resolution limit thus imposed is illustrated in Fig. 35. Fastie also showed that the resolution limitation imposed by straight slits can be removed by circularly curving the entrance and exit slits with a radius equal to the distance of the slit from the axis of the system, as shown in Fig. 35.

Once the line-curvature limitation is removed, the limitation to the system resolution may, in the absence of diffraction limit, be imposed by the coma of the spherical mirror. The comatic limit is shown in Fig. 36. as a function of the aperture ratio of the system. This figure gives a rather conservative estimate of the resolving power, but is nonetheless useful for preliminary system evaluation.

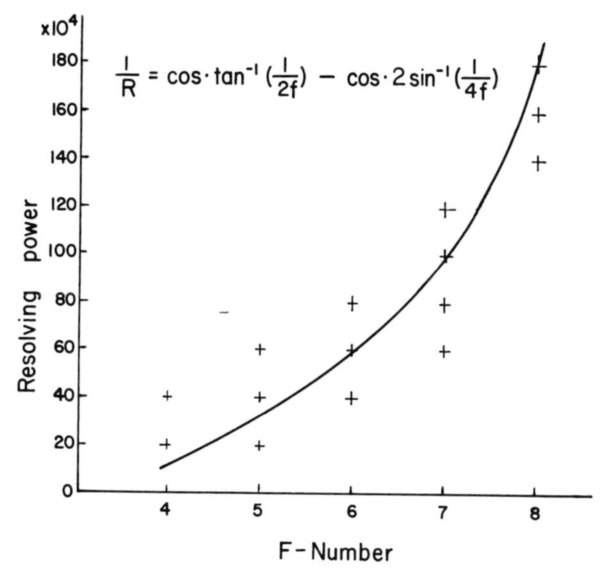

FIG. 36. Resolving power of the Ebert–Fastie monochromator.

2. *The Plane Grating in Nonparallel Light*

If we consider a plane grating to be the equivalent of a concave grating of infinite radius, we can see that a focused spectrum can be formed using a plane grating in conjunction with some other focusing means. The focal

---

[18] W. G. Fastie, *J. Opt. Soc. Am.* **42**, 641–647 (1952).

distance from the grating ρ is given by $\rho = q(\cos^2 \beta)/(\cos^2 \alpha)$, where q is the zero-order focal distance.

In a zero-deviation situation, $\alpha = \beta$, and hence this system can easily be converted to a monochromator. As it happens, the resolution is limited by a sort of pseudocoma which arises because of the convergence in the plane of the grating groove. The light not perpendicular to the grating groove is apparently diffracted by a groove whose width decreases as the cosine of the angle off perpendicular. The resolving power of a grating used in converging light is thus limited to $8q/L$, where L is the slit height.

3. *The Concave Grating As a Monochromator*

Most of the mountings described so far are useful for either photographic or photoelectric recording of the spectrum. The concave grating is also useful in monochromators. Almost any mounting can be used for a monochromator by moving an exit slit along the focal plane, but those in which the wavelength can be varied across a slit fixed in position are most useful.

The concave-grating monochromator mount currently in most common use is that described by Seya[19]. Seya found that if the angular deviation and focal distances are judiciously chosen (Fig. 37), the only adjustment needed is to rotate the grating about its vertex, and only small resolution loss will result over wide wavelength ranges. The deviation needed is large, 70°15′, which is a great convenience because equipment at the entrance and exit slits are well separated.

Other arrangements, notably those of Tousey *et al.*,[20] have also been used (see Fig. 38).

4. *Second-Order Problems*

The existence of multiple orders of diffraction in grating monochromators can sometimes cause difficulty by reducing the purity of the spectrum. There are several methods by which such difficulties can be avoided.

Most straightforward is the introduction of appropriate filters into the system. Generally speaking, the multiple-order problem requires that shorter wavelengths be removed in the presence of longer ones. Absorbing filters are usually quite capable of accomplishing the necessary attenuation of shorter wavelengths.

In some systems, however, the same grating is used in several different diffraction orders. The grating is always used on the blaze, but one order

---

[19] M. Seya, *Sci. Light* (*Tokyo*) **2**, 8–17 (1952); T. Namioka, *ibid.* **3**, 15–24); H. Grenier and E. Schaffer, *Optik* **14**, 263–276 (1957).

[20] R. Tousey, F. S. Johnson, J. Richardson, and N. Toran, *J. Opt. Soc. Am.* **41**, 696–698 (1951).

at a time is selected by an auxiliary system. Such monochromators will require for the auxiliary system either a carefully chosen set of interference filters or a prism monochromator working in series with the grating monochromator.

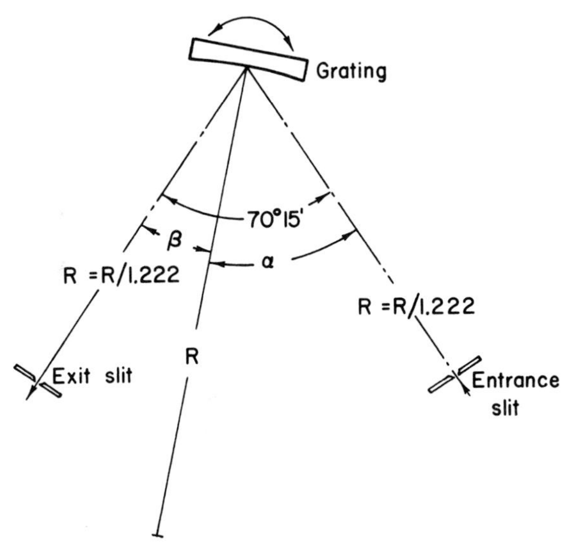

Fig. 37. The Seya monochromator.

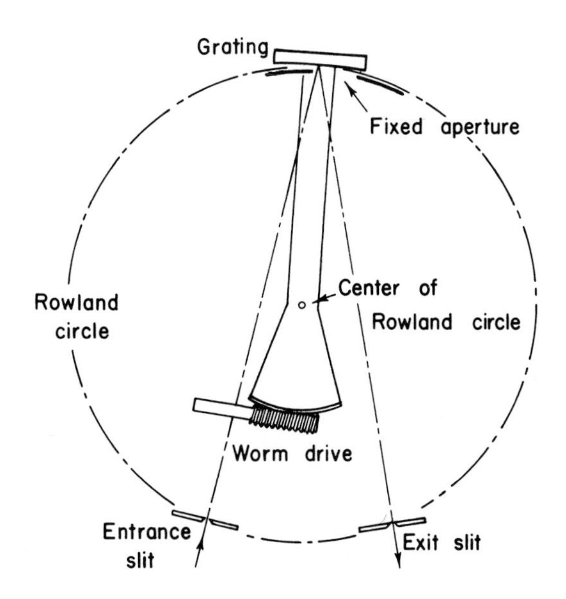

Fig. 38. The monochromator of R. Tousey, F. S. Johnson, J. Richardson, and N. Toran, *J. Opt. Soc. Am.* **41**, 696–698 (1951).

VII. DOUBLE MONOCHROMATORS

Two monochromators may be used in series to achieve higher dispersion or greater spectral purity.

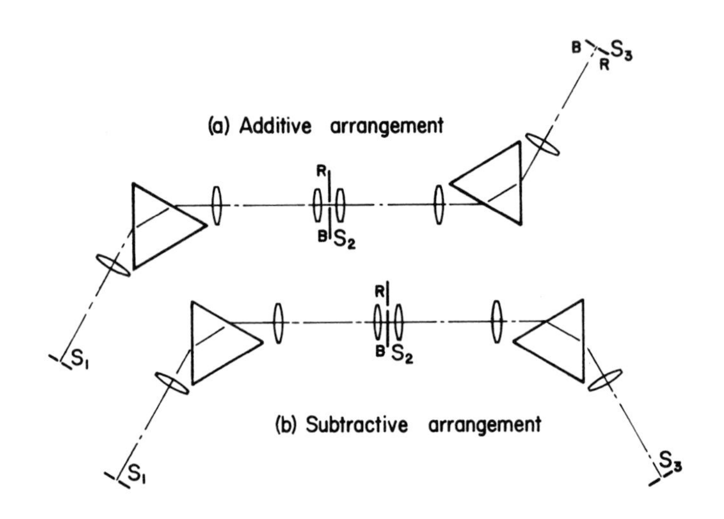

FIG. 39. The two types of double monochromator.

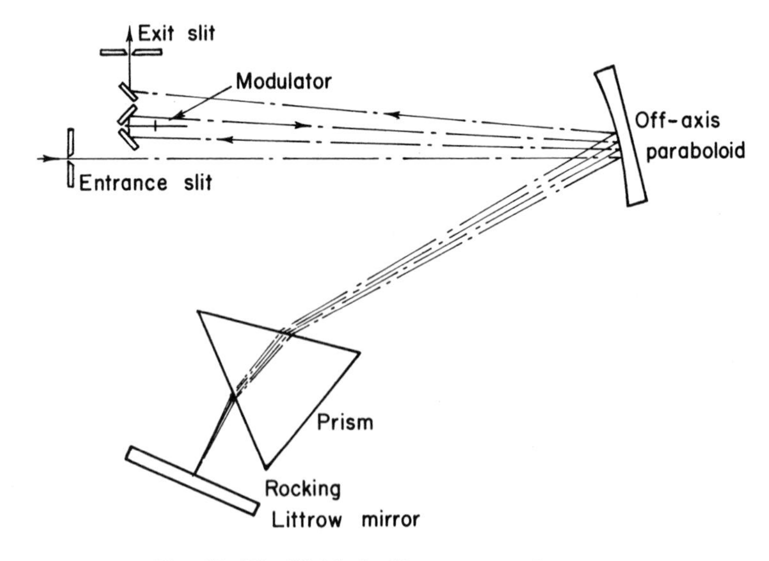

FIG. 40. The Walsh double-pass monochromator.

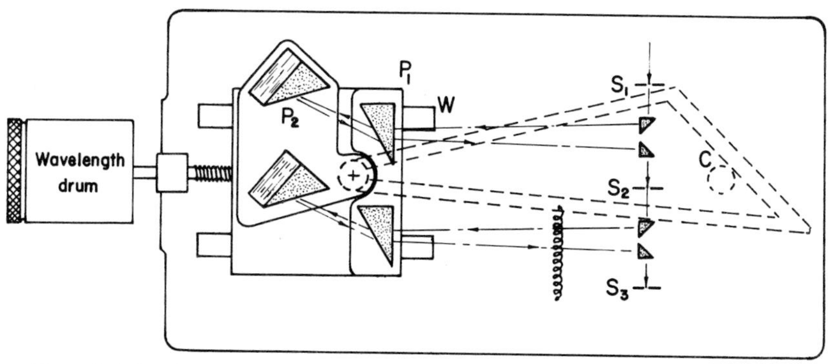

Fig. 41. The Hilger–Müller quartz monochromator. To change wavelength, the prism table is moved on ways (W), and the rear of the table carrying the second half-prisms equipped with roof mirrors (P_2), is suitably rotated by the cam-and-roller (C). The three slits (S_1, S_2, and S_3,) are in a linear array. The instrument covers a wavelength range from 0.2μ to 3.2μ.

1. *Double Dispersion*

Two monochromators may be used in series as in Fig. 39a in such a way that the dispersions of the systems add. From the efficiency equation of Section VIB, it is easy to see that we may expect greater efficiency, provided that the transmittance is high. However, more important, we can expect a great reduction in stray light. It is not necessary that both monochromators be prism systems, or that both be grating systems. Double-dispersion monochromators have been constructed with one prism monochromator and one grating monochromator. Indeed, it is not neccesary that two different systems be used. Figure 40 shows a double monochromator, in which the light is passed twice through the same system. To assure that the light detected has indeed passed the system twice, the light is modulated at the center slit and the detector system is arranged so that only modulated light is recorded. Another classical additive double monochromator is the Hilger–Müller monochromator shown in Fig. 41.

2. *Zero Dispersion*

In Fig. 40, a right-angle corner mirror is used at the midpoint of the system. If only a single mirror were used, we would not have a doubly-dispersed spectrum, but rather one which was entirely nondispersed—a zero-dispersion system.

Consider the zero-dispersion double monochromator shown in Fig. 39b. At the middle focal plane of such a monochromator, we have an entire spectrum displayed. If we did not isolate any particular portion of that spectrum, the second monochromator would act to collapse the spectrum into a single slit image of white light. It is easy to see how this happens by imagining that white light is introduced into what was the exit slit. We see then that the spectrum produced by this inverted procedure is identical to that produced by the direct procedure, the letters B and R at the middle slit indicating blue and red, respectively.

There are a number of advantages of the zero-dispersion monochromator in addition to very low stray light. Scanning the spectrum can be rapid because only a lightweight slit need be moved. In addition, a spectrum of any desired energy distribution can be produced at a fixed exit slit by an appropriately chosen masks or set of apertures at the middle focal plane. In any double monochromator, a field lens must, of course, be placed at the middle focal plane to image the aperture of one monochromator into the aperture of the other.

VIII. TIME-RESOLVED SPECTROSCOPY

In connection with the spectroscopic analysis of some problems of chemical kinetics, it is useful to scan the spectrum rapidly. The zero-dispersion double monochromator is useful here because the spectrum can be scanned rapidly by moving only a lightweight slit. In this geometry, the wavelength range is easily scanned with an oscillating mechanical system. In general, however, for high-speed applications, rotating mechanical parts are preferred. One geometry by which this high-speed scan can be achieved is shown in Fig. 42. The Littrow mirror is rotated about the axis shown. The reflection from the right-angle mirror removes the vertical component of the beam limitation imposed by the rotation of the Littrow mirror.

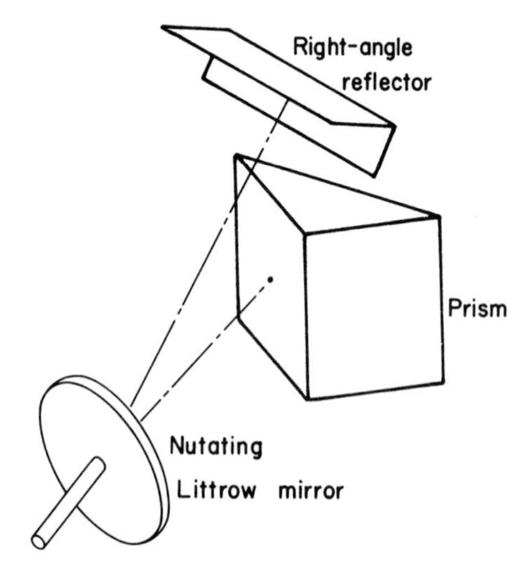

FIG. 42. Time-resolving spectrometer.

CHAPTER 4

Spectrophotometers

WALTER G. DRISCOLL
St. Vincent Hospital, Worcester, Massachusetts

I. INTRODUCTION

Histories[1] of scientific growth relate that around 1672 Sir Isaac Newton presented the first public accounting of his experiments with a glass prism, sunlight, and other radiant sources. These initial experiments gave birth to spectroscopy and spectrophotometry, and they stimulated interest in instruments constructed to explore the significance of the visible spectrum and also of the invisible regions adjacent to it.

---

[1] J. H. Jeans, "The Growth of Physical Science," pp. 185–187. Premier, New York, 1958.

For many years, spectrophotometry was performed almost entirely by visual matching of two monochromatic comparison fields, the wavelength progressing by narrow steps until the whole visual spectrum had been covered. Excellent summaries of visual spectrophotometry have been given by the Optical Society of America Committee on Spectrophotometry for 1922–23[2] and later by Gibson.[3] In the ultraviolet region, a quartz spectrograph was generally used as a detector, with a rotating logarithmic sector in front of the slit to act as a beam attenuator; by this means a photometric match could be secured at each wavelength throughout the spectrum.[4] For spectrophotometry in the infrared, electrical heat detectors were employed, although their low sensitivity was a severe handicap.

Contemporary spectrophotometers are integrated systems for examining or analyzing luminous sources and materials as a function of spectral information. Generally, such systems consist of a source of radiant energy, a monochromator, a detector, and a recorder. Furthermore, provisions usually exist, either before or after the monochromator, for inserting a sample for analysis into the beam of radiant energy.

Early spectrophotometric configurations, associated with such names as Lummer-Brodhun, König-Martens, and Hilger-Nutting,[5] have certainly influenced and stimulated the development of the sophisticated modern systems. Photoelectric recording has largely replaced visual detection, primarily because of the restricted sensitivity of the eye and the time and fatigue associated with the visual approach. Although the newer adaptations may take on various forms and geometries, very often, features of the visual designs continue to exist. One of the earliest automatic spectrophotometers was the well-known General Electric Corp. instrument designed by Hardy.[6]

An ideal source of radiant energy for use in spectrophotometers would be one with uniform intensity throughout the entire region of its application, or perhaps with supplementary intensity at those wavelengths where the transmittance of the optical elements or the sensitivity of the detector is reduced. Generally speaking, tungsten-filament sources have proved quite adequate in the visible region of the spectrum insofar as continuous radiation is concerned. If a line or a band source is desired in the visible

---

[2] *J. Opt. Soc. Am.* **10**, 169–241 (1925).

[3] K. S. Gibson, *J. Opt. Soc. Am.* **24**, 234–249 (1934).

[4] G. R. Harrison, R. C. Lord, and J. R. Loofbourow, "Practical Spectroscopy," pp. 326–361. Prentice-Hall, Englewood Cliffs, New Jersey, 1954.

[5] J. W. T. Walsh, "Photometry," pp. 355–361. Constable Press, London, 1963.

[6] A. C. Hardy, *J. Opt. Soc. Am.* **28**, 360–364 (1938). The construction of the instrument has been described by J. L. Michaelson, *J. Opt. Soc. Am.* **28**, 365–371 (1938). See also Harrison, Lord, and Loofbourow,[4] pp. 406–410.

region, gas discharge sources such as mercury, xenon, or neon can be resorted to. Gas sources are also quite adaptable for use in the ultraviolet region of the spectrum,[7] when tungsten sources become inadequate.

Tungsten sources find further utility in the near-infrared region of the spectrum, but in the middle and far infrared regions, they give way to globar elements and Nernst glowers,[8,9] which are rich in these energies. Furthermore, these latter infrared sources need not be housed in enclosures which might show total or selective absorption of the energy being discharged.

It should be noted that all sources of radiant energy are inefficient generators of luminous energy because a large portion of their input energy is released thermally. Since this heat cannot readily be utilized in the photometric system, it must be appropriately eliminated either by conduction to adjoining heat sinks or circulating coolants, such as water, or by convection in a forced air stream.

The radiant energy from the source, interacting with material samples or the instrumentation or the environment associated with both, may produce unwanted conditions of scattering, fluorescence or phosphorescence, electrical or mechanical noise, etc. Their elimination is one of the major engineering tasks associated with spectrophotometer design.

II. DISPERSIVE AND NONDISPERSIVE SYSTEMS

The heart of a spectrophotometer, its monochromator, is the means by which the source radiation is sorted out into discrete wavelength bands. This can be accomplished by using the selective absorption or the transmission properties of filters, or by employing the dispersive properties of prisms and gratings. Each method has its advantages and disadvantages. Analytical requirements should therefore dictate the suitability of a particular choice.

A. Filter Spectrophotometers

Optical filters, like electrical filters, can be low-pass, high-pass or bandpass. Furthermore, the bandpass type may have broad characteristics or quite narrow band properties. All of these types are occasionally used in spectrophotometers. For example, the low- and high-pass properties find application in grating instruments as order-sorters. They may also be used

---

[7] L. R. Koller, "Ultraviolet Radiation." Wiley, New York, 1952.

[8] H. L. Hackforth, "Infrared Radiation," p. 31. McGraw-Hill, New York, 1960.

[9] W. Brügel, "An Introduction to Infrared Spectroscopy," p. 104. Methuen, London, and Wiley, New York, 1962.

in fluorescent or phosphorescent measuring instruments to isolate the ultraviolet exciting radiation from the reemitted radiation of a different wavelength. These applications, however, are ancillary to our main interest in this section.

Narrowband filters are necessary for wavelength isolation unless very simple spectra are being considered. Simple spectra, such as are encountered in the determination of sodium, calcium, potassium, etc., in the blood and body fluids for clinical purposes, can be adequately separated using optical filters of reasonably broad bandwidths ranging from 50 to 500 Å.

Filters (see Volume I, Chapter 3 of this series) are available in the form of natural or artificial glasses, plastics, or other transparent media. Alternately, they may be provided by depositing thin films or a series of such thin films of different compositions and physical properties (i.e., indices of refraction) onto a transparent substrate (see Volume I, Chapter 8 of this series). The complexity of the serial film structure generally determines the narrowness of the filter bandpass. Filters of this type depend upon the reenforcement or cancellation of various wavelengths of light in passing through them, and, accordingly, they are referred to as interference filters. The current state of the art in evaporating multiple layers of alternately high and low index materials yields spectral bands as narrow as 1 or 2 Å wide with a transmittance of 50–80%. Again, depending upon the spectra being analyzed, this may be a very respectable value, particularly when one realizes that a large radiant flux can be available through a filter 1 or 2 in. on a side or in diameter. A spectrophotometer designed with one or more such filters in a fixed optical array, or mounted on a rotatable filter wheel to interrupt the source radiation sequentially, may be small in size and possess simplicity of construction.

B. Prism Spectrophotometers

For many years, prism instruments (see Chapter 3) monopolized the analytical scene, since original or replica reflecting or transmission gratings were not generally available at a reasonable cost.

Prisms made from a variety of materials are being used extensively even today. Considerable spectral purity can be realized with an instrument using them. If care is exercised in their selection so that they are large enough, properly chosen for the wavelength range of interest, adequately protected from the ill effects of their environment, etc., they may be the most appropriate selection for a given family of applications. Since the index of refraction of materials used for refracting optics such as lenses and prisms is generally not linear, this property adds complexity to a prism spectrometer system in the form of mechanical devices to provide output readings which are linear in wavelength and slit-width adjustments, to afford adequate analytical energy of known spectral quality. The transmission properties and the suitability of readily available materials for use in the ultraviolet, visible, and infrared regions of the spectrum leave a great deal to be desired. Broad wavelength coverage often requires a selection of several prism materials. Figure 1 shows the monochromator section of such

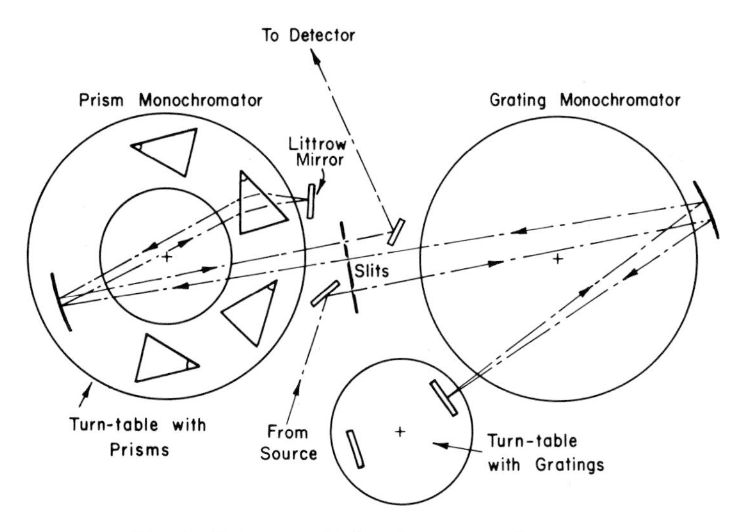

FIG. 1. Unicam multiple-prism spectrophotometer.

a multiple-prism spectrophotometer manufactured by Unicam Instruments, Cambridge, England. The prism monochromator, which has four interchangeable prisms, is ganged with a grating monochromator having two interchangeable gratings by cams linear in wave number and driven by a common shaft.

C. GRATING SPECTROPHOTOMETERS

Wide-range grating units are now available commercially, usually as modified versions of prism instruments. Advantage is taken of the fact that reflection gratings, both plane and concave, are now being manufactured to a variety of specifications, and that, when used in conjunction with reflecting optics, the shortcomings of refracting systems due to absorption losses, aberrations, etc., can be traded off for high transmission, high optical speed, high dispersion, and consistently high resolution. Due to the high dispersion of gratings, slit widths can be larger. This makes the slit design less critical with regard to quality and alignment of the slit jaws. Moreover, the effect of small amounts of optical aberration in the grating collimating system is smaller with wide slits, so that high-aperture spherical mirrors may often be used, even with gratings of large area.

The wavelength scale associated with a prism instrument, as previously mentioned, is dependent upon the index of the prism materials and the associated optics. Hence the calibration is a function of the temperature. This situation is not as signficant in the case of a grating instrument. Moreover, the angular rotation of the grating to cover a given wavelength range is considerably higher than that of a prism, so that the wavelength measurements on a grating instrument may be made with consistently higher precision.

Grating instruments are not without their shortcomings, however. Dispersion of radiation by a grating results in multiple orders of spectra, and hence order-sorters in the form of prisms or filters are required if wide wavelength coverage is anticipated. Such coverage often dictates the use of multiple gratings, which again introduces mechanical complexity into the array. See Figs. 1 and 2 for examples of these situations.

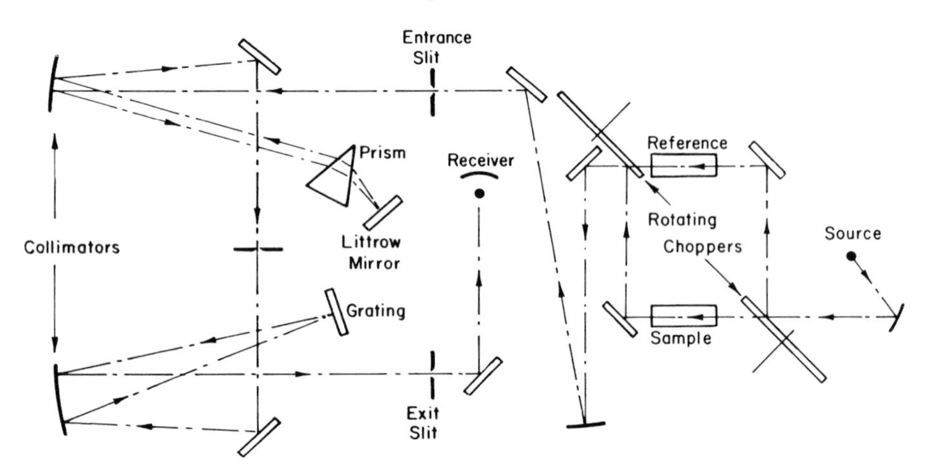

FIG. 2. Beckman prism-grating spectrophotometer.

D. NONDISPERSIVE SYSTEMS

Filter systems, discussed in Section IIA, fall into the classification of nondispersive systems. There is an alternate approach to the application of nondispersive instruments, in which such filters, prisms or gratings are not used; instead, total radiation is passed through a sample and a reference, and variations in the transmitted radiation are correlated with a change in sample concentration. Analysis is possible because the presence of a material component is evidenced by the presence of a specific absorption band, whose amplitude is related to its concentration at a particular wavelength. Systems have been developed to make continuous measurements at two or more wavelengths, and such instrumentation finds its place in process control or where continuous analyzers are applicable.[10] The mechanical and optical simplicity of these units leads to low cost and the maximum in reliability and dependability.

III. OPTICAL, ELECTRICAL, AND MECHANICAL SYSTEMS

Although they may not be universally recognized as such, spectrophotometers currently in use or available commercially are sophisticated engineering systems. In general, a comparative analysis of all available spectrophotometers reveals that they may be categorized into a limited number of types, such as open-loop systems, closed-loop systems, single- or double-beam systems, and single- or multiple-pass systems. These variations can be illustrated by block diagrams which differ in only a few aspects.

A. Open-Loop, or Single-Beam, Systems

The subsystems of single-beam, or open-loop systems can be represented by the block diagram in Fig. 3. Such a system is typical of several that may be employed to obtain absorption, reflection, transmission, or fluorescent spectra as meter readings or as graphical recordings. Because of their design and organization, the analytical spectrum which is of

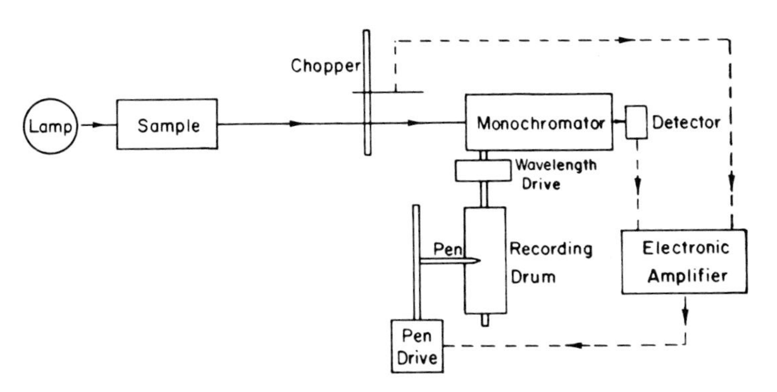

Fig. 3. Single-beam, or open-loop, system.

[10] J. W. O'Laughlin and C. V. Banks, *in* "The Encyclopedia of Spectroscopy" (G. L. Clark, ed.), pp. 19–33. Reinhold, New York, 1960.

primary interest has superimposed upon it the energy and wavelength distribution characteristics of the instrument source and detector as well as the atmospheric and environmental factors which influence the light path. In principle, a single-beam spectrophotometer, used properly, can provide analytical results equivalent to those of the more convenient double-beam spectrophotometer to be discussed in the next section. Such use requires that a control spectrum, which represents the source, detector, environment, etc., but not the influence of the sample being analyzed, be obtained or provided.

When such a control spectrum is compared with the control-plus-sample spectrum, the spectral contributions due to the sample alone may be discerned. The equivalence of analytical data obtained in this way with data from double-beam systems depends on many factors, which for the most part are influenced by time-variant conditions in the system. This comes about because the spectra in a single-beam instrument are run serially. If the source of exciting energy is not properly regulated and driven from a stable supply, time variations in its intensity distribution and, even more seriously, its energy-wavelength distribution will distort the final calculations and spectral data derived therefrom. Similarly, detector response, which again may be dependent upon power-source stability, fatigue characteristics, temperature, etc., can lead to erroneous interpretations. Previous sections and chapters of these volumes have detailed the properties of sources and detectors, so suffice it to say here that a single-beam mode of operation requires understanding and some engineering refinements which can be set aside in double-beam arrangements.

The "uvispek" being offered commercially by Hilger and Watts, Ltd. of London, England, is a typical single-beam spectrophotometer which is offered with a tungsten-filament light source or a hydrogen-discharge source or both. In the wavelength region from 1850 Å to 10,000 Å, a silica prism is employed; in the region from 3900 Å to 10,000 Å, a glass prism may be selected. Two photomultipliers are specified to cover this total spectral range.

Commercially available instruments are cited here to exemplify the material being discussed. A particular choice is not intended to imply that other manufacturers do not offer equivalent or similar equipment.

B. Closed-Loop, or Double-Beam, Systems

A block diagram of a double-beam instrument, or a closed-loop system, is shown in Fig. 4. This configuration automatically refers all of its spectral data to a zero curve and yields meter readings or spectrograms on a chart of absorbance or transmittance *vs* wavelength.

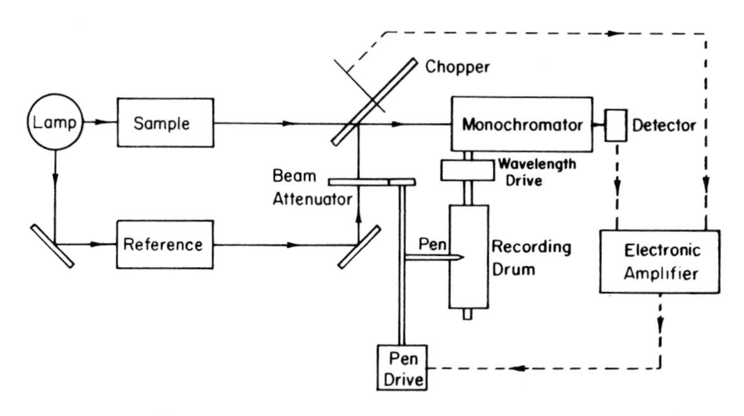

FIG. 4. Double-beam, or closed-loop, system.

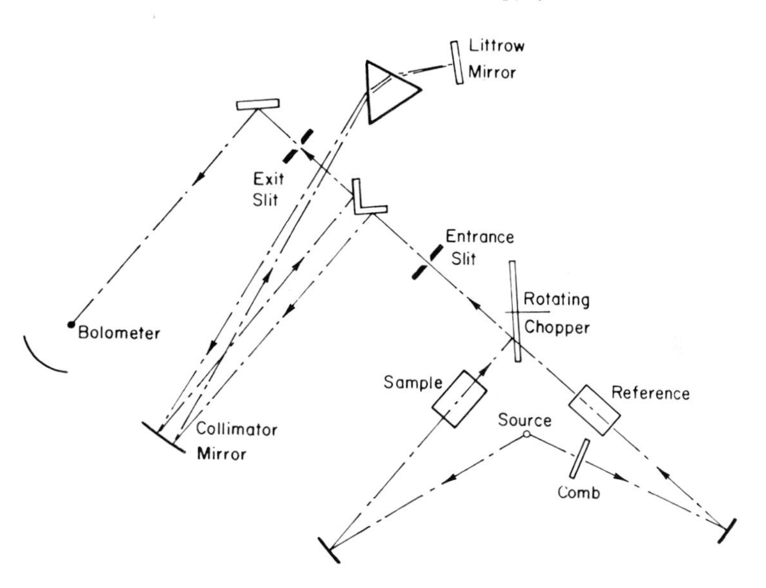

FIG. 5. Baird-Atomic double-beam spectrophotometer.

The optical diagram presented in Fig. 5 illustrates the arrangement of components in Baird-Atomic's[11] Model NK-1 infrared prism spectrophotometer. It is noted that the radiation from the globar source (a low-resistance rod of silicon carbide with a blackbody peak at 1.8–2.0 μ at 1200°C) is divided into two beams by two source mirrors. One beam passes through the sample under study and the other serves as a reference or

[11] Baird-Atomic, Inc., 33 University Road, Cambridge, Massachusetts.

standard. At certain points in the infrared spectrum, the sample absorbs energy from the sample beam while the reference beam remains unchanged. When this inequality in the two beams occurs, it is detected by the system, and a shutter, or "comb," is moved into the reference beam. The "comb" blocks a portion of the reference beam and decreases its intensity to match the intensity of the sample beam. When the "comb" has moved to a point where the two beam intensities are equal, the detector output becomes zero—this point is a null. Instruments employing this technique are called "null-indicating" systems. The amount of "comb" movement necessary to equalize the two beams is a measure of the transmittance of the sample under study, and its movement is accordingly linked with the recording-pen movement.

A double-beam instrument employing the optical "null" principle tends to limit analytical errors caused by variations in the performance of the source, detector, and electronic components. Furthermore, differences in successive recordings are minimized because changes in ambient or instrument temperature, and environmental absorptions in the light paths between the source and detector, affect both beams in a similar way.

Double-beam systems are not restricted to use in the infrared, for by selection of a suitable source, maximum radiation of a desired type (i.e., ultraviolet, visible, or infrared) can be provided.

The scanning of the spectrum is usually accomplished by moving some component of the system such as an optical grating, prism, mirror, or exit slit and detector; while the dispersion and the programming of the radiation through the system is provided by filters, or by a prism or a combination of prisms, and, in some cases, by a grating, combinations of gratings, or combinations of prisms, gratings, and filters, or possibly using even more esoteric combinations or mirrors, cams, and lever arms.[12] The dispersed radiation is then directed to a suitable detector, and the resulting electronically-amplified signals presented to a meter or to a recorder.

Figure 2 is an optical diagram of Beckman's[13] IR-7 prism-grating spectrophotometer. It is illustrative of an advanced system which incorporates a fore-prism and a grating in a double-beam system. The grating provides greater dispersion and greater energy than is generally available using a prism alone. The fore-prism serves as an order-sorter for the grating, and contributes to the initial dispersion in all orders of the grating. It is materially effective in reducing stray light in the system even at wide slit widths. In this particular array, three precision cams drive the prism, the grating, and the triple-slit system. It should be apparent that such

---

[12] See, for instance, W. Brower, U.S. Patent 2,856,531 (Oct. 1958).

[13] Beckman Instruments, Inc., Fullerton, California.

coordinated mechanical movement requires the application of the most modern engineering skills to provide reproducible results without excessive noise from overdamped or underdamped servo systems, backlash in gear trains, distortion of mechanical parts, or some combinations of these.

Another unusual system for use in the ultraviolet, visible, and infrared is the Cary[14] model 14, which is reportedly designed with an engineering philosophy which recognizes that precision bearings will wear, electronic tubes will deteriorate, and optical components will decline in transmission or reflection with age and use.

Figure 6 shows the path of ultraviolet and visible radiation through the

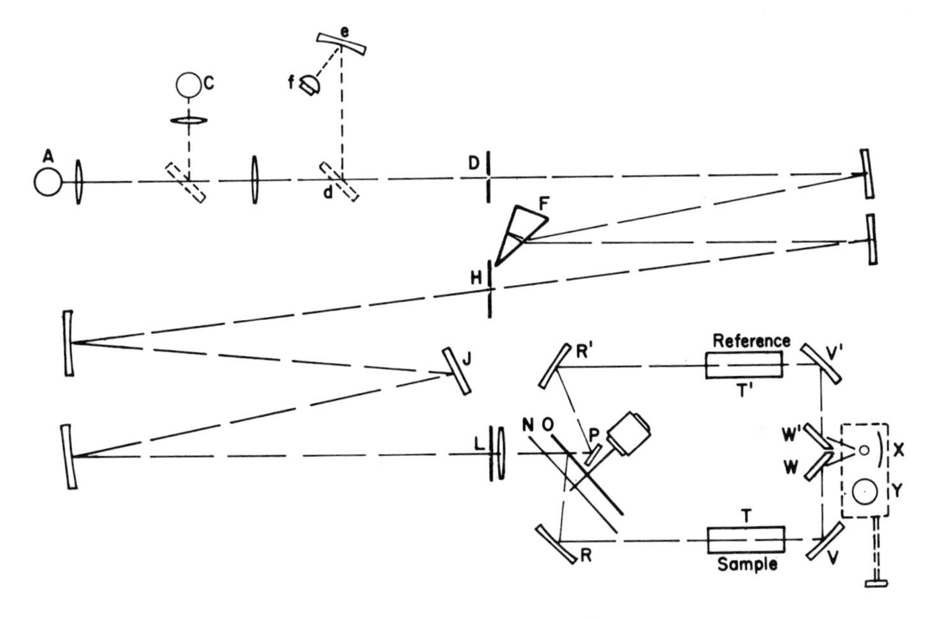

FIG. 6. Cary prism-grating spectrophotometer.

instrument. Radiation from the hydrogen lamp (*A*) or the tungsten lamp (*C*) enters the double monochromator through slit (*D*). It is dispersed by the prism (*F*) and the grating (*J*); (*H*) is an intermediate slit of variable width. Monochromatic radiation leaves through slit (*L*) and is sent at 30 cps alternately through the reference cell (*T'*) and the sample cell (*T*) by means of a rotating semicircular mirror (*O*) which is driven by a synchronous motor. It alternately reflects the light beam to the mirror (*R*), or allows it to pass through to the mirrors (*P*) and (*R'*). On the shaft of the

[14] Applied Physics Corporation, Monrovia, California.

synchronous motor is also a chopper disk (N) which produces a dark interval between successive half cycles of the alternation. The beam through the reference cell, consisting of pulses of monochromatic light at 30 cps, and a similar beam through the sample cell, are directed to the photocell(X) by the mirrors (V, V' and W, W'). The light pulses of the two beams are out of phase with each other, so that the photocell receives light from only one beam at a time. The photocell signals are measured and compared with a "flicker-beam photometric system."

The flicker-beam system uses a single photomultiplier. Synchronized with the alternate pulses of sample and reference radiation seen by a receiver is a train of photoelectric timing signals. These signals introduce the attenuation of the recorder slidewire during the reference interval, but allow the sample signal to be passed to the comparison circuit unattenuated. The signals are then compared, and any difference causes the pen motor to readjust the slidewire until equality is achieved.

When the instrument is used in the infrared region, the tungsten light source (Y) is slid into the place of the phototube (X). Radiation from the source is divided between the sample and the reference cells, chopped, and sent into the monochromator through the slit (L) by the chopper and mirror. Alternate pulses of sample and reference light are dispersed by the monochromator and directed to the lead sulfide cell (f) by mirrors (d) and (e). By this means, any nonmonochromatic infrared radiation thermally emitted by surfaces in the chopper unit, which would appear in the light beams reaching the lead sulfie cell if it were placed at (X), is eliminated by the monochromator from the light to be measured.

Another well-known null-type infrared spectrophotometer is that of the Perkin-Elmer Corporation.[15]

C. Derivative Spectrophotometry

The preceding discussion relates primarily to standard configurations of spectrophotometers. Giese and French[16] and French *et al.*[17] showed that in the case of overlapping bands in a spectrum, more discerning methods of measurements than transmission and absorption may be of great advantage. They suggested the measurement of the derivative of the transmittance, etc., and called these techniques "derivative spectrophotometry."

The techniques for determining the derivative fall into two general

[15] J. U. White and M. D. Liston, *J. Opt. Soc. Am.* **40**, 29–41, 93–101 (1950). See also "The Encyclopedia of Spectroscopy,"[10] p. 459.

[16] A. T. Giese and C. S. French, *Appl. Spectry*. **9**, 78–96 (1955).

[17] C. S. French, A. B. Church, and R. W. Eppley, Carnegie Inst. of Washington, Yearbook, No. 53, p. 182, 1954.

classes; those which operate on the output signal of the spectrophotometer, and those in which the derivative is obtained through operations within the spectrophotometer itself. Examples of operations on the output signal are: electronic differentiation, mechanical differentiation of a recorder-pen movement, and measurements of the slope of a spectrophotometric curve. Examples of operations within the optical system yielding derivative measurements are: vibration of entrance or exit slits, spectrum shift by beam-position modulation-chopper selection of wavelengths, or two separate monochromators. Gunders and Kaplan[18] have demonstrated the equivalence of the two classes, and they have pointed out the added system complexity that results from such designs. Mechanical, optical, and electronic engineering problems are much more significant than for the more conventional configurations.

D. The Signal-to-Noise Concept

The signal-to-noise concept, particularly in recent years, has taken on new and significant meaning. The signal is the information that one desires and hopefully can conserve within a system; the noise is the unwanted, yet persistently present information which tends to degrade or limit the availability of the signal. The signal is selectively introduced or generated for analytical purposes; noise invariably finds its way into the analytical system in spite of care and selectivity.

Each subsystem in any type of instrumentation may contribute noise in a variety of forms, i.e., mechanical, optical, electrical, etc. An illuminating source may fluctuate in brightness or spectral distribution as a function of time. Likewise, false radiation in the form of stray, scattered, diffracted, or fluorescent light may degrade the performance of a system. Furthermore, previous chapters (e.g., Electrooptical Devices and Infrared Detectors) and many familiar texts on electronic circuitry and systems[19–22] stress the importance of minimizing the inherent noise in each subsystem by engineering around the problem by using tried and tested engineering practices.

As an example of an inherent problem in null-indicating double-beam systems, one should be aware of the relationship between amplifier gain,

---

[18] E. Gunders and B. Kaplan, *J. Opt. Soc. Am.* **55**, 1094–1097 (1965).

[19] J. D. Strong, "Concepts of Classical Optics," pp. 470–474. Freeman, San Francisco, California, 1958.

[20] M. J. E. Golay, *J. Opt. Soc. Am.* **46**, 422–427 (1956).

[21] R. C. Jones, *Advan. Electron. Electron Phys.* **5**, 1–96 (1953).

[22] K. M. van Vliet, J. R. Whinnery, S. Rice, and D. C. Forster, *in* "The Encyclopedia of Electronics" (C. Susskind, ed.), pp. 546–555. Reinhold, New York, 1962.

speed or recorder-pen response, and noise. The pen-balancing system is a closed servo loop operating through a light beam, a light-beam attenuator such as a "comb" or wedge, a detector, amplifier, and balancing motor. It has the general properties of a servo loop whose response characteristics depend upon the loop gain, which is often adjustable by the amplifier gain control and by the monochrometer slit width.

Although it is not appropriate here to discuss in detail the ways and the means of limiting undesirable noise, it is significant to point out that good design practice in spectrophotometers, particularly if cost is not a serious design limitation, calls for: sources, electronic amplifiers, and read-outs that are driven by stable and well-regulated power supplies, well-baffled optical paths, vibration-free or chatter-free mechanical or motor assemblies, and possibly even cooled detector systems.

Little need be said of the importance, too, of the proper selection of optical materials, suitably coated if necessary; of appropriate slit widths; of electrical circuitry and compatible detector time-constants; and of realistic spectral scanning times, in order to obtain analytical satisfaction. These factors also contribute to the total noise, and ultimately to lowering the signal-to-noise ratio of the system.

A signal-to-noise ratio in excess of one ($S/N > 1$) is desirable in high-performance spectrophotometers. Signals which must be extracted from data with $S/N < 1$ require costly and unusual techniques which are not generally practical in analytical spectrophotometers for routine or even research analysis. Until the cost and the complexity of low-signal, high-noise instrumentation is reduced, it is preferable to seek an alternate and more appropriate technique.

Fortunately, commercial spectrophotometers of superior design are not without merit. Many industrial suppliers boast, and for the most part deliver, instrumentation with very respectable specifications including S/N values well in excess of 1 for certain prescribed standards and concentration minimums.

E. System Environment

Somewhat related to the previous discussion of signals and noise is the environment in which a spectrophotometer must perform. For example, if water vapor or some other undesirable gas (e.g., CO_2) in the instrument environment limits its performance, suitably dried air, or an instrument purged with a noninterfering dry gas (e.g., argon or nitrogen), or even evacuated systems, are available and realistically engineered for cost which is not extreme in the light of the acquired new capability.

F. Reflection, Fluorescence, Phosphorescence, and Raman Scattering

Spectrophotometers are used for measurements other than absorption and transmission. In particular, one should consider the measurement of specular and diffuse reflectance, emissivity, fluorescence, phosphorescence, and Raman scattering, at least briefly, because each adaptation to a spectrometer system requires particular engineering attention.

Reflection measurements have been made for many years in the visible and near-infrared regions of the spectrum, typical instrumentation being the Hardy recording spectrophotometer manufactured by the General Electric Company.[6] In this instrument, a light-integrating sphere provides a ready means of making specular and diffuse reflectance measurements of samples of a large variety of shapes and sizes. Measurements are made in comparison with a white, calibrated reflection standard such as magnesium carbonate or a white vitrolite tile. Subsequently, other manufacturers either built this feature into their units or offered it as a reflectivity attachment. In fact, and where applicable, the reflectivity attachments often incorporate heater arrangements, so that the measurements can be made over a relatively large range of temperatures (i.e., 400–1200°C), thus making it possible to measure the emissivity of a sample. Since the engineering complexity and details of related reflectivity attachments are beyond the scope of this chapter, it is suggested that the interested reader consult several of the prominent manufacturers of spectrophotometers for details on their individual offerings.

Suffice it to say that such measurements are not wholly satisfactory throughout the entire light spectrum, partly because of the fact that the proper materials are not available (i.e., a perfectly-white reflecting source for infrared specular and diffuse measurements), and partly because the introduction of controlled temperatures and heat sources into a complex optical–mechanical array is not without its own difficulties.

Fluorescent and phosphorescent attachments have also been made for many of the commercially available spectrophotometers. Furthermore, several companies offer special instrumentation, which generally consist of a double spectrometer, one to produce suitable narrow bands of ultraviolet light for excitation and the other to analyze the fluorescent or phosphorescent light emitted from the excited sample. The problems inherent in these adaptations are not distinct from those found in engineering other types of spectrophotometers. Most commercial systems for the measurement of fluorescence and/or phosphorescence operate as single-beam instruments rather than as double-beam instruments. This becomes about because of the difficulties inherent in attempting to correlate the output of the ultraviolet-emitting source with the level of emission of the fluorescence as a function of wavelength. Consequently, one feature of a double-beam system, namely, the removal of the influence of the exciting source, is not readily accomplished, although several roundabout approaches to double-beam operation have been attempted here and in England.

It probably should be mentioned, however, before leaving the discussion of fluorescence, that care must be exercised to eliminate the interfering effects of fluorescence from normal absorption, transmission, and reflection measurements, since a luminous contribution of this type can affect analytical results seriously if it is present. The judicious use of filters or light baffling may be necessary expediencies.

Although there are not many spectrophotometers available commercially for making measurements of Raman scattering, the Cary[14] instrument Model 81 is an excellent example. Raman instrumentation might be considered as probably the ultimate in spectrophotometer engineering. The Raman effect depends upon an inherently inefficient physical process, perhaps one thousandfold less efficient than the Rayleigh molecular-scattering process. Thus a high-intensity light source and an expensive monochromator of exceptional light-gathering power must be coupled to an efficient optical train.

IV. SCANNING

In spectrophotometers, as in other information-yielding systems, assuming that they are not energy-limited, optically inferior, or noise-limited, the spectral information that can be collected is proportional to the time taken to accumulate and/or scan it. For this reason, several engineering tradeoffs may be realistically considered.

A. Continuous Scanning as a Function of Time and/or Wavelength

These scanning modes are perhaps the simplest and the most readily designed, since a mechanical drive can be made to function reliably and reproducibly with a minimum of complexity. A one-to-one relationship between time and wavelength is quite readily effected by a simple gear transmission, lever systems, or, at most, calibrated cam linkages which retain calibration for extended periods.

B. Stepwise and/or Differential Rate Scanning

These modes generally introduce greater mechanical and electrical complexity into a spectrophotometer. Their presence is often desirable, however, to afford extended time intervals for scanning portions of spectra containing numerous inflections while quickly passing over spectral structure of little information content. Mechanical and, particularly, electrical tolerances inherent in the servo subsystems which are employed for such purposes contribute to the complexity and the cost of variable-velocity scanning. Considerably more attention must be applied to their original design, to their analytical use, and also to their calibration and maintenance.

C. Ratio Detection and/or Continuous Analysis

A third, though only a pseudo method of scanning also merits consideration if an application or a design is limited to comparing analytical results at two, or a limited number, of wavelengths. This is the direct· ratio system, wherein the response to a sample is detected at different wavelengths and reported as some function (e.g., difference, ratio, or log of ratio, etc.) of the detected signals. This technique or mode is relevant in relating the concentration of ingredients in mixtures, or ions in solutions. Process-control spectrophotometers and continuous plant stream analyzers are often designed in this way. If nondispersive subsystems are combined with ratio measurements and readout, extremely simple yet effective control instrumentation may result.

V. OUTPUT DATA AND RECORDING

A. Analog Processing

The scanning modes previously discussed generally yield a continuous flow of transmittance or absorbance data as some function of time or wavelength. Although the data or information may not be forthcoming at a constant rate, it appears near the output of a spectrophotometer in continuous or analog form. In much of the instrumentation in existence today, it is ultimately presented or readout in continuous graphical or analog form.[15, 23–25]

B. Digital Processing

An analog presentation of information need not be employed, however, since there is some, if not considerable, merit in intercepting the analog information at some intermediate subsystem in the spectrophotometer and

---

[23] W. S. Baird, H. M. O'Bryan, G. Ogden, and D. Lee, *J. Opt. Soc. Am.* **37**, 754–761 (1947).

[24] N. Wright, and L. W. Herscher, *J. Opt. Soc. Am.* **37**, 211–216 (1947).

[25] A. Savitzky and R. S. Halford, *Rev. Sci. Inst.* **21**, 203–212 (1950).

sampling it or digitizing it prior to ultimate signal processing and readout.[20, 26] Ease of signal amplification, improvement in subsystem S/N characteristics, information coding, and, possibly, the convenience of ultimately feeding spectral data to a process control computer or an automatic sample matrix analyzer, or an alternate closed-loop servo system, could well justify its use.

VI. PERFORMANCE

The performance of a spectrophotometer[27] is a function of numerous instrumental parameters. The wavelength repeatability, the wavelength accuracy, the photometric repeatability, spectral slit widths, and cell or sample constants are particularly significant. Many of these factors are interrelated. Furthermore, they may be influenced by the resolution, the linearity, the stray energy of the system, etc. In addition, the verification of instrumental performance implies that tests or analyses are made using proper operating conditions and that factors, such as the ambient temperature, the response time of the system, the signal-to-noise ratio, and the scanning speeds are compatible with their original design concepts.

A. Wavelength Repeatability and Accuracy

The wavelength of radiant energy being used in a spectrophotometer at a particular instant is generally given by a dial reading in a manual unit, or by a chart presentation in a recording unit. Pure compounds or prepared mixtures that have accurately known wavelength characteristics should be employed as reliable standards. If a spectrophotometer consistently provides the same dial readings or chart indications for a specific spectral position, as evidenced by an absorption or emission band or line, then it possesses the quality of wavelength repeatability. The wavelength accuracy is related to the deviation of the average wavelength reading at a test band or line from its known wavelength.

There are numerous possible reasons for a lack of wavelength repeatability. Changes in the temperature of various parts of the system or its subsystems resulting from not allowing adequate start-up time to assure thermal equilibrium, or from the failure to adequately dispose of the heat generated internally, may be the fault. The backlash between gears and in gear trains, looseness in mechanical linkages or eccentricities in rotating

---

[26] D. F. Hornig, G. E. Hyde, and W. A. Adcock, *J. Opt. Soc. Am.* **40**, 497–503 (1950).

[27] ASTM Preprint, Report of Committee E-13 on Absorption Spectroscopy, 66th Annual Meeting, June 23–28, 1963.

motions, and other conditions such as worn parts, frictional variations resulting from inadequate lubrication or accumulations of dirt, are often contributors.

These troubles often evidence themselves quite seriously when scanning speeds are higher than good engineering practices would dictate. Slow or modest scanning speeds may not contribute in a major way to a lack of performance and repeatability.

B. Photometric Repeatability and Linearity

The photometric data obtained by an analyst are of significance only when complete knowledge of the applicability of the physical and chemical laws being employed is known in conjunction with the photometric performance characteristics of his instrumentation.

C. Spectral Resolution and Slit Widths

Resolution is defined as the ratio $\lambda/\Delta\lambda$, where λ is the wavelength of the region in question and $\Delta\lambda$ is the separation of two absorption bands that can just be distinguished.

It should be pointed out again that all operating parameters must be set to the same values in both the resolution test and the analytical methods described previously. Such parameters include the wavelength scanning speed, the response time, the amplifier gain, detector bias, servosystem gain settings, slit widths, and other relevant variables. On meter-reading instruments, one should take adequate time to assure that needle deflections have reached equilibrium. Since resolution and slit-width settings are closely related, the engineer or analyst should take considerable care in selecting these settings. The optimum slit width is determined by the spectral characteristics of the sample and the dispersion of the instrument being designed or used. The narrowest slit width should be applied that will yield an acceptable signal-to-noise ratio. Where instrument resolution is more than adequate, the signal-to-noise ratio should be maximized.

The spectral slit width of an instrument may be calculated by multiplying the mechanical slit widths (measured in millimeters) at each analytical wavelength being considered by the reciprocal linear dispersion (in wavelength units per millimeter) as obtained from the manufacturer's literature. For most instruments, we should mention again, the linear dispersion is a function of the wavelength, and it should be treated accordingly. If the mechanical slit width is less than about 0.1 mm, the value obtained for the spectral slit width may not be significant unless slit curvature and diffraction effects are considered.

D. Other Considerations Affecting Performance

Absorption cells, reflection attachments, and any other fixtures relating to the use of the instrument should neither interfere with nor contribute to the problems of analytical work. Differences in absorption cell length, leakage, etc., should of necessity be avoided. Similarly, effects of polarization resulting from the use of optically-active or birefringent components or samples should be viewed with suspicion. Effects of fluorescence due to samples and materials may be troublesome. Stray radiant energy reaching the detector at wavelengths different from the nominal wavelength band at which an instrument is set must be sorted out and rejected if possible. This last item may take on significant meaning in a design study, and, accordingly, should be considered in detail.

The stray radiant energy is commonly expressed as a percentage of the total energy or power reaching the detector in the absence of an absorbing sample. "Near" stray energy is at wavelengths near that of the nominal wavelength. Its presence may become very significant near sharp, narrow bands, at any wavelength. Stray radiant energy causes observed absorbance values to differ from true values. Usually, the observed values will be lower. It is one of the factors causing curves of concentration vs absorbance to be nonlinear at high values of absorbance. There is normally a definite relationship between the stray radiant energy at one wavelength and the values at all other wavelengths if the same instrument, source, and detector are used; this makes it possible to characterize this difficulty for the instrument in general. The determination is best done at the extreme point in the range where the percentage of stray radiant energy is the greatest.

E. Equipment Maintenance and Reliability

Faulty performance in an instrument which cannot be readily diagnosed demands particular instrument knowledge usually available only to the manufacturers. It is difficult to expect that equipment operating and maintenance manuals will provide answers and repair instructions which are adequate. They can, at most, reflect accumulated experience and anticipated trouble centers. Very often, however, difficulties not previously experienced will occur and require specialized attention. The importance of an awareness of this on the part of a user cannot be overstressed. Individuals with only a cursory familiarity with a unit and all of its intricacies should not undertake even minor repairs without realizing that they may further degrade its performance or contribute to a major service problem. Fortunately, most manufacturers are now offering service and maintenance contracts which provide timely relief at a reasonable cost.

CHAPTER 5

Colorimeters

HARRY K. HAMMOND III

*Optics Metrology Branch, National Bureau of Standards,**
Washington, D.C.

I. INTRODUCTION

In order to discuss colorimeters as optical instruments, it will be desirable to outline briefly the basic principles on which colorimetry is based, and to understand the procedure by which spectrophotometric data are reduced to colorimetric parameters.

Colorimetry, like photometry, requires that radiant flux be evaluated in accord with the spectral sensitivity of the human eye. Instrumental measurements must correlate with what the eye sees. In photometry, only one spectral weighting function is required to evaluate the luminous intensity of a light source. Colorimetry, on the other hand, requires three weighting functions in order to evaluate luminous intensity and chromaticity.

---

* The views of the author are his own and do not necessarily reflect the opinion of the National Bureau of Standards.

The fundamental measurement of any form of radiant flux generally involves geometrical as well as spectral considerations, and colorimetry is no exception. Geometrical considerations are extremely important in the colorimetry of many types of reflecting materials, particularly where the color may change with angle of illumination or view. Fluorescence may add an additional consideration to the choice of illuminating and viewing conditions for measurement.

The designation "colorimeter" is used for several classes of instrument that differ widely in their design and use. Data from a spectrophotometer provide the most complete colorimetric information on a specimen; thus this instrument could be classed as a colorimeter, but usually it is not so considered. The word colorimeter is most frequenctly applied to a class of instruments that is more accurately described by the term color-difference meter because these instruments measure color differences more accurately than they measure colors. In the materials-appearance control laboratory, photoelectric tristimulus colorimeters are rapidly replacing visual instruments.

In the chemical laboratory, we find instruments designated as colorimeters that are used to evaluate concentrations of chemical constituents. These instruments are not colorimeters in the usual sense because they are not designed to evaluate color as such; instead, they provide an index of the amount of some material in solution by measuring the absorption of a characteristic band.

The color temperature meter is a special class of colorimeter designed to provide an index of the temperature of an incandescent body by indicating the temperature of a complete radiator having the same chromaticity.

II. PRINCIPLES OF COLOR MEASUREMENT

A. Basic Premises and the CIE System

Colorimetry is based on several simple premises. These are most easily visualized by thinking in terms of colored lights. The first premise is that the color of any specimen light can be matched visually by adding together appropriate amounts of any three colored lights that one designates as standards, provided that each standard light is truly independent, i.e., one standard light cannot be matched by some combination of the two remaining standards. The second premise is that negative amounts of one of the standard lights may be required, i.e., it may be necessary to mix one of the standard lights with the specimen light to obtain a match with the remaining standards. This procedure of matching colored lights was used by

Wright[1] and others to determine experimentally the amounts of three spectrum lights required to match all spectrum colors. As long as the choice of standard lights is arbitrary, many color systems could be devised.

In 1931, the International Commission on Illumination (formerly abbreviated to ICI, but now more usually designated CIE from its French name, Commission Internationale de l'Éclairage) recommended that color data be expressed in terms of a standard system.[2] In effecting standardization, two additional restrictions were added to the two premises already stated. One of the standard colors was chosen to have a spectral distribution identical to that of the relative spectral sensitivity of the average human eye. The spectral distributions of the two remaining standards were transformed mathematically so that no negative amounts would be required to match any test color. The spectral distributions of the three CIE observer functions $\bar{x}_\lambda$, $\bar{y}_\lambda$, and $\bar{z}_\lambda$ standardized in 1931 for a 2° field of view, are shown as solid curves in Fig. 1. In 1964, the CIE recommended a set of supplementary color-matching functions that apply to a 10° field of view.[3] These are shown in Fig. 1 by dotted lines. The larger visual field is more representative of the viewing conditions encountered in commercial use. To avoid confusion over which set of functions is being used, one speaks of the 1931 CIE Standard Observer Functions or the 1964 CIE Supplementary Observer Functions. In symbolic notation, the subscript 10 is used to designate the larger field.

The colors of reflecting or transmitting objects depend on the spectral quality of the light falling on the objects as well as the spectral character of the objects themselves. For this reason, relative spectral energy distributions for three light sources were standardized by the CIE about 1931.[2] Source A was intended to be representative of incandescent lamp light, source B of average sunlight, and source C of average daylight. Source A was defined as a gas-filled tungsten-filament lamp operated at a color temperature of 2854 K. Source B was defined as source A modified by a two-cell liquid filter of specified composition. Source C was defined in the same manner as source B, but with a filter of different composition. In practice, spectral distributions provide more useful definitions, and in 1964, the CIE adopted this position.[3] Source C has a chromaticity close to that of a Planckian radiator at 6750 K. However, its spectral distribution is not Planckian, nor does it approximate any phase of daylight as closely as desired for present-day colorimetry. For this reason, a series of D

---

[1] W. D. Wright, A redetermination of the trichromatic coefficients of the spectral colors, *Trans. Opt. Soc.* (*London*) **30**, 141–164 (1928–9).

[2] CIE *Proc. 8th Session* Cambridge. pp. 19–24 (1931).

[3] CIE, *Proc. 15th Session* Vienna, 1963, Vol. A, p. 35 (1964).

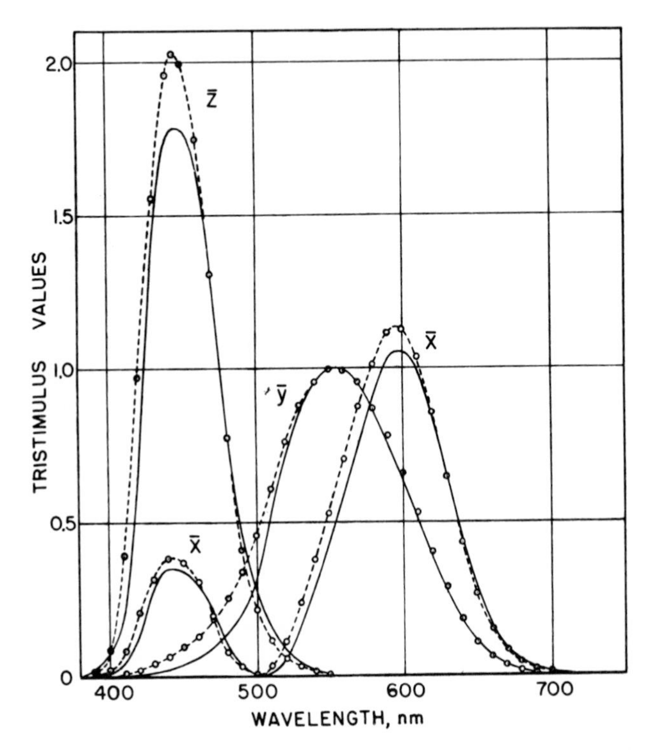

FIG. 1. Comparison of spectral tristimulus values for 1931 CIE standard observer (2° field; solid curves) and 1964 supplementary observer (10° field; dashed curves with open dots). Comparison is based on equal areas under the $\bar{x}$, $\bar{y}$, and $\bar{z}$ curves.

sources representing various phases of daylight has been proposed.[4] These D sources are designated by a subscript representing the correlated color temperature. The CIE has adopted D_{6500} as the preferred average daylight. Where a bluer daylight is desired D_{7500} should be used; similarly, if a redder source is desired, D_{5500} should be selected. Tables giving the spectral distributions of these sources are now available.[5] but physical sources with these spectral distributions have yet to be developed. Standardization of the spectral distributions from other types of sources such as fluorescent lamps has been proposed, but has not yet been adopted by by the CIE.

---

[4] D. B. Judd, D. L. MacAdam, and G. W. Wyszecki, Spectral distribution of typical daylight as a function of correlated color temperature, *J. Opt. Soc. Am.* **54**, 1031–1040, 1382 (1964).

[5] G. W. Wyszecki and W. S. Stiles, "Color Science," p. 9, Table 1.4. Wiley, New York, 1967.

B. The Spectrophotometer as a Colorimeter

The spectral distribution of radiant flux, H_λ, obtained by use of a spectrophotometer may be used to compute the color of a light source by making use of the CIE standard observer functions $\bar{x}_\lambda$, $\bar{y}_\lambda$, and $\bar{z}_\lambda$. The tristimulus values X, Y and Z of the source are obtained by integrating the products of relative spectral flux distribution H_λ and each of the three observer functions. In theory, this is done for a range of wavelengths from zero to infinity. In practice, the range from 0.38 to 0.77μm is usually considered adequate for the visible spectrum. The appropriate equations are

$$X = \int H_\lambda \bar{x}_\lambda \, d\lambda, \qquad Y = \int H_\lambda \bar{y}_\lambda \, d\lambda, \qquad Z = \int H_\lambda \bar{z}_\lambda \, d\lambda.$$

For light sources, the units of H_λ determine the numerical magnitudes of the tristimulus values. In the case of transmitting or reflecting specimens, the integrations are similar except that they involve triple products comprising: (1) the spectral distribution of radiant flux H_λ of the light source for which the color of the specimen is to be evaluated, (2) the spectral transmittance or reflectance B_λ of the specimen, and (3) the standard observer functions $\bar{x}_\lambda$, $\bar{y}_\lambda$, and $\bar{z}_\lambda$. These integrated products are normalized by dividing each by the integrated product of the spectral flux distribution H_λ and the luminous efficacy function $\bar{y}_\lambda$: thus

$$X = \frac{\int B_\lambda H_\lambda \bar{x}_\lambda \, d\lambda}{\int H_\lambda \bar{y}_\lambda \, d\lambda}, \qquad Y = \frac{\int B_\lambda H_\lambda \bar{y}_\lambda \, d\lambda}{\int H_\lambda \bar{y}_\lambda \, d\lambda}, \qquad Z = \frac{\int B_\lambda H_\lambda \bar{z}_\lambda \, d\lambda}{\int H_\lambda \bar{y}_\lambda \, d\lambda}.$$

These tristimulus values are usually expressed on a numerical scale such that the Y value is the luminous transmittance or reflectance in per cent. The CIE chromaticity coordinates x, y, and z are computed from the tristimulus values X, Y, and Z by determining the fraction of each tristimulus value with respect to the sum of the three tristimulus values,

$$x = \frac{X}{X+Y+Z}, \qquad y = \frac{Y}{X+Y+Z}, \qquad z = \frac{Z}{X+Y+Z}.$$

Since $x + y + z = 1$, only two chromaticity coordinates need be given, usually x and y.

For a nonrecording spectrophotometer, the integration is performed numerically by summing the triple products for a given small wavelength interval. In the day of desk calculators, 0.01-μm intervals were usually deemed small enough. If a mechanical integrating device is attached to a recording spectrophotometer, the summation interval can be infinitesimally small and the integration can be performed as the spectrum is scanned. Alternatively, the spectral data can be recorded on punched-paper or magnetic tape and then fed to a high-speed computer for integration. This latter system has the advantage of permitting the spectral transmittance or reflectance data to be processed so as to obtain the color of the object for several different light sources. In addition, the color coordinates can be computed simultaneously for several types of color space. Because equal distances in different parts of the 1931 CIE color space do not represent equally-perceptible color differences,[6] many modifications of this space have been proposed, but the ideally uniform space has not yet been devised. However, a uniform color space was recommended by the CIE in 1964.[3] The rectangular coordinates are designated U^*, V^*, and W^*. They are related to the tristimulus values X, Y, and Z, as follows:

$$W^* = 25Y^{1/3} - 17, \qquad 1 < Y < 100$$
$$U^* = 13W^*(u - u_0), \qquad V^* = 13W^*(v - v_0)$$
$$u = 4X/(X + 15Y + 3Z) = 4x/(-2x + 12y + 3)$$
$$v = 6Y/(X + 15Y + 3Z) = 6y/(-2x + 12y + 3)$$

where u_0 and v_0 are the values of u and v for the nominally achromatic color placed at the origin of the u,v system.

C. Illuminating and Viewing Conditions

In the case of reflecting specimens, the directions of illumination and viewing can have an important bearing on the colorimetric result.[7] The CIE standard conditions are 45° illumination and 0° viewing.[2] However, only one spectrophotometer and a few colorimeters are currently constructed with this geometry. Most spectrophotometers and several colorimeters utilize unidirectional illumination at an angle different from 45° and view the specimen hemispherically, or use the converse conditions. Why should this be so? Two reasons for the use of hemispherical viewing are: (1) specimen orientation is not critical, and (2) nearly all the reflected flux is collected for measurement, so that the instrument is operated under the optimum geometrical condition for maximum sensitivity. When unidirectional viewing is used, the flux accepted for measurement from a perfectly diffusing specimen will be a fraction determined by the ratio of the solid angle subtended by the detector at the specimen divided by 2π.

In the case of instruments utilizing hemispherical viewing, there is no measurement problem for matte or low-gloss specimens. Glossy specimens, however, will be incorrectly evaluated for color if the surface-

---

[6] G. A. Fry, this series Vol. II, p. 49, Fig. 37, 1965.

[7] L. Mori, H. Sugiyama, and N. Kambe, Influences of illuminating and viewing condition and gloss of surface upon colorimetric results, *Acta Chromatica* **1**, 6–11 (1952).

reflected light is retained within the instrument sphere, because a visual observer seldom appraises the color of a glossy specimen with a strong highlight reflected in the surface. If the specimen is illuminated perpendicularly, some of the surface-reflected light will leave the sphere through the entrance port, the exact amount depending not only on the degree to which the specimen surface approaches mirror quality, but on the relative diameters of entrance beam, entrance port, and sphere, as well as on the divergence of the beam.

To provide a degree of control of the mirror-reflected flux, the specimen is illuminated at some angle other than perpendicular and an absorber or black cap is placed at the position of mirror reflection of the incident beam. This procedure works well for plane, mirrorlike specimens. However, when the gloss of the specimen is reduced, or if the specimen surface is curved or wavy, some surface-reflected flux is likely to be included in the measurement, the amount depending on the instrument geometry and the surface condition of the specimen. For these reasons, some workers prefer to make color-control measurements without attempting to eliminate surface-reflected light.

Many commercial specimens fluoresce when exposed to short-wave radiation. Many colorimeters and most spectrophotometers are not suitable for evaluation of the color of a fluorescent specimen because the specimen is irradiated after spectral selection of the energy by the monochromator or filter. Accurate evaluation of a fluorescent specimen requires irradiation with the complete spectrum of the source for which the color measurement is desired. However, compact, high-intensity sources having spectral distributions duplicating daylight have not been available. Because of these shortcomings, the statement is frequently made that "this method" or "this instrument" is "not applicable to the measurement of fluorescent specimens." The statement is true, but it is not very helpful, because many specimens are at least mildly fluorescent. When this kind of instrument is used for measurement of supposedly nonfluorescent specimens, a good procedure is to make a qualitative estimate of fluorescence by examining each specimen under the radiation from a mercury lamp equipped with a visible-absorbing, ultraviolet-transmitting filter.

D. Color-Difference Measurement

In many industrial applications of color control, evaluation of the color difference between a production specimen and a control standard is most important. This kind of evaluation can be performed by using a spectrophotometer, but it can also be done with an instrument that determines the tristimulus values of the specimen directly in terms of the tristimulus values of the standard. From the standpoint of the man in charge of color control, however, comparison of tristimulus values is not an easy method by which to judge whether a specimen matches a standard within an acceptable tolerance. For this purpose, the tristimulus values are usually modified and combined in such a way as to provide an index of color difference that will have some degree of correlation with the perceived color differences.[8] The modification is necessary in part to minimize the effect of the nonuniformity of the CIE color space. Various modifications of the CIE space have been proposed, and have been used with various degrees of success. The modifications most frequently used are those proposed by Hunter,[9] Adams,[10] MacAdam,[11] and the CIE.[3]

A unit of color difference was first proposed by Judd at the National Bureau of Standards, and for this reason he called it the NBS unit.[12] Other investigators have referred to this unit as the judd. The judd is defined by a complex color-difference equation,[13] but the magnitude of this unit of color difference was chosen such that smaller differences would be unimportant in most commercial transactions, although the unit is about four times as large as the smallest difference observable at the 5% level when colors are viewed in fields of large angular extent separated by a fine dividing line.

Specification of the number of judds within which a specimen must match a standard is very useful for procurement purposes. For production control, however, one needs to know not only the magnitude of the total color difference, but also the magnitude and direction of the components that make up the difference, so that appropriate action can be taken where necessary to reduce the difference between specimen and standard.

E. Color Standards

Material color standards are often used as a means of checking the calibration or performance of colorimeters[14] and spectrophotometers.[15]

---

[8] D. B. Judd, A. Maxwell triangle yielding uniform chromaticity scales, *J. Res. Natl. Bur. Std.* **14**, 41–57 (1935); *J. Opt. Soc. Am.* **25**, 24–35 (1935).

[9] R. S. Hunter, Photoelectric tristimulus colorimetry with three filters, *Natl. Bur. Std. (U.S.) Circ.* C-429 (1942).

[10] E. Q. Adams, *X-Z* planes in the 1931 ICI system of colorimetry, *J. Opt. Soc. Am.* **32**, 168–173 (1942).

[11] D. L. MacAdam, Projective transformations of ICI color specifications, *J. Opt. Soc. Am.* **27**, 294–299 (1937).

[12] D. B. Judd, Specification of uniform color tolerances for textiles, *Textile Res.* **9**, 253–263, 292–307 (1939).

[13] R. S. Hunter,[9] pp. 17–18.

[14] R. S. Hunter, Tests for the uniformity of 45°–0° reflectometers, *J. Opt. Soc. Am.* **53**, 390–393 (1963).

[15] H. J. Keegan, J. C. Schleter, and D. B. Judd, Glass filters for checking performance of spectrophotometer-integrator systems of color measurement, *J. Res. Natl. Bur. Std.* **A66**, 203–211 (1962).

Color standards are also used to provide a simple means for specification of colors, when desired, in preference to a more fundamental specification in the CIE system. Sets of standards have been assembled to cover commercially important regions of color space. Lovibond glass standards[16] were originally developed in the nineteenth century to aid in the color control of beer, but they have been widely used for evaluating the colors of all types of transparent materials. Munsell painted papers have filled a parallel need for color standards for opaque materials.[17]

Standards for special purposes have been made from transparent glass and plastic as well as from opaque glass, plastic, paper, painted panels, ceramic tile, and porcelain enamel on steel. There are two important points to consider in the selection of material for use as a standard: (1) the standard and the specimen should have the same spectral character, and (2) the standard should have a high degree of permanence. These two requirements are often contradictory. The first requirement is most satisfactorily met by making a standard from the same material as the specimens for which it will serve as a standard. This often means sacrificing permanence. The second requirement is best met by making the standard of glass or vitreous enamel. This often means that the standard and the specimen will be of different materials and thus may differ importantly in spectral character. The drift in color of standard painted panels has been an important problem in the paint industry, but Huey[18] showed that color change in painted panels can be retarded appreciably by storing them at low temperature, such as 0°C, a temperature obtainable with a home freezer.

Fundamental transmittance measurements are made by obtaining the ratio of the transmitted to incident flux. In the case of reflectance measurements, however, the measurement is always made relative to a reference white standard. For many years, the reference standard has been a freshly smoked layer of magnesium oxide. Recently, the CIE recommended replacement of smoked MgO with the ideally-reflecting, perfectly-diffusing standard.[19] A practical method for absolute calibration of reflectance standards for hemispherical viewing has recently been described.[20–22] In

---

[16] Tintometer Ltd., Milford, Salisbury, England.

[17] Munsell Color Co., 2441, N. Calvert St., Baltimore, Maryland 21218.

[18] S. J. Huey, Low-temperature storage of color standard panels, *Color Eng.* **3**, No. 5, 24–27 (1965).

[19] CIE, *Proc. 14th Session* Brussels, p. 36 (1959).

[20] J. A. Van den Akker, L. R. Dearth, and W. M. Shillcox, Evaluation of absolute reflectance for standardization purposes, *J. Opt. Soc. Am.* **56**, 250–252 (1966).

[21] D. G. Goebel, B. P. Caldwell, and H. K. Hammond III, Use of an auxiliary sphere with a spectroreflectometer to obtain absolute reflectance, *J. Opt. Soc. Am.* **56**, 783–788 (1966).

[22] Standard Method for Absolute Calibration of Reflectance Standards, ASTM E-306. Book of Standards Part 30, Am. Soc. for Testing and Mater. Philadelphia, Pennsylvania, 1967.

spite of progress on absolute calibration techniques, industry still needs durable, easily-reproducible instrument standards for daily use, such as pressed powder standards of reagent grade MgO or $BaSO_4$.

Most color standards are calibrated by measuring them on a spectrophotometer. However, it has also been found desirable to develop standards to check the performance of spectrophotometer–integrator systems.[15]

III. VISUAL COLORIMETERS

Because the eye is a sensitive and readily available detector of color difference, several designs of visual colorimeters have been used, and some are still in use. These instruments have the advantage of simplicity of design, but they have two important disadvantages: (1) much time is required for measurement, and (2) the result depends in part on the spectral sensitivity of the eye of the operator. Depending on whether the instrument is designed to add reference lights to match the test light, or to subtract color from a reference light, visual colorimeters are classed as "additive" or "subtractive."

In the basic visual colorimeter, the operator observes a field with one part illuminated by the unknown color and the other part by measured amounts of reference colors. The simplest field is circular and split down the middle. For precise results, the dividing line must be very sharp. The reference colors are adjusted by trial until the best match for the test color is obtained. Provision must be made for independently varying the amounts and chromaticities of the reference colors. The amounts of each reference color required provide a measure of the unknown color.

A. Additive Colorimeters

When colors from different lights or from filtered portions of the same light are added, we have an "additive" type of colorimeter. In an early design of this type of colorimeter, such as that described by Nutting,[23] monochromatic light was added to white light to match the test color, thereby obtaining measured parameters that correlated directly with the psychological attributes hue and saturation. These attributes, however, are not fundamental to color measurement or specification, and Guild[24] severely criticized this method of colorimetry.

---

[23] P. G. Nutting, A new precision colorimeter, *Bull. Bur. Std.* **9**, 1–5 (1913).

[24] J. Guild, A criticism of the monochromatic-plus-white method of colorimetry, *Trans. Opt. Soc.* (*London*) **27**, 130–158 (1926).

Visual tristimulus colorimeters have been constructed by a number of experimenters over the years. Except for the early work of Ives[25] in 1908, the published reports of these instruments fall into two periods, 1924–1936[26–31] and 1949–1958.[32–36] These instruments are not suited to product control; instead, they provide useful visual research tools. The visual tristimulus design has a number of shortcomings: (1) the restricted chromaticity range fails to include many important colors, (2) construction is expensive, (3) data are not easily reduced to desired form, and (4) sensitivity is low.

In an attempt to overcome the first shortcoming, Donaldson[37] added three more primaries to his colorimeter. The six primaries are produced by passing incandescent-lamp light through colored filters (Fig. 2). The transmitted light is then mixed in a small white-lined sphere. Light from this sphere illuminates the reference field, and light from the test specimen illuminates the test field. A set of identical filters is provided at the eyepiece of the instrument to facilitate initial adjustment. The observer adjusts the intensity of each primary by turning a micrometer screw connected to the shutter that controls the transmitting area of the source filter. The shutter positions are indicated on carefully ruled scales, and these scale readings constitute the color specification for the test color.

---

[25] H. E. Ives, The Ives colorimeter in illuminating engineering, *Trans. Illum. Eng. Soc.* (*N.Y.*) **3**, 627–644 (1908).

[26] F. Allen, A new tri-color mixing spectrometer, *J. Opt. Soc. Am.* **8**, 339–341 (1924).

[27] J. Guild, A trichromatic colorimeter suitable for standardization work, *Trans. Opt. Soc.* **27**, 106–129 (1925–26).

[28] W. D. Wright, A trichromatic colorimeter with spectral primaries, *Trans. Opt. Soc.* **29**, 225–242 (1927–28).

[29] H. P. J. Verbeek, Een trichromatische Colorimeter, *Physica* **13**, 77–82 (1933); **14**, 1082–1084 (1934).

[30] R. Donaldson, A trichromatic colorimeter, *Proc. Phys. Soc.* (*London*) **47**, 1068–1073 (1935).

[31] S. M. Newhall, An instrument for color stimulation and measurement, *Psychol. Monographs* **47**, 199 (1936).

[32] W. R. J. Brown and D. L. MacAdam, Visual sensitivities to combined chromaticity and luminance differences, *J. Opt. Soc. Am.* **39**, 808–834 (1949).

[33] D. L. MacAdam, Loci of constant hue and brightness determined with various surrounding colors, *J. Opt. Soc. Am.* **40**, 589–595 (1950).

[34] W. S. Stiles, 18th Thomas Young Oration: The Basic Data of Colour Matching, *Phys. Soc.* (*London*) *Year Book*, p. 44 (1955).

[35] I. Hennicke and W. Münch, Ein visuelles Farbmessgerät nach dem Gleichheitsverfahren, *Farbe* **6**, 189–194 (1957).

[36] H. Beck and M. Richter, Neukonstruktion des dreifarben-Messgerätes nach Guild-Bechstein, *Farbe* **7**, 141–152 (1958).

[37] R. Donaldson, A colorimeter with six matching stimuli, *Proc. Phys. Soc.* (*London*) **59**, 554–560 (1947).

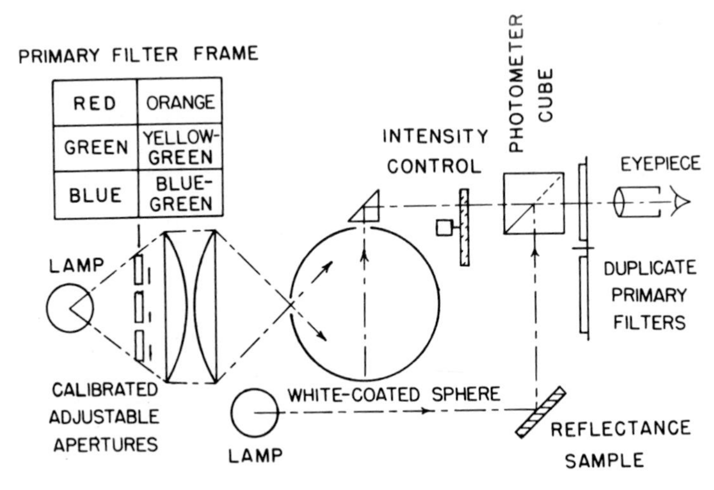

FIG. 2. Optical system of the Donaldson six-primary visual colorimeter. The amount of each primary is indicated by the position of the calibrated adjustable aperture associated with each filter. Constancy of calibration is dependent on maintaining beam geometry. Primary colors are mixed in white-coated sphere before being compared with test color. Intensity of mixture is controlled by adjustable rotating sector. Duplicate primary filters in eyepiece assist observer to obtain initial aperture settings.

Another visual colorimeter, of much simpler design, utilizes colored papers, such as those supplied by the Munsell Color Company, to form sectors on a rotating disk, generally referred to as a Maxwell disk.[38] Four papers, three chromatic and one black, provide the necessary degrees of freedom to match a test color. If the tristimulus values of each paper are known, the tristimulus values of the mixture can be computed as the area-weighted mean of the components. This colorimeter has had considerable use in evaluating the colors of agricultural products such as hay, cotton, tomatoes, and the like, where the test material is nonuniform and the color of the test specimen is conveniently averaged by spinning on another disk.[39]

B. Subtractive Colorimeters

In this type of colorimeter, filters are introduced in tandem into a single reference beam in order to subtract out the colors not needed to match the test color. Probably the most widely used subtractive colori-

---

[38] D. Nickerson, A colorimeter for use with disc mixture, *J. Opt. Soc. Am.* **21**, 640–642 (1931).

[39] D. Nickerson, Color Measurement and its Application to the Grading of Agricultural Products, U.S. Dept. Agr., Misc. Publ. 580 (1946).

meter is the Lovibond tintometer.[40] It has been used for three-quarters of a century in various forms for measuring colors of all kinds of commercial materials that transmit light, such as beer, oil, sugar solutions, etc. The color of the comparison field is determined by Lovibond red, yellow, and blue glasses. The unit of each of these colored glasses is arbitrary, but the units are related by the fact that one unit of each color of glass taken together in daylight produces a neutral color. Glasses of each color are provided in a series of increasing saturation so that combinations can be obtained in steps of one-tenth unit. Although designed long before the CIE system of color specification, the instrument can be adapted to it.[41]

Subtractive colorimeters have also been made by using wedges of colored material calibrated from spectrophotometric measurements. Jones[42] constructed an instrument from wedges of colored gelatine. His wedges were of such density gradient that colors of high saturation could be read directly. Judd[43] designed a subtractive colorimeter using wedges of pale-green and pale-yellow glass. The wedges are constructed in pairs. They are mounted on separate rack strips with the thick end of one over the thin end of the other. Movement of the rack pinion moves each wedge of a pair into or out of the beam simultaneously, thereby providing a relatively rapid rate of change of density without producing a gradient across the field of view. This instrument has been extremely useful for small-difference colorimetry at the National Bureau of Standards, and it is still in use there.

A specially-designed subtractive colorimeter using a xenon lamp and cuvettes to hold dyes has been used by Friele[44] in colorant mixture formulation.

IV. PHOTOELECTRIC TRISTIMULUS COLORIMETERS

A. General Considerations

Ever since the invention of the photoelectric cell about 30 years ago, work has been directed at replacing the human eye in colorimetry with

---

[40] J. W. Lovibond, The Tintometer—a new instrument for the analysis, synthesis, matching, and measurement of colour, *J. Soc. Dyers Colourists* **3**, 186 (1887).

[41] R. K. Schofield, The Lovibond Tintometer adapted by means of the Rothamsted device to measure colours on the CIE system, *J. Sci. Instr.* **16**, 74–80 (1939).

[42] L. A. Jones, A colorimeter operating on the subtractive principle, *J. Opt. Soc. Am.* **4**, 420–431 (1920).

[43] D. B. Judd.[12] See also D. B. Judd and G. Wyszecki, "Color in Business, Science, and Industry," 2nd ed. p. 191. Wiley, New York, 1963.

[44] L. F. C. Friele, A subtractive colorimeter for the formulation of colorant mixtures, *Color Eng.* **2**, No. 7–8, 12–17 (1964).

photocells and filters, with gradually increasing success. The basic idea is to determine the spectral sensitivity of the cell and then to design filters so that filter–photocell combinations will duplicate the tristimulus functions of the CIE standard observer.[9] This can be done easily to a first approximation, but duplication to a high degree of accuracy is exceedingly difficult (see Fig. 3). Even if the filter–photocell combination is made very

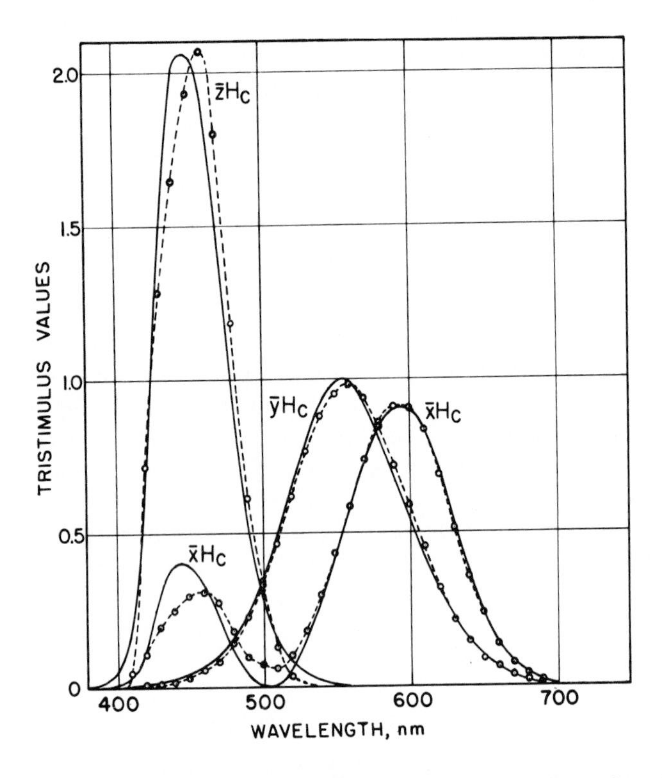

FIG. 3. Comparison of spectral tristimulus values obtained in Hunter's early colorimeter design[9] with target values for CIE 2° observer and source C; solid curves (daylight). Areas under CIE $\bar{x}H_c$, $\bar{y}H_c$, and $\bar{z}H_c$ curves are in relation 98 : 100 : 118. Colorimeter curves (dashed curves with open dots) are adjusted to have corresponding areas. Note that the disparity between pairs of these curves is only slightly larger than that for CIE 2° and 10° observer functions in Fig. 1.

accurate for one instrument, the same filters may not serve equally well in a second instrument, because of variation in spectral sensitivities among photocells, even from the same production lot. Nevertheless, photoelectric tristimulus colorimeters have found wide application in the field of industrial color control, where they are used primarily as color-difference meters to evaluate the difference in color between a production specimen and a standard of similar spectral character.

In its simplest form, the tristimulus colorimeter contains a light source for illuminating the specimen, a photocell for viewing it, and three tristimulus filters that are inserted in turn in either the illuminating or viewing beam. If the filters are placed in the illuminating beam, the specimen receives less heat from the light source, but the filters receive more heat. On the other hand, if the specimen exhibits fluorescence, a more accurate evaluation of color will be obtained by placing the tristimulus filters in the viewing optics. If photocells are associated with each of the tristimulus filters and are so positioned that each cell views the specimen simultaneously, then it becomes possible to combine the electrical outputs and thereby effect analog-type computation of data, as has been done in the instruments designed by Hunter.[45]

Every photoelectric tristimulus colorimeter on the market today is provided with an incandescent-lamp source, though some manufacturers are making provision for use of xenon or mercury lamps as auxiliary sources for evaluating fluorescent materials. Photovoltaic cells are used as detectors in many instruments, but photoemissive types are also used. Photovoltaic cells have been favored for use as detectors in colorimeters because: (1) their spectral sensitivity is maximum in the green region of the spectrum, where the eye is most sensitive, (2) the spectral sensitivity varies less from one cell to another than for photoemissive cells, and (3) they convert light into an electric current that can be measured directly with a microammeter without amplification. For these reasons, photovoltaic cells are invariably used in the design of simple, low-cost instruments. They are also used, however, in some sophisticated instrument designs. Because the spectral sensitivity peaks near the middle of the visible spectrum, modification by means of filters to duplicate approximately the desired spectral functions is relatively easy. Photoemissive cells, on the other hand, usually have high blue sensitivity with corresponding low red sensitivity, and thus filter design for long-wave functions is more difficult. Both types of detectors are temperature sensitive; the spectral response curves shift toward longer wavelengths as the photocell temperature is raised. Improved temperature stability can be provided by operating photocells at a thermostatically controlled temperature above that of the laboratory.

Filters are usually made of glass for permanence, but gelatine filters and sealed liquid filters[46] have been used. Filters are usually designed with

---

[45] R. S. Hunter, Photoelectric color difference meter, *J. Opt. Soc. Am.* **48**, 985–995 (1958).

[46] S. H. Emara and R. P. Teele, Development of filters for a thermo-electric colorimeter, *J. Res. Natl. Bur. Std.* **67C**, 319–325 (1963).

uniformly thick layers to completely cover the sensitive area of the photocell, but some filters have been designed with segments of different colored material to obtain a better fit of the cell response to the desired spectral function.[47]

B. Commercial Instruments

A detailed discussion of all the features of each commercial instrument is beyond the scope of this chapter. In 1967, photoelectric tristimulus colorimeters made by 11 different manufacturers were available in the United States, and most manufacturers produced several models. Each instrument has certain appealing features in design, operation, or price.

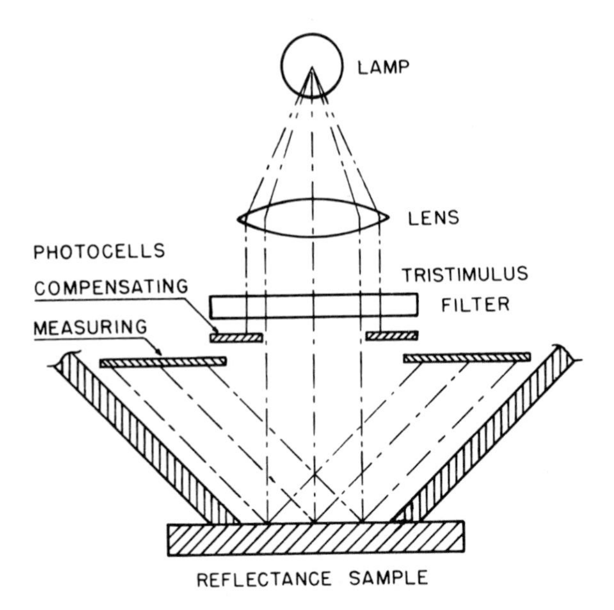

Fig. 4. Optical design of the Photovolt reflection meter. Incandescent lamp light passes through a tristimulus filter with added blue component so that the photovoltaic cell responds as if CIE source C were used. A portion of incident light strikes the annular compensating photocell; the remainder passes through hole in cell to reach the reflectance sample. The measuring photovoltaic cell receives light reflected in an annular cone centering on 45°. The annular design eliminates the azimuthal orientation problem encountered with linearly-textured specimens. Tristimulus filters are inserted in turn or mounted in rotating turret. The optical head is separated from the rest of the photometer, providing a lightweight unit that can be placed in any position required by the sample.

---

[47] I. Nimeroff and S. W. Wilson, A colorimeter for pyrotechnic smokes, *J. Res. Natl. Bur. Std.* **52**, 195–199 (1954).

In general, the more sophisticated the design, the higher the price. Many instruments use the illuminating and viewing geometry prescribed by the CIE, namely, 45° illumination and 0° viewing, or the converse. Some popular instruments, however, use hemispherical illumination and unidirectional viewing geometry; others use the converse of this geometry. Most instruments are designed for the colorimetry of reflecting specimens, but many can be used with transmitting specimens.

Some instruments are designed to be portable so that the instrument can be taken to large specimens when desired. Instruments in this class are the Photovolt reflection meter, the Lange colorimeter, and the Hilger and Watts colourmeter. Each of these instruments utilizes 45°–0° geometry (or the converse), a photovoltaic cell as detector, tristimulus filters inserted in turn, and a galvanometer or microammeter to indicate the photocell output (Fig. 4).

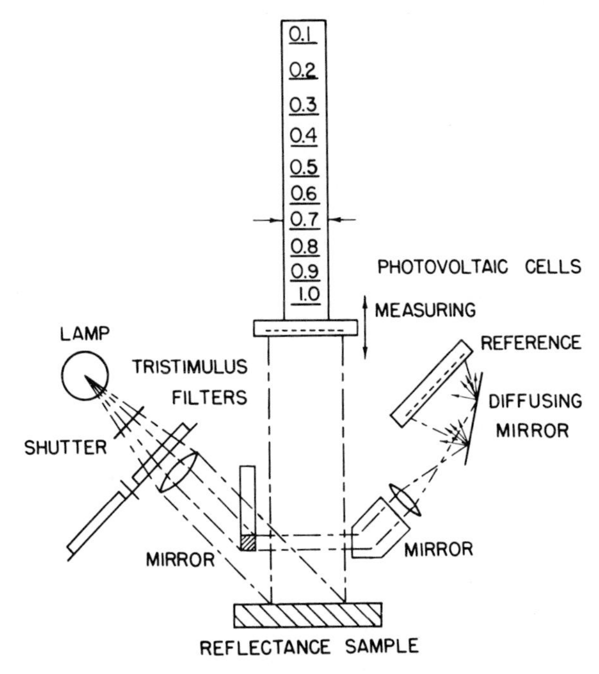

FIG. 5. Optical design of the Hunter multipurpose reflectometer. Opening shutter allows light from the incandescent lamp to pass through the tristimulus filter and reach the sample at a 45° angle. The added blue component in filters causes the photovoltaic cell to respond as if CIE source C were used. A portion of the incident beam is reflected by mirrors to the reference photocell. The measuring photocell receives light reflected by the specimen in the perpendicular direction. The cell position is adjusted to obtain a current balance with the reference photocell; the position is indicated on an approximate inverse-square scale.

Less portable instruments like the Hunter multipurpose reflectometer are usually somewhat more precise.[48] This instrument utilizes the same basic geometry and photovoltaic detector as the portable type, but the reflectance reading is made on an inverse-square scale determined by the distance required from the specimen to balance the photocurrent of the measuring cell with the current from a comparison cell that monitors the light coming directly from the source (Fig. 5). A somewhat similar instrument that utilizes a cam to reduce the cross section of the comparison beam to achieve photocell balance is known as the Colormaster differential colorimeter.[49] This instrument splits the incident bean in two so that light is reflected from the test specimen and a comparison specimen onto separate photoemissive cells. The illuminance of the comparison cell is attenuated by means of a cam in the comparison beam until the test- and comparison-cell currents are equal as determined by means of a microammeter in an electronic bridge circuit. The cam position is the measured parameter (see Fig. 6). This design has high measurement sensitivity, and thus the ability to evaluate small color differences precisely.

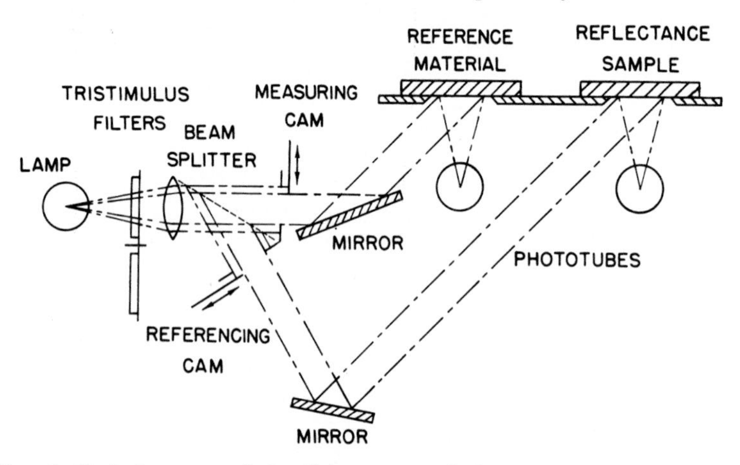

FIG. 6. Optical system of the Colormaster colorimeter. Incandescent-lamp light, after passing through a combination tristimulus and source C filter, is divided into two beams by the semitransparent beam splitter to provide continuous illumination of the reference material as well as a reflectance sample. The parameter measured is the position of the measuring cam required to reduce the area of reference beam to equalize the illumination on reference and measuring photocells. Uniform flux density across the cam window is required.

---

[48] R. S. Hunter, A multipurpose photoelectric reflectometer, *J. Res. Natl. Bur. Std.* **25**, 581–618 (1940); also *J. Opt. Soc. Am.* **30**, 536–559 (1940).

[49] L. G. Glasser and D. J. Troy, A new high-sensitivity differential colorimeter, *J. Opt. Soc. Am.* **42**, 652–660 (1952).

All of the early designs of colorimeters used a slow and tedious manual means of balancing the comparison- and test-photocell currents. The Gardner Laboratory was the first to recognize the increased speed and ease of operation obtainable by making colorimeters with servodriven, self-balancing potentiometers. For some years, their complete line of instruments has been offered with the automatic balancing feature. The cost of automation is substantial, but in production control, it can be easily justified. The Martin Sweets color-brightness tester,[50] used extensively in the paper industry, is now available in both manual and automatic balancing models.

A few colorimeters are designed with hemispherical illumination and unidirectional viewing by positioning the specimen at the port of a uniformly illuminated white-lined sphere. One advantage to the user of this type of geometry is that the measurement is not dependent on sample orientation. Hemispherical illumination is used in the Zeiss Elrepho[51] and the IDL Color Eye.[52] The Elrepho uses two photoemissive cells, one for the sample and one for the comparison specimen. The Color Eye uses one multiplier phototube and a rotating decentered lens to provide a mechanical beam switch for alternately viewing the test and comparison specimens (see Fig. 7). Both instruments use mechanical beam attenuation as the

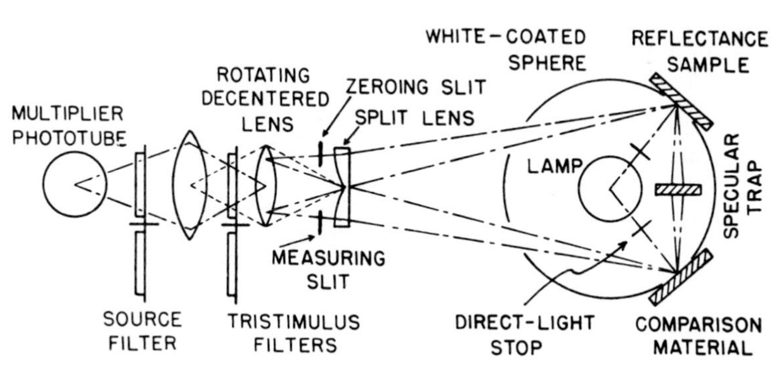

FIG. 7. Optical system of the Color-Eye colorimeter. Note that the sample is illuminated diffusely by incandescent-lamp light and that the multiplier phototube alternately views the comparison material and the reflectance sample many times per second through a rotating decentered lens. Tristimulus filters placed in the viewing beam provide more accurate data on fluorescent samples.

---

[50] L. R. Dearth, W. M. Shillcox, and J. A. Van den Akker, The standard brightness tester, as a four-filter colorimeter, *Tappi* **46**, No. 1, 179A–188A (1963).

[51] H. J. Höfert, Ein Filterphotometer zur Remissionsmessung, *Z. Instrumentenk.* **67**, 118–124 (1959).

[52] G. P. Bentley, Industrial tristimulus color matcher, *Electronics* **24**, 102–103 (Aug. 1951).

measured parameter. The Color Eye can be obtained with a series of narrowband interference filters in a turret to make the instrument serve as an abridged spectrophotometer. Several other colorimeter manufacturers now offer or are contemplating this additional feature.

The converse geometry, unidirectional illumination and hemispherical viewing, is used in designs by Hunterlab; Joyce, Loebl and Co.[53]; and by Jobin and Yvon.[54] Joyce, Loebl and Co. provide their instruments with sectored filters individually masked to produce the best possible spectral fits of the filter–photocell combinations to the CIE tristimulus functions (Fig. 8). Hunterlab uses supplemental trimming filters to adjust the spectral response of production photocells to the design values for the tristimulus filters. The use of separate detectors permits the combination of the electrical outputs so as to provide built-in analog computation of Hunter's chromaticity coordinates *a* and *b*, as well as the square root of reflectance as the correlate for lightness. These unique features are also found in instruments with 45°–0° geometry designed by Hunter.[45]

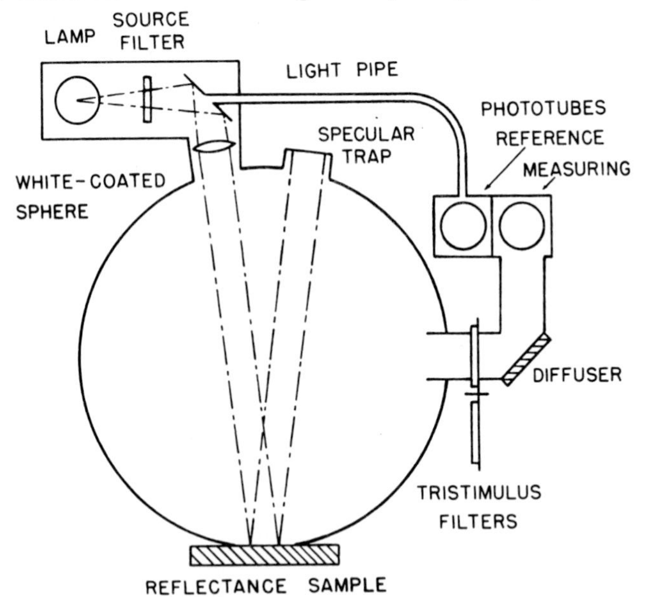

FIG. 8. Optical system of the Joyce–Loebl Colorcord colorimeter. Note that the reflectance sample is viewed diffusely. Sector-type tristimulus filters are used to improve the duplication of CIE tristimulus functions. Note that a diffuser is placed between the filters and the phototube. Use of a light pipe permits the reference phototube to be placed adjacent to the measuring phototube, thereby minimizing drift due to differential heating of the phototubes.

[53] J. Hambleton, A direct-reading tristimulus colorimeter, *Brit. Ink Maker* **9**, No. 1, 31–35 (1966).

[54] Manufactured by Jobin and Yvon, 26 rue Berthollet, Arcueil (Seine), France.

Instruments usually view a specimen area from 1 to 2 in. in diameter. In many instruments, lenses can be inserted to illuminate smaller areas. For some materials, such as bulk cotton and acoustical tile, larger illuminated and viewed areas are desirable. A special colorimeter has been designed for cotton.[55] In addition to illuminating and viewing a 4-in. diameter area, this instrument is provided with servo controls that automatically plot the chromaticity and indicate the lightness. Another instrument that has been adapted to large-area specimens is the Color Eye. An LS model of this instrument is available in which the usual 4-in.-diameter sphere has been replaced with one 18 in. in diameter, so that instead of measuring a 1-in.-diameter specimen area, a 4-in. area is evaluated.

A summary of some of the important features of tristimulus colorimeters currently being manufactured is contained in Table I. The relative merit of these various instruments will not be discussed here; however, members of the staff of the Institute of Paper Chemistry have investigated many instruments over the years and have published their findings.[56]

---

[55] D. Nickerson, R. S. Hunter, and M. G. Powell, New automatic colorimeter for cotton, *J. Opt. Soc. Am.* **40**, 446–449 (1950).

[56] Staff of the Institute of Paper Chemistry, A study of photoelectric instruments for the meaurement of color: reflectance and transmittance.

Part I, A general discussion of color and color measurements, *Paper Trade J.* **105**, No. 18, 135–141 (Oct. 28, 1937), and No. 19, 27–39 (Nov. 4, 1937).

Part II, The General Radio color comparator, type 725A, *Paper Trade J.* **105**, No. 25, 46–50 (Dec. 16, 1937).

Part III, The Lange photoelectric reflection meter, *Paper Trade J.* **105**, No. 27, 42–46 (Dec. 30, 1937).

Part IV, The Eimer and Amend improved reflection meter, *Paper Trade J.* **107**, No. 20, 33–37 (Nov. 17, 1938).

Part V, The Hunter multipurpose reflectometer, *Paper Trade J.* **107**, No. 25, 29–40 (Dec. 22, 1938).

Part VI, The Higgins reflection meter, *Paper Trade J.* **112**, No. 1, 13–22 (Jan. 2, 1941).

Part VII, The Photovolt photoelectric reflection meter, Model 610, *Tappi* **33**, No. 10, 85A–89A (Oct. 1950).

Part VIII, The Densichron as a reflection meter, *Tappi* **34**, No. 8, 126A–132A. (Aug. 1951).

Part IX, The Hunter color and color-difference meter, *Tappi* **34**, No. 9, 134A–158A (Sept. 1951).

Part X, The "Color-Eye" (P.P.G.-I.D.L.), *Tappi* **35**, No. 11, 141A–158A (Nov. 1952).

Part XI, The Alinco brightness tester, Model 50, *Tappi* **41**, No. 9, 196A–204A (Sept. 1958).

Part XII, The Elrepho photoelectric reflectance photometer, *Tappi* **43**, No. 2, 230A–239A (Feb. 1960).

Part XIII, The Zeiss-Electrophotometer Elko II, *Tappi* **43**, No. 5, 253A–260A (May 1960).

Part XIV.[50]

TABLE I

Summary of Characteristics of Some Commercially Available Photoelectric Tristimulus Colorimeters for Laboratory Measurement of Reflecting Specimens

Instrument	Geometry[aa]		Light on specimen[bb]	Specimen position[cc]	Detector(s)[dd]	Measurement scales[ee]	Readout[ff]	Approx. Price[gg]
	Source	Receptor						
—[a]	0°	45°(C)	W,TF	Any	SePV(2)	R,G,B	Meter	0.5–0.6
—[b]	0°	45°(C)	W,TF	→|	SePV(2)	R,G,B;R_λ	Meter	0.6
—[c]	0°	45°(C)	W,TF	Any	SePV(1)	R,G,B	Galvanometer	0.4
—[d]	0°	45°(C)	W,TF	Any	SePV(1)	R,G,B;R_λ	Galvanometer	3.0
—[e]	45°	0°	W,TF	$\overline{\uparrow}$	PE,S–4(2)	R,G,B	Counter, M	1.5
—[f]	45°	0°	W	$\overline{\uparrow}$	PE,S–4(2)	X_B, X_R, Y, Z	Counter, M	1.9–2.2
—[g]	45°	0°	W	$\overline{\uparrow}$	PE,S–4(2)	X_B, X_R, Y, Z	Counter, A	4.6–5.8
—[h]	45°	0°	W,TF	→|	SePV(2)	R,G,B	Scale, M/A	1.0/1.9
—[i]	45°(2)	0°	W	$\overline{\uparrow}$	SePV(5)	R,G,B	Dial, A	1.9
—[j]	45°(4)	0°	W	$\overline{\uparrow}$	SePV(5)	R,G,B	Dial, A	2.5
—[k]	0°	45°	W	$\overline{\uparrow}$	SePV(7)	L,a,b	Counter, M	2.0
—[l]	45°(2)	0°	W	$\overline{\uparrow}$	SePV(4)	L,a,b	Counter, A	3.3
—[m]	45°(4)	0°	W	$\overline{\uparrow}$	SePV(5)	L,a,b	Counter, A	3.9
—[n]	45°(4)	0°	W	$\overline{\uparrow}$	SePV(5)	L,a,b	Counter, A	4.3
—[o]	45°	0°	W	$\underline{\downarrow}$	PE,S–10(3)	L,a,b	Counter, M	3.3
—[p]	45°	0°	W	$\underline{\downarrow}$	PE,S–10(3)	L,a,b	Counter, M	3.4
—[q]	0°/8°	H; 8	W	→|	PE,S–10(3)	L,a,b	Counter, M	4.0
—[r]	45°	0°	W	$\underline{\downarrow}$	PE,S–10(4)	L,a,b	Counter, M	4.0
—[s]	45° (C)	0°	W	$\underline{\downarrow}$	PE,S–10(3)	$L.a.b$	Counter, M	4.2
—[t]	H; 4	40°	W	→|	PM,S–4(1)	$X_R, Y, Z; R_\lambda$	Counter, M	4.3–5.5
—[u]	H; 18	6°	W/Hg	→|	PM,S–4(1)	$X_R, Y, Z; R_\lambda$	Counter, M	6.8–9.0

—[v]	45°	0°	W	↓	PE,S–4(2)	R,G,B; R_λ	Counter, *M*	3.5
—[w]	H	0°	W/Xe	↓	PE,S–4(2)	R,G,B; R_λ	Counter, *M*	4.5
—[x]	7°	H; 12	W	↓	PM,S–4(2)	X, Y, Z	Counter, *M*	3.8
—[y]	7°	H; 12	W	↓	PM,S–4(2)	X, Y, Z; R_λ	Counter, *M*	4.8
—[z]	15°	H	W,TF	↓	PM,S–5(1)	R,G,B	Counter, *M*	2.0

[a] Reflectometer, Model 610, Photovolt Corp., 1115 Broadway, New York, N.Y. 10010.

[b] Lumetron, Model 402E, Photovolt Corp.

[c] Colorimeter, Dr. Bruno Lange, Berlin, Germany. In USA, International Scientific and Precision Instrument Co., Inc., 910 17th Street, N.W. Washington, D.C., 20006.

[d] Colorimeter, Model J-40, Hilger and Watts Ltd., 98 St. Pancras Way, Camden Road, London, N.W.1, England. In USA, Engis Equipment Co., 8035 Austin Ave., Morton Grove, Illinois 60053.

[e] Colormaster, Model V, Manufacturers Engineering and Equipment Corp., 250 Titus Avenue, Warrington, Pennsylvania 18976.

[f] Color Brightness Tester Model S-1, The Martin Sweets Co., Inc., 3131 West Market St., Louisville, Kentucky 40212. Available with calibration service supplied by the Institute of Paper Chemistry, Appleton, Wisconsin.

[g] Automatic Color Brightness Tester, Model S-2, Martin Sweets.

[h] Reflectometer, Model 100M-1, Gardner Laboratory, Inc., 5521 Landy Lane (P.O. Box 5728) Bethesda, Maryland 20014. This is the original "multipurpose reflectometer."

[i] Reflectometer, Model AUX-2, Gardner Laboratory.

[j] Reflectometer, Model AUX-HS-2, Gardner Laboratory.

[k] Colorimeter, Model C-4, Gardner Laboratory.

[l] Colorimeter, Model AC-1, Gardner Laboratory.

[m] Colorimeter, Model AC-2a, Gardner Laboratory.

[n] Colorimeter, Model AC-3, Gardner Laboratory.

[o] Colorimeter, Model D25A, Hunter Associates Laboratory, Inc., 9529 Lee Highway, Fairfax, Virginia 22030.

[p] Colorimeter, Model D25AF, Hunter Associates Laboratory.

[q] Colorimeter, Model D25P, Hunter Associates Laboratory.

[r] Colorimeter, Model D25L, Hunter Associates Laboratory.

[s] Colorimeter, Model D25M, Hunter Associates Laboratory.

[t] Color Eye, Model 0–1/Sig, Instrument Development Laboratories Division, Kollmorgen Color Systems, 67 Mechanic Street, Attleboro, Massachusetts 02703.

[u] Color Eye, Model LS/Sig, Instrument Development Laboratories.

[v] Elko II, Carl Zeiss, Oberkochen, Federal Republic of Germany. In USA, Carl Zeiss, Inc., 444 Fifth Avenue, New York, N.Y. 10018.

[w] Elrepho, Carl Zeiss.

[x] Colorcord, Model IIa, Joyce, Loebl, and Co., Ltd., Gateshead, England. In USA, Joyce Loebl, Inc., Building 3, South Avenue, Burlington, Massachusetts 01803.

[y] Colorcord, Model IIb, Joyce, Loebl, and Co.

[z] Reflectometer, A. Jobin and G. Yvon, 26, rue Berthollet, Arcueil (Seine), France.

[aa] Principal beam directions only are given. Divergence of beams from principal directions varies from a few degrees to many degrees, depending on the instrument. Where more than one beam is used for 45° illumination, the number is given in parentheses. *H* stands for hemispherical illumination or viewing, and the number following indicates the sphere diameter in inches. *C* stands for conical.

[bb] *W* stands for incandescent lamp light, usually about 3100°K. *TF* means tristimulus filtered. Hg and Xe indicate auxiliary mercury and xenon sources.

[cc] Some photometer heads can be moved to measure a specimen in any position; others are designed to illuminate the specimen primarily in the direction of the arrow.

[dd] SePV: Selenium photovoltaic; PE: Photoemissive, with spectral sensitivity given by the *S*-number; *PM*: photomultiplier.

[ee] Measurement scales of reflectance are usually designated red (R) (amber), green (G), and blue (B). R_λ indicates that narrow-band filters are also available, usually as an accessory. X_B and X_R are the blue and red lobes of the CIE *X* function. Instruments that read *L*, *a*, *b* directly can also be obtained to read *X*, *Y*, *Z*.

[ff] Readout is on a meter, galvanometer, counter, scale, or dial, as indicated. Here M indicates that balance is obtained manually; A automatically by means of a servodriven motor.

[gg] Approximate price is given in kilodollars as of September 1967. The price is given primarily as an index of the complexity of the instrument design.

The Inter-Society Color Council has recognized the need for a catalog of color-measuring instruments, and, in 1965, established a committee to compile one.

V. SPECIALIZED COLORIMETERS

Not all instruments that are called colorimeters actually measure color. In the "color comparator," the color of an unknown is compared with that of a series of standard samples usually assigned values on an arbitrary numerical scale. The "chemical colorimeter" is used by the analytical chemist to evaluate the concentration of a material in solution.

The optical design of a color comparator is usually quite elementary. In its simplest form, a test tube containing the sample and another containing the standard are viewed side by side against the same background (usually white) and judged for equality of color. The chemical colorimeter, on the other hand, requires many of the elements of the conventional colorimeter designed for light-transmitting specimens.

A. Color Comparators

Although the colors of liquids can be judged in cylindrical tubes by viewing perpendicular to the axis of the tube, an improved method utilizes a flat-bottomed tube in which the solution is viewed along the axis of the tube, thereby providing a longer optical path for nearly transparent solutions and doing away with the lens effect of the cylinder. Other designs eliminate the lens effect by using sample cells of square cross section. By adding a flat-bottomed, cylindrical glass plunger, the depth of liquid can be adjusted for sample or standard to provide a color match. The eye is most frequently used to determine when the colors are equal, but a photocell with a filter isolating a principal absorption band of the material being tested can also be used. Because standard colored solutions are difficult to reproduce, some color comparators have been provided with standards of colored glass.[57] The use of color comparators is discussed in books on analytical chemistry.[58]

---

[57] Method of Test for ASTM Color of Petroleum Products, ASTM D 1500, Book of ASTM Std. Part 17 (1967). Method of Test for Color of Transparent Liquids (Gardner Color Scale), ASTM D 1544, Book of ASTM Std. Part 21 (1967).

[58] W. B. Fortune, Color comparimeters, *in* "Analytical Absorption Spectroscopy" (M. G. Mellon, ed.), pp. 116–160. Wiley, New York, 1950.

B. Chemical Colorimeters

These instruments are not intended to measure color in the usual sense; instead, they are designed to measure the transmittance of a solution at a wavelength that is characteristic of the concentration of a component of the solution. The light source is usually an incandescent lamp, but a vapor arc may also be used. The spectrum selector may consist of a series of glass "monochromatic" filters (interference filters) or an inexpensive grating monochromator. The detector and measurement mechanism are some form of simple photoelectric photometer. The manufacturer provides a manual that lists the various types of materials that can be analyzed, together with a table or graph to convert photometer readings into concentration of material in solution.

One form of instrument that is still in use was developed by Evelyn[59] soon after the invention of the photovoltaic cell. In consists of a flashlight-type lamp and reflector on one side of a test tube and a photovoltaic cell on the other. Glass monochromatic filters provide the spectrum selection. The availability of inexpensive replica gratings has permitted the manufacture of an instrument with a small spectrometer as the spectrum-selection device.

Chemical colorimeters are sometimes called filter photometers; their use is discussed in books on analytical chemistry.[60]

VI. COLOR TEMPERATURE METERS

A. Basic Principles

Because the color of an incandescent light source changes as the temperature of the source is altered, color has long been used as an index of temperature. Since a substantial change in temperature produces only a small change in color, a tristimulus colorimeter is not the most sensitive instrument for making this kind of color measurement. A more sensitive method for Planckian-type radiators is to measure the ratios of source intensities at wavelengths in the far-blue and far-red portions of the spectrum. At the extremes of the visible spectrum, the eye becomes insensitive, but the precision of photocells at these wavelengths can be quite high. On the other hand, the photometric ratios need not be determined at the extreme ends of the visible spectrum; instead broad regions of red and blue isolated by filters are preferable, so that detectors of moderate sensitivity

---

[59] K. A. Evelyn, A Stabilized photoelectric colorimeter with light filters, *J. Biol. Chem.* **115**, 63–75 (1936).

[60] R. H. Muller, Filter photometers.[58] pp. 161–185.

can be used. If photometer readings are made with high precision, color-temperature differences as small as 1 K can be detected.

B. Design Considerations

Any stable photoelectric photometer of high precision can be adapted to read color temperature. The procedure is to obtain two colored filters, usually a red and a blue, that will provide nearly equal photometer readings in some region of the color-temperature scale for which measurement is desired. For incandescent lamps, the desired range is usually from 2000 to 3000°K. A series of lamp standards of color temperature is used to obtain a plot of ratios of red/blue readings as a function of color temperature.

In actual instruments designed for the purpose, it is customary to adjust the illuminance of the photometer so as to obtain a predetermined reading for one of the filters, say the red one. The instrument reading for the blue filter can now be calibrated to indicate color temperature directly. This system is used in an accessory to a commercial photometer.[61]

Increased precision for laboratory purposes can be obtained by alternating the filter readings rapidly. In some designs, this is done by placing them on a rotating disk[62] or on a vibrating relay.[63] When the filters are interchanged rapidly, the detector pulses must be separated for the respective filters and then integrated. Precision laboratory color-temperature meters of this type have not been built commercially, but a photoelectric pyrometer of high precision that can be utilized is available commercially.[64]

GENERAL REFERENCES

Supplementary material on color, colorimeters, and colorimetry will be found in the following books:

F. W. Billmeyer and M. Saltzman, "Principles of Color Technology." Wiley (Interscience), New York, 1966.

D. B. Judd and G. Wyszecki "Color in Business, Science, and Industry," 2nd ed. Wiley, New York, 1963.

W. D. Wright, "The Measurement of Colour," 3rd ed. Hilger, London, 1964.

O. S. A. Committee on Colorimetry, "The Science of Color." Opt. Soc. of Am., Washington, D.C., 1953.

G. Wyszecki and W. S. Stiles, "Color Science: Concepts and Methods, Quantitative Data and Formulas." Wiley, New York, 1967.

---

[61] Manufactured by Photo-Research Corp., 837 N. Cahuenga Blvd., Hollywood, California.

[62] W. J. Brown, A physical color temperature comparator, *J. Sci. Inst.* **31**, 469–471 (1954).

[63] P. Hariharan and M. S. Bhalla, A precision direct-reading colour temperature meter, *J. Sci. Inst.* **35**, 449–452 (1958).

[64] Manufactured by Milletron, Inc., 540 Alpha Drive, Pittsburgh, Pennsylvania 15238.

CHAPTER 6

Astronomical Telescopes

A. B. MEINEL

Unversity of Arizona, Tucson, Arizona

I. GENERAL SPECIFICATIONS

The design and construction of large telescope systems present one of the major challenges in the field of optical instrumentation. In large telescopes, one encounters the requirement of high optical image quality in structures so large that the flexural deformation of the optical and mechanical systems are serious limiting factors. For the past three decades, the frontier of telescope art was defined by the 200-in. Palomar telescope. At this time, it appears likely that another period of advance in the state of the art of telescopes is about to occur. There are plans under way throughout the world for ten telescopes larger than 120-in. aperture, led by the 238-in. telescope of the USSR. It is therefore appropriate at this time to review the problems of large telescope contruction.

In the last three decades, there have been tremendous strides taken in many technological fields, yet telescope art has remained relatively stationary, perhaps as a consequence of the fact that present telescopes perform quite well under the limitation imposed by the atmosphere. As space and balloon telescopes become important new tools, the boundary conditions will change and new capabilities will be required. In this chapter, we will not discuss the Palomar 200-in. or Lick 120-in. telescopes, since they are extensively described elsewhere.[1]

A. Atmospheric Limitations

In the case of a telescope located on the ground, the principal limiting factor is the atmosphere.[2] The magnitude of the effect of the atmosphere varies considerably with the use of the telescope. For example, the atmosphere at the best astronomical sites rarely permits a large telescope to give images less than 0″.5 for observation integration times exceeding 1 sec. The average size of the resulting photographic image is closer to 1″.2. On the other hand, for short exposures with moderate-sized instruments (of the order of 20 in. in aperture and 0.05 sec of exposure), one can get images as small as 0″.15–0″.20 under the best conditions.

The atmospheric limitation is set by the modulation transfer function for the atmosphere, which is characterized by an average "cell-size" of 3–10 in. This cell-like pattern is constantly in motion and varies in detailed structure with time. As a consequence, a small telescope will time-average to exactly the same distribution of intensity in the image as a large one, but the large telescope usually integrates over enough spatial distance that it never has as good an image size with a short exposure time as will a smaller telescope.

The relationship between the image profile and the seeing disturbance is given in Fig. 1. If the only variable is the amplitude of the disturbance,

---

[1] I. S. Bowen, The 200-inch Hale Telescope, p. 1, and W. W. Baustein, The Lick observatory 120-inch telescope, p. 16, *in* "Telescopes" (G. P. Kuiper and B. M. Middlehurst, eds.). Univ. of Chicago Press, Chicago, Illinois, 1960.

[2] J. Stock and G. Keller, Astronomical Seeing,[1] p. 138.

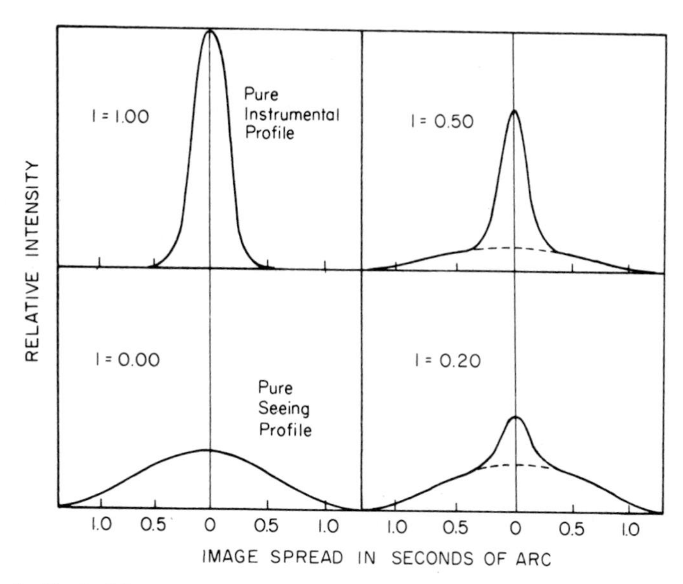

FIG. 1. Typical instrumental profile of a star image for a telescope under different seeing conditions. In each case, the telescope aperture is large compared to the characteristic mean wavelength of the seeing disturbance. The quantity I represents the fraction of the total light that is disturbed by the modulation transfer function into the instrumental profile.

then the instrumental profile is changed only by the subtraction of energy out of the profile and its redistribution in the profile determined by the modulation transfer function for the disturbance. When the rms value of the seeing disturbance is small, the original image sharpness is not significantly altered.

Additional boundary conditions are imposed by the atmosphere. At the ground, the ultraviolet $\lambda < 3000$ Å is absorbed for all astronomical objects by ozone between 3000 Å and 2000 Å and by molecular oxygen below a wavelength of 2000 Å. Most optical glasses are transparent for $\lambda > 3500$ Å, but telescope systems that must reach the atmospheric cutoff and involve transmission elements must use special materials like fused silica ($\lambda >$ 1800 Å), sapphire ($\lambda > 2000$ Å), or other crystalline materials. Above the atmosphere, the entire spectral region is accessible for solar observations and, for $\lambda > 900$ Å, for stellar observations, but new problems arise from loss of reflectivity of mirrors and the lack of transmitting materials. Now that balloon and space telescopes have removed this limitation, the problem of telescope design has opened into a new area where the current state of the art in optical, thermal, and mechanical areas constitute the apparent limiting factors.

B. Optical Specifications

The specifications pertinent to a telescope depend upon the perfection expected of the optical system. The most critical component in the performance of the optical system is the primary mirror, because of several factors. First, the size of the mirror presents a problem in the attainment of the desired accuracy and smoothness of the optical figure. Second, the size means that the thermal relaxation time is large and that optical deformations due to mechanical stresses caused by thermal gradients can occur. Third, the size means that mechanical constraint forces can cause deformations of the mirror surface.

By optical specifications, we mean the optical performance of the mirror under ideal conditions in the laboratory, where the errors in optical figure are the residuals remaining after the optician has concluded his efforts. For a large astronomical mirror, the specifications in terms of energy spread that are accepted as excellent are:

50%	within	0″.3 (dia)
80%	within	0″.6 (dia)
100%	within	1″.0 (dia)

The image profile represented by these figures is similar to that shown in Fig. 1. The energy distribution of an astronomical mirror is usually determined from a spot diagram using a Hartmann test screen with the order of 50 holes. While there are theoretical objections to the accuracy of a Hartmann spot diagram for near-diffraction-limited optical systems, all large telescope mirrors are so for from diffraction-limited performance (i.e., $>80\%$ of the energy within the central maximum) that the Hartmann test constitutes a practical means of evaluation.

In the case of large, diffraction-limited mirrors, as may be required in the performance of space-astronomy tasks in the next decade, one must use other means of defining the optical specifications for the mirror. One parameter that is becoming useful is a value of the root-mean-square value (rms) for the deviation of the surface, since this quantity can be statistically related to the amount of energy distributed outside the diffraction image. The exact relation has been given by Shack.[3] The resultant specifications are:

2%	outside diffraction pattern $= 1/50\lambda$	rms
10%	outside diffraction pattern $= 1/20\lambda$	rms
50%	outside diffraction pattern $= 1/8\lambda$	rms

[3] R. V. Shack, *Appl. Opt.* **3**, 1171–1181 (1964).

The acceptance of a telescope mirror should only be made on the basis of more than one type of test. Current astronomical practice is to use the Hartmann test to verify the absolute shape of the surface (done without any auxiliary test elements) and a Foucault knife-edge photograph, with necessary null corrective elements, to verify smoothness. These two tests should be further augmented, in the case of a near-diffraction-limited mirror, by an interferogram test to establish the rms value. It must be remembered that for an interferogram to be definitive of the absolute shape of the surface, it must be obtained without auxiliary corrective optics, except perhaps for a test flat.

The specifications placed upon the optical performance of a telescope depend to a considerable degree upon the application. A telescope to be used for photometry or spectroscopy need not have the resolution required for high-resolution photography. In the case of a terrestrial telescope, both the seeing and the emulsion, e.g., determine the maximum useful focal length. The astronomer always uses a relatively grainy photographic emulsion because the quantum efficiency for developed grain is highest for large grains. A large-grained emulsion, however, has a rapid blackening rate, and hence can handle less information in terms of net density, but a fine-grained emulsion with large information-handling potential has a lower quantum efficiency. Tests with Eastman Kodak emulsions, e.g., show that a plate having a resolution of 1 μ has about half the quantum efficiency of a plate having a resolution of 15 μ. In some cases, the additional data-handling capability of the 1-μ emulsion may override its lower quantum efficiency. A photographic telescope is generally designed for a 15-μ resolution emulsion, and the image size for maximum rate of information acceptance is approximately 15 μ. This means that for $0''.25$ seeing blur, a telescope must have a focal length of approximately 4.0 ft. On the other hand, when good photographic photometry is to be done, one wants a larger image, i.e., more grains in the image, and focal lengths larger than 4.0 ft are still quite useful in astronomy.

Most image tubes, intensifiers and orthicons, have poorer resolution, more like 20–40 μ, so they match longer focal length systems. In the absence of higher-resolution image tubes or high-efficiency, fine-grained photographic emulsions, one must contemplate very long focal lengths for a large diffraction-limited space telescope. The small cathode sizes of present image tubes, moreover, means that one can record only very small fields of view.

The specifications to be assigned to the rest of the system are closely tied to the optical specifications that are met in operation. These are quite different from the specifications for the optician, since they add two new important effects: (1) mounting deformations and (2) thermal deformations. One generally sets the mounting tolerances to be of approximately the same magnitude as that of the mirror. For example, for a large terrestrial telescope, one would assign errors in the following manner: (1) mirror supports, 0″.3; (2) decollimation, 0″.3; (3) thermal deformation, 0″.5; and (4) drive smoothness, 0″.3 wander/min; which, when added to (5) seeing, 0″.3–1″.5; and (6) mirror figure, 0.″3; means that it really is remarkable when all factors combine favorably and a telescope yields 0″.5 images or better. It is on these occasions that the astronomical yield can be great; every effort must be made to maximize these occasions.

II. OPTICAL CONFIGURATIONS

The optical system of a telescope consists of a large primary mirror to gather the radiation, and secondary optics, when required, to assist in focusing the light in accordance with the desired application. Secondary optical trains can consist of as few as a single reflecting element or many mirrors and transmitting elements.

The design of a telescope optical system must recognize the potential need to adapt to many uses and detectors. It is therefore important to have available at least one focal position of good resolution that has a reasonable field of view and no transmitting elements in the optical path. Current experience with the 200-in. Palomar telescope indicates that it is important to be able to do photometry from a wavelength of 3000 Å to as far in the infrared as 1.0 mm. A telescope needs, in addition, a multiplicity of observing stations, both to accommodate different basic *f*-ratios and complex instrumentation.

A. Prime- or Newtonian-Focus Configurations

The optical configuration for the prime focus and the Newtonian focus is quite simple. In the first case, one only has the primary mirror to provide the focal properties. A parabolic primary provides a focus, and telescopes prior to 1962 have uniformly been provided with such a mirror figure. The Newtonian configuration is achieved by adding a flat mirror near the focus to relocate the image at a convenient location for access by the astronomer. Telescopes larger than 100 in. are often large enough to permit the observer to ride inside a cage at the top of the telescope tube. The field of good definition is very small at the prime- and Newtonian-focal positions, so that corrective optical elements must be added to permit effective use of these two focal configurations for direct photography.

B. Prime-Focus Wide-Field Correctors

The Ross corrector concept developed from Ross's suggestion[4] that an air-spaced doublet could be used to control the field curvature, coma, and astigmatism of a paraboloid. He subsequently added a meniscus ahead of the doublet to permit the correction of spherical aberration. The useful field of a Ross corrector is limited to 10–15′ of arc as a result of a combination of fifth-order aberrations and chromatic variation of the third-order aberrations. Ross correctors generally have poor lateral color performance.

The art of prime-focus correctors of the Ross type has been extensively explored in recent years by Wynne (UK)[5], Schulte (US), Baranne (France), and Köhler (Germany). A summary of the work by each of these designers was recently presented at a symposium sponsored by the International Astronomical Union.[6]

Ultraviolet transmission presents a special problem in Ross correctors, since flint glass is used. One corrector design by Baranne,[7] however, uses one glass type, and while chromatic variation of the focal position is large, the image quality is good from 3400 to 7000 Å.

A new approach to a prime-focus corrector was proposed by Meinel,[8] in which the aberrations of the paraboloid are corrected near the focus by an array of thin aspheric plates. A design by Schulte and Meinel for a Meinel corrector for 1° field for a 150-in. $f/2.7$ Ritchey–Chrétien primary is shown in Table I.[9] A similar design has recently been made by Köhler.[10]

It is perhaps disappointing to the optical designer, now that a large improvement has been made in prime-focus correctors, that the properties of photographic plates are shifting the emphasis in direct photography to the Cassegrain focus; however, the introduction of new detectors can have a profound influence upon telescope design. In particular, image tubes can be efficiently used at prime focus where the linear size of the field is small. In any event, the prime-focus position on a telescope and suitable correctors must still be considered as essential for modern research with a large telescope.

---

[4] F. E. Ross, *Astrophys. J.* **81**, 2, 156 (1935).

[5] C. G. Wynne, *Appl. Opt.* **4**, 1185–1192 (1965).

[6] D. L. Crawford (ed.), *IAU Symp. No. 27, Construction Large Telescopes, Tucson,* 1965. Academic Press. New York, 1966.

[7] A. Baranne, *Rep. Europe South. Obs. Comm.* (1964, 1965).

[8] A. B. Meinel, *Astrophys. J.* **118**, 335 (1953).

[9] D. H. Schulte, *Appl. Opt.* **2**, 141–151 (1963).

[10] H. Köhler, *Appl. Opt.* **7**, 241 (1968).

TABLE I

MEINEL–SCHULTE ASPHERIC CORRECTOR

	Primary	1	2	3	4	5	6	
C	-2.337×10^{-4}	-1.340×10^{-4}	Plano	$+3.740 \times 10^{-4}$	Plano	-5.20×10^{-4}	Plano	Focus
AD	$+8.643 \times 10^{-13}$	$+1.058 \times 10^{-7}$		-7.27×10^{-7}		$+1.94 \times 10^{-6}$		
AE		$+5.91 \times 10^{-12}$						
Separation	−914.0	−1.52	−50.80	−1.39	−50.80	−1.27	−50.38	
Material	Air	Silica	Air	Silica	Air	Silica	Air	

Sag(cm) $= Cr^2 + ADr^4 + AEr^6$ (r in cm)

C. CASSEGRAIN-FOCUS CONFIGURATIONS

The telescope consisting of a parabolic primary mirror and a hyperbolic secondary is traditionally used, since the system produces a sharp focus with either the paraboloid alone or with both mirrors. This arrangement is an advantage where one needs the prime focus or Newtonian focus as well as a Cassegrain or coudé focus. A second advantage is that the mirrors can be optically figured by conventional methods.

As long as the astronomer is interested in only a small field close to the axis, this system is satisfactory. The principal aberration is coma, which varies with the primary f-ratio, F, the linear distance off-axis, l, and the secondary magnification, m:

$$\text{coma}_T = 3l/16m^2F^2$$

For a standard Cassegrain telescope, mF is the effective f-value of the system at the Cassegrain focus. Astigmatism is the principal third-order aberration that depends upon the magnification of the secondary.

The Ritchey–Chrétien design is currently popular for large telescopes because it yields a Cassegrain configuration with zero coma. The hyperbolic figure of the Ritchey–Chrétien primary, with $e = 1.05$–1.15, is compatible with the design of prime-focus correctors.

In this system, the principal working foci are the Cassegrain and the coudé; no transmitting elements are required for good field performance, so that any detector from 3000 Å to 1 mm in wavelength can be used. The principal remaining aberrations are astigmatism and field curvature, but these can easily be corrected by a single-element transmitting corrector near the focus. The properties of a group of Ritchey–Chrétien configurations have been summarized by Dessey.[11]

D. Cassegrain Wide-Field Correctors

The changing emphasis to direct photography at the Cassegrain focus has led optical designers to try to improve the Cassegrain field of view. Designs for Cassegrain correctors have been proposed by a number of persons. Rosin[12] and Wynne[6] have developed some exceedingly good correctors by using a doublet of zero power about half-way from the focus to the secondary, plus, when needed, several more elements near the focus. One can, in fact, obtain near-diffraction-limited performance over fields of view as large as 30′. Another design developed by Schulte and Meinel[5] uses a weak aspheric corrector plate plus a field-flattening lens near the focus.

Another type of Cassegrain corrector is represented by the focal reducer as designed by Meinel[13] and Courtes.[14] In this case, the telescope f-ratio is changed for fast photography with narrow-band filters. The normal field aberrations of the Cassegrain focus can be corrected in the transformation. An optical diagram of a focal reducer for the 82-in. McDonald telescope is shown in Fig. 2. A recent study of the families of two-mirror systems has been given by Shack.[15]

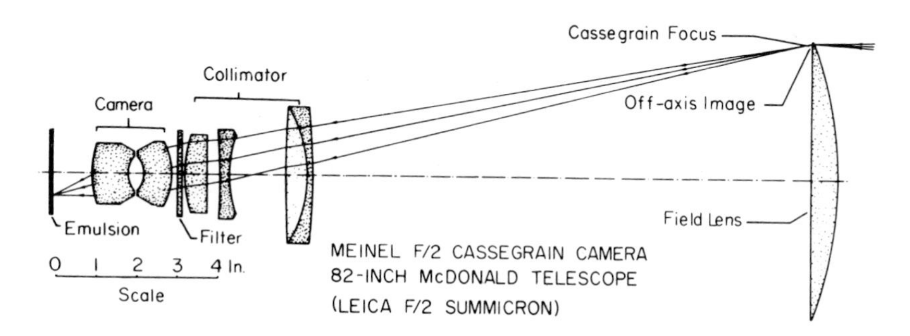

FIG. 2. Optical diagram for an $f/2$ focal reducer designed by Meinel for use with the 82-in. McDonald and the 40-in. Yerkes ($f/3$ effective) telescopes.

---

[11] J. L. Dessey, *Rep. Obs. Astron., Cordoba* **100**, (1963).
[12] S. Rosin, *J. Opt. Soc. Am.* **51**, 331–335 (1961).
[13] A. B. Meinel, *Astrophys. J.* **124**, 652 (1956).
[14] G. Courtes, *Astron. J.* **69**, 317 (1964).
[15] R. V. Shack, *Opt. Sci. Center. Newsletter* **3**, 64 (1969).

E. Hybrid Configurations

Two configurations that have been occasionally used for Cassegrain telescopes are the spherical primary and the elliptical primary (Dall–Kirkham) systems. The former has an advantage of a spherical primary at the expense of a very aspheric convex secondary, a small useable field due to coma, and high sensitivity to decollimation. The Dall–Kirkham system has the advantage of a spherical secondary and a primary for which both conjugate foci are available for direct test by the optician during figuring. It too has a small field due to coma, being usually three to four times worse than for a normal Cassegrain system.

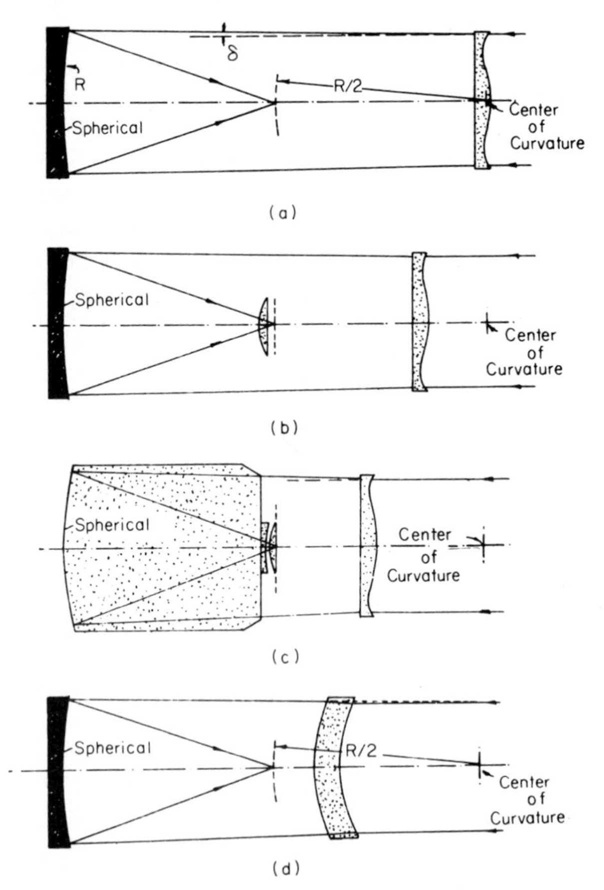

Fig. 3. Four catadioptric configurations. (a) A normal Schmidt with corrector at the center of curvature of the mirror. (b) A flat-field modification using a spaced field-flattener and displaced corrector. (c) A flat-field solid Schmidt for spectrograph camera use. (d) A normal Maksutov configuration. The asphericity of the corrector is greatly exaggerated in (a), (b), and (c).

F. Catadioptric Configurations

The requirement for telescopes with useful fields exceeding 1° can be met only by catadioptric systems. Here, the power of the system is provided by a concave spherical mirror and a full-diameter corrective element (Fig. 3). The Schmidt configuration achieves a wide field by the use of a planoaspheric transmitting corrector plate located at the center of curvature of the mirror. The Maksutov,[16] or Bouwers,[17] system achieves the same correction by use of a concentric or nearly-concentric transmitting meniscus, the mean center of curvature being located at the center of curvature of the spherical mirror. These configurations have a curved focal surface with a radius about half that of the primary mirror. All third-order image aberrations are negligible, and the maximum size that can be used is set by oblique spherical aberration (Schmidt), by axial chromatic aberration (Maksutov), and by chromatic variation of spherical aberration (both). These problems, coupled with the difficulty of obtaining a large glass disk of optical quality, set the maximum practical size for these telescopes at about 1.5 m.

Flat-field configurations of the Schmidt or Maksutov type can be obtained, but the aberrations are considerably larger, and these designs are generally used for small systems such as for cameras in spectrographs.

G. Coudé Configurations

The third focal position of a modern telescope is the coudé focus, which is growing in importance as auxiliary instruments become larger and more complex. The coudé focus is generally placed at the extremity of the south pier, although designs have been proposed in which this focus is located at other positions, depending on the structural design of the telescope. The coudé focus in all designs, however, is characterized by a fixed final position with respect to the building rather than to the moving telescope, as are the prime and Cassegrain focal positions.

The usual range of *f*-ratios at the coudé focus is from 30 to 36, since the beam must be piped a considerable distance from the telescope. The two-pier asymmetrical design, as in 82-in. McDonald telescopes, is compact enough for the use of an *f*/24 coudé beam. The coudé *f*-ratio can be selected arbitrarily, but some new telescope designs use a single reversible secondary with one side figured for Cassegrain use and the other for coudé use. Since the axial positions of the "Cass" and coudé reflecting

---

[16] D. D. Maksutov, *J. Opt. Soc. Am.* **34**, 270–284 (1944).

[17] A. Bouwers, "Achievements in Optics," pp. 16–65. Elsevier, Amsterdam, 1946.

surfaces then are approximately the same, the coudé f-ratio is fixed by the mechanical design.

a. Mirror Configurations. The number of mirrors required in a coudé arrangement varies with telescope configuration. In the case of a symmetrical mounting, one needs either three or five mirrors. The 100-in. Mt. Wilson, 120-in. Lick, and 200-in. Palomar telescopes use a configuration in which the No. 3 mirror must be moved in declination for each star through half the angle that the telescope is moved. For telescope positions to the south of the zenith, the three-mirror configuration is used (Fig. 4a). For angles appreciably north of the zenith, the primary mirror intercepts the beam, and an auxiliary No. 4 and No. 5 mirrors must be used (Fig. 4b).

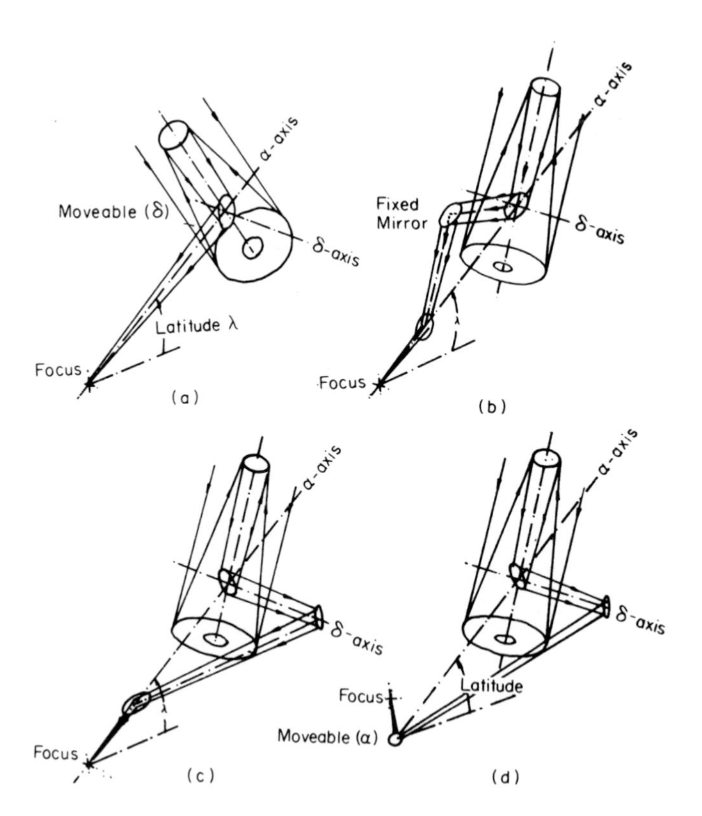

FIG. 4. Four Coudé folding mirror configurations. (a) Three-mirror, used for southerly declinations. (b) Five-mirror alternative to (a) when northerly declinations cause interference by the primary mirror. (c) Fixed 5-mirror Coudé. (d) Horizontal 5-mirror Coudé. In both (a) and (b), the No. 3 flat must be tilted. In (c), Nos. 3, 4, and 5 flats are fixed. In (d), the No. 5 mirror must be tracked with the polar axis motion.

The 84-in. and 150-in. Kitt Peak telescopes use a fixed five-mirror coudé. This arrangement uses the No. 3 mirror to deflect the beam out through the declination axis (Fig. 4c). The No. 4 mirror is at the end of the declination axis and reflects the beam to the No. 5 mirror located in the polar axis housing.

The 150-in. Kitt Peak and 60-in. Cerro Tololo (Chile) telescopes use a new horizontal coudé configuration.[18] In these telescopes, the coudé beam is rendered horizontal by the use of either the No. 5 mirror or a small No. 6 mirror (Fig. 4d). The trade off in this case is having simplified construction of the dome and access to the coudé spectrograph.

b. Coudé Flats. Coudé optical flats present two special problems in construction because of their elongated shape and the large angle of incidence that often is required. The first problem requires special handling in the optical shop, since the mirror must either be cut from a finished round mirror, or temporary "wings" must be added to make the rectangular blank circular for polishing.

The second problem is that the coudé mirror has a rather tight tolerance on residual curvature. In a near-normal usage, any residual curvature of a mirror can be focused out of the system, but in a highly inclined mirror, the residual curvature produces astigmatism. The amount of residual astigmatism that one can tolerate can then be computed by the following approximate expression:

$$\Delta s = \frac{\Delta \varepsilon}{8F} \frac{\cos \phi}{1 - \cos \phi}$$

where the quantities are as defined in Fig. 5. It is interesting to note that the error in the focal plane is apparently independent of the distance of the mirror from the focus. This results because we have defined Δs over the actual beam size intercepting the mirror; hence the absolute curvature of the mirror must decrease linearly with the distance L. In other words, the inclined flat must have a greater accuracy the farther it is from focus. The tolerance for a coudé flat inclined at an angle of $\phi = 60°$ for an image tolerance of $\Delta\varepsilon = 50\ \mu$ in the focal plane is $\Delta s = 0.4$ waves ($\lambda = 5000$ Å) for an $f/30$ beam.

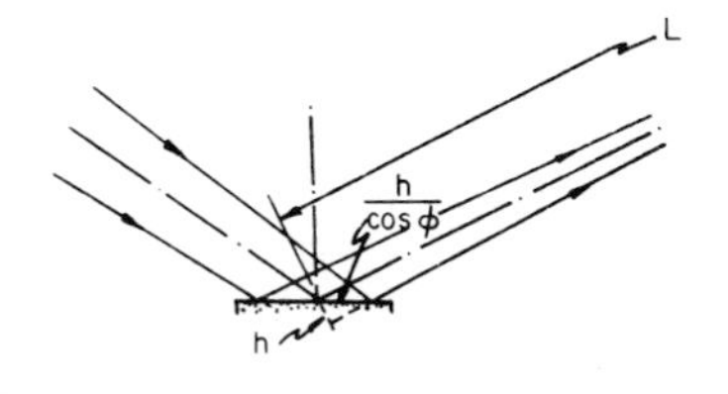

FIG. 5. Diagram for determination of the flatness tolerance for coudé folding flat mirrors.

[18] *Sky Telescope* **35**, 72–76 (1968).

c. Coudé Field of View. The field of view at the coudé focus is quite small, of the order of 5′, so that natural optical aberrations are negligible. On the other hand, the number of reflections required to get to the coudé focus is from one to three more than for the Cassegrain, and the surface-error contributions add up to a point where the image quality is generally poorer than at the other focal positions.

d. Field Rotation. The property of field rotation at the coudé focus adds another optical complication to some types of observation. One can, however, use a derotator consisting of either a Dove or a Schmidt prism of fused silica. The Dove prism will produce some lateral color, while the Schmidt (Pechan) prism has no dispersive power, as the incident light is not far from parallel.

e. Coudé Spectrographs. A coudé spectrograph consists of a collimating tilted spherical mirror, a grating assembly of one or more interchangeable gratings, and an array of cameras. Coudé spectrographs are large instruments, since the effective speed of such a spectrograph for high-dispersion work on stars is dependent upon the beam size of the spectrograph and independent of the telescope aperture. For a complete analysis of spectrograph performance, see Bowen.[19] Coudé spectrograph camera arrays usually cover speeds of $f/1.5$ to $f/12$. Extreme apertures, greater than $f/1.0$, can be achieved by the addition of an aplanatic sphere to the standard coudé camera.

The choice of the plane of the spectrograph dispersion of a coudé spectrograph is one in which convenience must be balanced against performance: The dispersion of a coudé spectrograph is preferred in a plane defined by the polar axis and a horizontal line normal to the polar axis. This configuration minimizes any optical effect from horizontally-stratified thermal air layers in the coudé room. A more convenient arrangement is for the plane of dispersion to lie in a plane defined by the polar axis and a vertical line through the polar axis, as in the case of the 82-in. McDonald telescope. In this case, however, different wavelengths traverse different layers of air, and wavelength errors and variation of image quality could arise from stratified air layers if there is a thermal gradient present.

III. OPTICAL COMPONENTS

The optical materials used in current telescopes are clearly divided according to function. Mirrors are generally used for the large-power elements because of their wideband frequency response, freedom from chromatic aberrations, the possibility of using opaque materials such as metals, and availability in large sizes. Transmitting elements are used as

---

[19] I. S. Bowen, *in* "Astronomical Techniques" (W. A. Hiltner, ed.), pp. 34–62. Univ. of Chicago Press, Chicago, Illinois, 1962.

infrequently as possible, since glass of high quality is available only in small sizes, generally in diameters less than 30 in., and the UV absorption and chromatic aberrations limit performance of the resulting system. The corrector plate for Schmidt telescopes is one exception, since it is possible to obtain large-diameter, thin plates of good optical quality from plate glass manufacturers.

A. Vitreous Materials

1. *Pyrex*

The customary material for large telescope mirrors has been low-expansion borosilicate glasses such as Corning 7160 Pyrex or Schott Duran 50 (see Vol. I, p. 153 of this series). This glass can be cast in large disks, up to 100 in. in diameter, and in ribbed structures. Pyrex materials have expansion coefficients from 2.4 to 3.2×10^{-6} per °C, and this limitation means that a large Pyrex mirror is subject to difficulties arising from thermal gradients within the material. Anneal specifications range form 30 mμ/cm to 60 mμ/cm, depending upon the manufacturer. Strains of this value do not appear to affect the resulting mirror as long as the disk is not cut after finishing, as is sometimes required for coudé mirrors. In the latter case, a strain of less than 25 mμ/cm must be obtained in a mirror blank.

Pyrex disks can be obtained either as solid or ribbed structures. There are many advantages to ribbing: (1) reduction in thermal relaxation time, (2) distribution of edge support loads throughout the disk, and (3) reduction in total weight. One problem of ribbed designs is that the support pockets interrupt the continuity of the ribs, thereby lowering the stiffness of the structure.

2. *Fused Silica*

The traditional "ideal" optical material has been fused silica, principally because of its low expansion coefficient. Until recently, however, fused silica could not be procured in sizes above a few tens of inches in diameter. The development effort by Corning, however, has made fused silica pf practical consideration for large telescope mirrors. At the present time, both Corning Glass Company and the General Electric Company are producing telescope disks 150 in. in diameter.

The low thermal expansion of fused silica is an advantage that makes a better telescope mirror in operation as well as during fabrication in the optical shop. The time that the optician spends waiting for a glass mirror to reach "equilibrium figure" is a significant fraction of the time of fabrication. The fact that fused silica disks are usually solid is, however, a disadvantage. Since their thickness is several times that for a ribbed glass disk, the thermal relaxation time is greatly increased. A solid disk of 100 in.

diameter and 12 in. in thickness would have a relaxation time of two to three days, while a 200-in. disk of the same t/D ratio would take two to three weeks. The effects of the larger thermal gradients within the disk offset the advantage of the lower expansion coefficient of fused silica. A ribbed Pyrex mirror of 2.5-in thick webs may therefore show less thermal distortion in astronomical use than a solid silica mirror of 12-in. thickness.

The method of fabrication of large, fused silica disks is quite different from that for glass. The deposition of SiO_2 in vapor from the combustion of silicon tetra-halogen or flame-fused dust yields a superior product for optical mirrors for disks of up to 60 in. in diameter and 4 in. in thickness. Both Corning and General Electric have solved the problem of disks as large as 150 in. in diameter by fabricating a large disk out of many selected smaller blanks. The Corning module is a hexagonal disk about 4 in. thick and 60 in. in diameter; the General Electric module is a hexagonal rod 6 in. across by about 20 in. in length. These modules are carefully cut and rough-polished, then stacked under pressure and fused together by heating the entire array. Silica becomes tacky at a temperature well below fluidity and makes a good bond between the parts.

Fused silica can be given a finer anneal than Pyrex, in large part due to the smaller thermal dimensional changes in the material when it passes through the annealing temperature cycle. Anneals of 10–15 $m\mu/cm$ are typical for fused silica disks.

Fused silica having a zero coefficient of expansion at room temperature is technically feasible by the addition of an impurity such as titanium. This type of fused silica has its zero crossing point at room temperature, or nominally 25°C, whereas normal fused silica has a zero expansion coefficient at approximately −80°C.

Corning has developed a material of very low thermal-expansion coefficient, ULE® titantium silicate, $\alpha = 1\text{–}2 \times 10^{-7}/°C$. This material is produced by vapor deposition in which approximately 8% of $Ti0_2$ is added to the $Si0_2$. The material has hardening characteristics and fabrication possibilities identical to fused silica, and the much lower expansion coefficient is causing this material to supercede ordinary silica for most large optical systems.

3. *Ceramic Glasses*

The general class of devitrified glasses known by trade names of Pyroceram and Cercor (Corning) and Cer-Vit (Owens-Illinois) offers to optics entirely new materials of great promise. While the early Pyroceram mirrors were opaque and grainy "ceramic" glasses, the new products are quite transparent and apparently do not exhibit the scattering of light from a polished surface that previously tended to dull interest in these materials.

The most spectacular property of these ceramic glasses is the very low coefficient of expansion and wide temperature range over which the coefficient is nearly zero. Typical values for the Cer-Vit mirror material and fused silica are shown in Fig. 6.

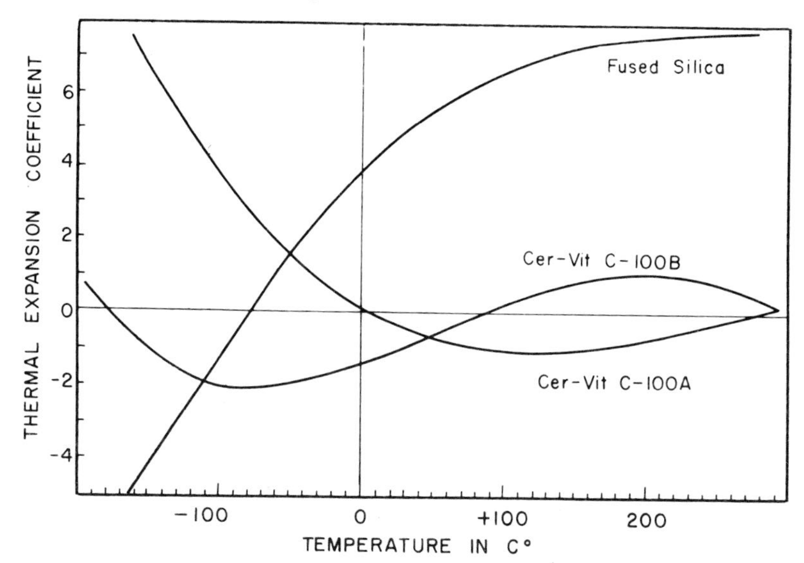

Fig. 6. Thermal expansion coefficient of fused silica compared to several ceramic glasses of the Cer-Vit type. ($1–2 \times 10^{-7}$/°C.)

The ceramic glasses are produced by the inclusion in the glass melt of nucleating agents that act as centers upon which crystalline phases develop upon heat treatment. The properties and compositions of several types of catalyzed crystalline glass are given by Porai-Koshits.[20] Cer-Vit glasses can be cast as a glass and processed before final heat treating for the desired thermal properties. At the present writing, these materials can be cast with section thicknesses of the order of 6–10 in., so that, in ribbed form, one could expect to see castings made up to the order of 100–200 in. in diameter.

B. Metal Mirrors

The use of metal mirrors is based primarily upon their high thermal conductivity and secondarily upon the wide variety of physical shapes and machining operations that one can employ. There are a number of materials that have been tested with varying degrees of success (see Vol. III, p. 46, of this series).

---

[20] E. A. Porai-Koshits, ed., Catalyzed crystallization of glass, *in* "All-Union Conference on the Glassy State," Vol. III. Consultants Bur., New York, 1964.

1. *Invar*

Invar is basically an iron–nickel alloy whose properties can be modified still further by heat treatment and mechanical processing. It is characterized by a very low expansion coefficient relative to other metals, which, in the best cases, is better than that of fused silica. Invars, however, have zero thermal expansion only over a narrow temperature range and satisfactory performance demands either knowledge of the thermal environment of the telescope or active control of the mirror temperature.

Invar has the disadvantage of high density ($\rho = 8.6$), and a solid mirror of satisfactory rigidity is very heavy. Attempts to make lightweight mirrors encounter the problem of machining and fabricating a tough material. If brazed or welded structures are attempted from sheet invar, one encounters the problem of the lack of dimensional stability typical of all lightweight, built-up metal structures. Epoxy-cemented metal structures have also had little success, apparently due to bonding stresses and differences in expansion coefficient between the metal and epoxy.

2. *Beryllium*

Beryllium has an expansion coefficient very much larger than either the vitreous materials or invar, but it has some properties that make this material of interest for mirrors. They are a low density ($\rho = 1.9$), and high modulus of elasticity (40×10^6 lb/in.^{2}), combined with a good ratio of the thermal expansion coefficient to the conductivity. Beryllium has one other fundamental characteristic that offers the possibility of high stability, in that it is hard enough to be used as a pure material. Alloys depend upon phase distribution for mechanical properties and these distributions tend to involve microstresses which can lead to changes with time. Commercial mirror-grade beryllium consists of metallic beryllium with a small amount of beryllium oxide. Even though beryllium should be a good choice for mirrors, the state of the art of mirror-blank forming is such that one cannot predict whether a given mirror blank will make a good mirror until it is finished. In order to achieve low-scattering optical performance, beryllium mirrors must be overcoated with Kanigen before optical polishing, since the pressed or forged beryllium material has appreciable voids remaining after working.

3. *Aluminum*

Aluminum has been successfully used for astronomical mirrors, the most notable case being in the Kitt Peak McMath Solar telescope. The choice of aluminum is attractive both from the standpoint of cost and with regard to an excellent thermal expansion/conductivity ratio. In the case of a solar telescope, the incident radiant flux is high enough to produce bad seeing by direct transfer from the mirror surface to the adjoining air, thereby ruining the mirror as effectively as if the actual surface of the mirror had been warped. Aluminum proves an attractive solution, in that the high conductivity of aluminum avoids a steep thermal gradient at the mirror surface.

C. Optical Fabrication

The time-honored method that the optician uses to arrive at the final optical figure of a surface is one of successive iterations, usually with a pitch-coated tool and a metallic oxide polishing agent. The application of this polishing action to a particular zonal error gradually results in a mirror in which the amplitude of the error is reduced to within the specified tolerance (see Fig. 8). In the case of a spherical surface, this process results in a good end product, since the nature of the polishing action is to bring all parts of the surface into contact, but in the case of a steeply aspheric mirror, no such intrinsic property exists, and the achievement of a smooth figure is solely dependent upon the skill of the optician.

Even in the case of spherical surfaces, the final mirror figure is not devoid of figure defects. When the regular circularly symmetrical zones have been removed, the surface is seen to have random lumps and hollows of 5–10 in. average diameter. The amplitude of this structure is dependent upon the polisher, being smaller for a hard pitch lap. The amplitude also appears to be dependent upon the length of time that the mirror is polished, growing with time. This result may appear strange, but it is perhaps analogous to the growth of "washboard" on a dirt road subject to heavy traffic. The appearance of a near-diffraction-limited mirror when tested with a very fine pinhole is shown in Fig. 7. Note the irregular nature of this surface. The rms error here, however, is only about $\lambda/30$.

In the case of aspheric surfaces, the difficulty of figuring grows rapidly with the amplitude of the asphericity. Our experience would indicate that the degree of difficulty increases approximately as the $\frac{1}{3}$ to $\frac{1}{2}$ power of the amplitude, or, for a simple paraboloid, approximately inversely as the f-ratio. One would therefore always like the f-ratio of a telescope paraboloid to be as large as possible, subject to the operation of boundary conditions. Unfortunately, considerations of the total telescope cost and mechanical flexure of the mounting result in the usual telescope primary f-ratio being closer today to 2.5 or 3.0 than to the $f/5$ of the recent past. As a consequence, mirrors of $f/3.3$ to $f/2.5$ aperture are not as smooth, in spite of increased effort by the optician. When the circularly symmetrical zones

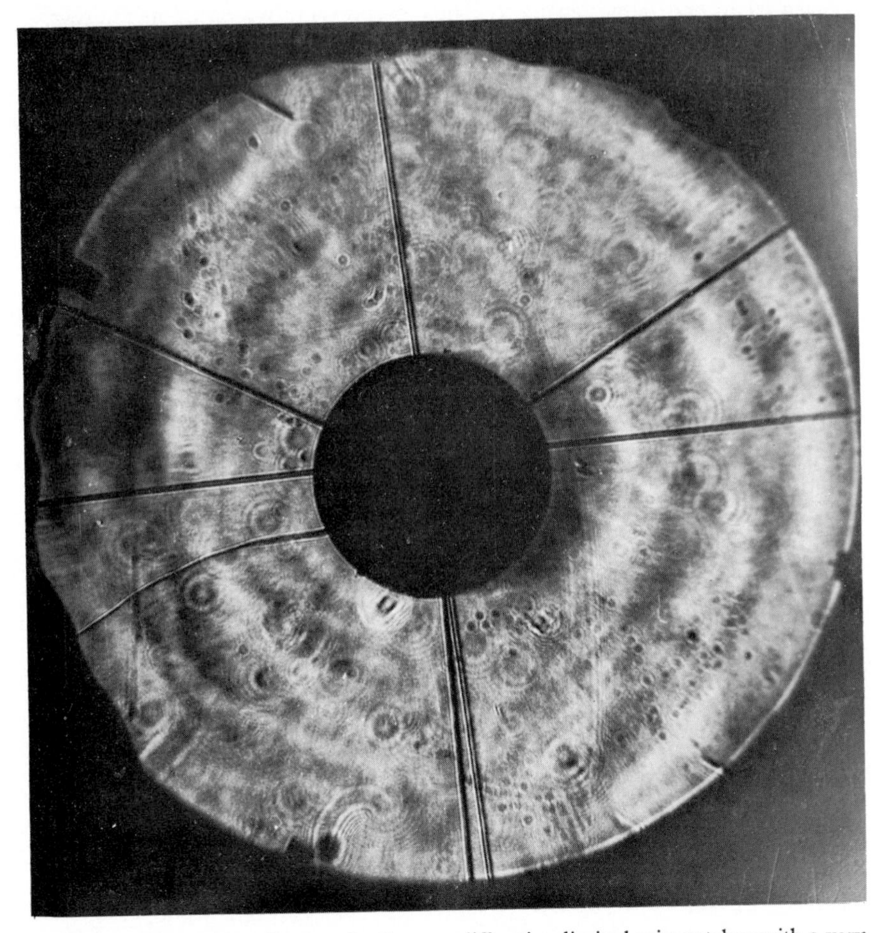

FIG. 7. Knife-edge photograph of a near-diffraction-limited mirror taken with a very small pinhole and monochromatic light. The rms value for this mirror is close to $\lambda/30$.

have finally been removed, the residual "lumpiness" of these high-aperture mirrors is noticeable. The slope errors of the residual zones may be of the order of 0".1 to 0".3, or a surface error of 0.05–0.15 waves for a 20-cm lump.

The polishing of aspheric plates for Schmidt cameras presents a special problem because of the amplitude of the asphericity, which can exceed 1 mm in depth over a 20-cm diameter. Optical working of a corrector is usually done either by means of a succession of ring tools, as developed by Hendrix, or a flexible sheet polisher of thin rubber on which facets are attached, as suggested by Hayward.[21] A close approximation to the final curve can be achieved using precision machined iron tools. Meinel has shown that one can machine an aspheric tool that statistically is very close to the necessary optical tolerances by cutting a series of zonal approximations to the curve. Radial steps are made fine enough so that the depths of successive zones differ by less than 5 μ. If the depth of each zone is cut with an error that is independent of that for any other zone, i.e., random errors, then the action of the Hayward polisher is such as to form an average over 10–20 zonal cuts. Residual errors from the grinding tool, of the order of one wave, can then be removed without serious difficulty by stacking small bar weights, or ring weights, as desired, upon the correct zone of the Hayward polisher.

In 1964, Texereau[22] demonstrated a new technique by hand-figuring the secondary optical elements of the McDonald 82-in. telescope until both the Cassegrain and coudé focal positions gave exceedingly fine optical performance. The Foucault knife-edge test of the system before and after figuring are shown in Fig. 8. The resultant star images at the coudé focus are shown in Fig. 9. It must be noted, however, that once this excellence of figure is achieved, it is necessary that the supporting systems for the mirror do not change with time.

The figure of the mirror during optical fabrication and test also depends upon thermal amd mounting deformations. In the case of a large mirror, the stiffness of the mirror decreases with size, for the same ratio of diameter to thickness, in proportion to $1/D^2$; hence a large mirror presents a problem of support in the optical shop as well as in the telescope. A large mirror should be mounted, if possible, on the same support system as it will later have in the telescope, and with the same precision. For a mirror of the size and stiffness of the 200-in. mirror, the support forces must balance the weight of the mirror to within 0.1%. Achievement of this accuracy together with sufficient ruggedness to withstand the weight of the polishing tool and the thrust of the polishing action requires careful engineering thought.

One customary way of mounting a mirror during polishing is on a pad or multiple pads of sponge rubber. In such an arrangement, one must be careful to pregrind the polishing table and carefully measure the force for a given deflection of each rubber pad. To fail to do these things is to risk astigmatism of the entire mirror or local deformation of the mirror about the support pads.

---

[21] R. Hayward, Unpubl. Rept. NDRC, ca (1944).

[22] J. Texereau, *Sky Telescope* **28**, 345–348 (1964).

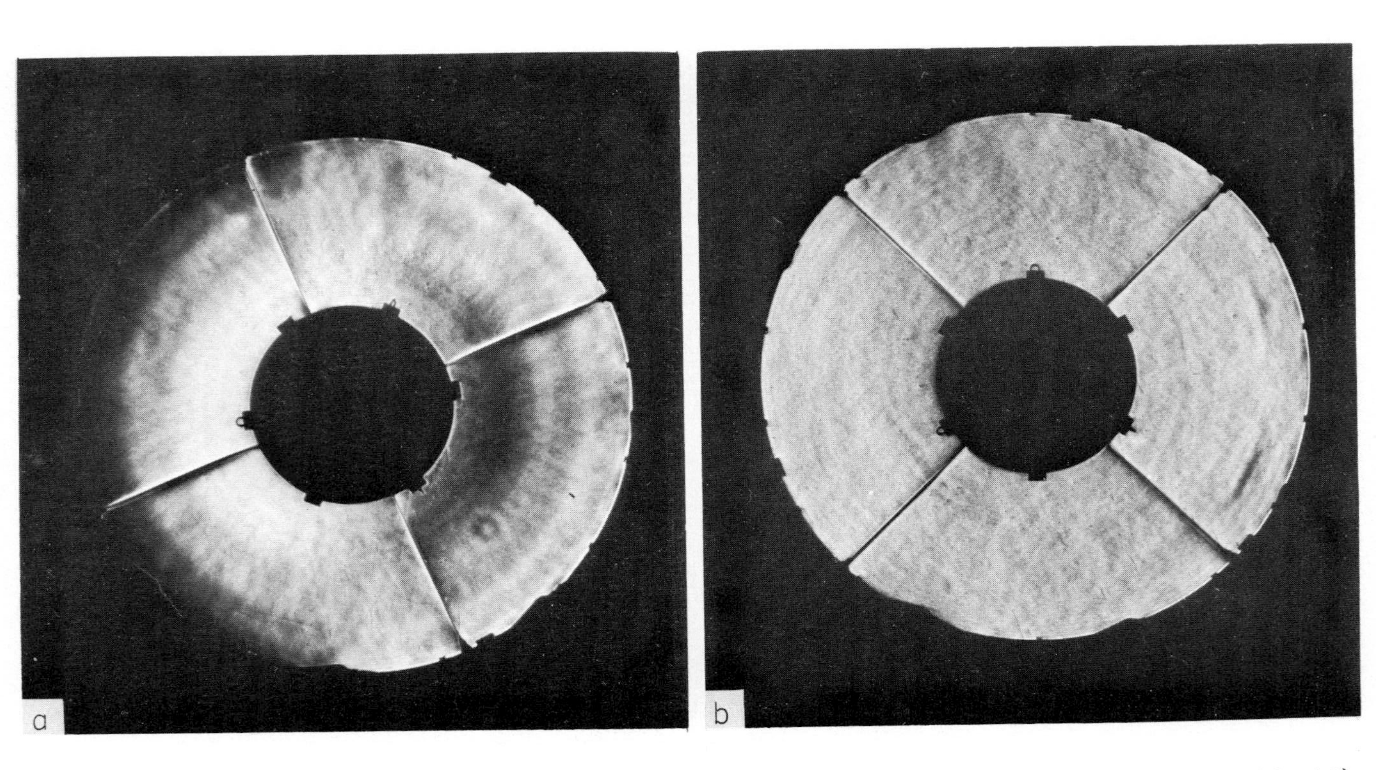

FIG. 8. Knife-edge photographs of the 82-in. McDonald Cassegrain system (a) before figuring, mirror overcorrected by 1.2λ; (b) after seventeenth refiguring run. (From Texereau.[22])

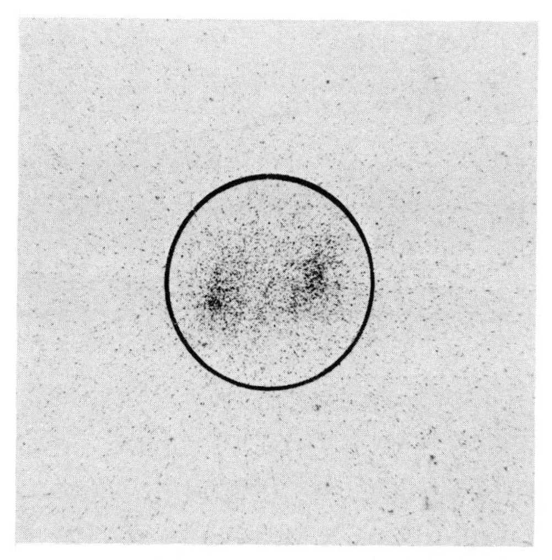

FIG. 9. Double star Σ359, 0.25-sec exposure, 82-in. McDonald telescope at the coudé focus, after refiguring. The circle has a diameter of 1 sec of arc.

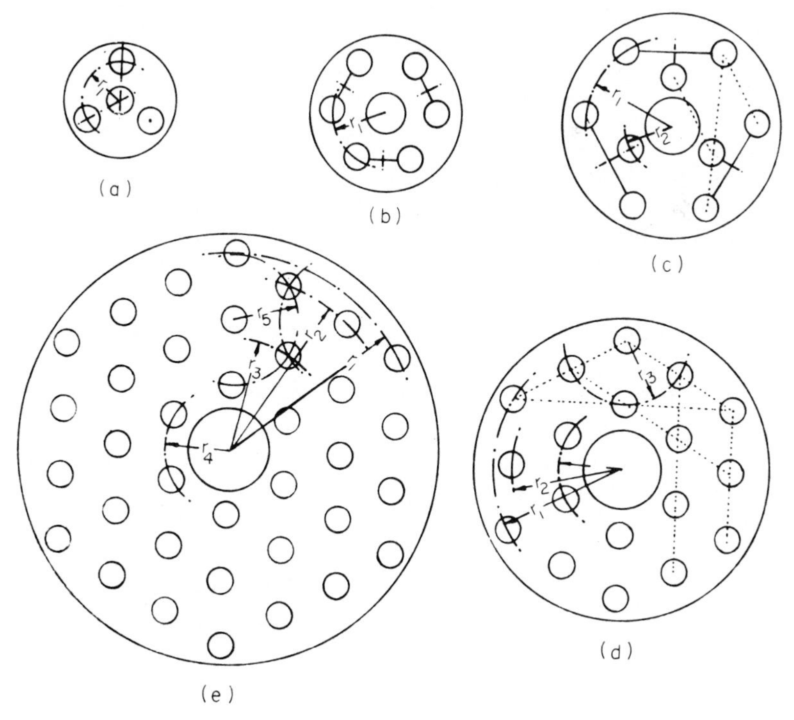

FIG. 10. Various mirror back-support layouts, using: (a) 3-point, (b), 6-point; (c) 9-point; (d) 18-point; (e) 36-point mountings. The values of the several radii are given in Table II.

The decrease in stiffness of a large mirror requires more support points. One cannot simply increase the thickness of a big mirror to achieve the necessary stiffness, since other problems become serious. Various support configurations for mirrors, ranging from 3- to 36-point layouts, are shown in Fig. 10 and Table II.

TABLE II

RADII OF THE VARIOUS SUPPORT-POINT GEOMETRIES OF FIG. 10[a]

Radius	Support configuration				
	3-point	6-point	9-point	18-point	36-point
r_1	0.67	0.71	0.80	0.87	0.91
r_2	—	—	0.46	0.75	0.81
r_3	—	—	—	0.43	0.60
r_4	—	—	—	—	0.52
r_5	—	—	—	—	0.30
Type[b]	*P*	*P*	*P*	*C*	*C*

[a] Radii are for unit mirror diameter.

[b] *P*, self-adjusting pivot pads usually employed, *C*, independent counterweighted pads usually employed.

The final correction of an optical surface beyond the current state of the art of pitch polishing presents a challenge that must be met if better astronomical mirrors are to be made. Several techniques, however, are now becoming available that offer the possibility of removing the random zones and lumps left by correctional polishing. These are (1) selective deposition of chromium, aluminum, or nonmetal by vacuum evaporation, and (2) ionic polishing.

The former is not a new technique, and was first demonstrated as feasible by Strong,[23] who corrected the residual circularly symmetrical zones of a mirror using a rotating mask between the source and the mirror. Efforts are currently being made to apply this technique to asymmetrical zonal errors. This requires the evaluation of the magnitude of the surface errors and the control of the deposition in both amount and location. While this approach is promising, the figure of the mirror is not "permanent," and subsequent realuminizing of the mirror requires a repetition of the entire process.

[23] J. D. Strong, "Procedures in Experimental Physics," p. 181. Prentice-Hall, Englewood Cliffs, New Jersey, 1938.

Ionic polishing, accidentally discovered by Meinel *et al.*,[24] involves the accurate removal of a surface by a beam of ions in the energy range under 1 MeV. Ions in this range apparently remove the surface without damage to the degree of polish or stress of the surface; in fact, the evidence indicated that the polish is improved significantly over a pitch polish. In case of ionic polishing, the surface is permanently changed. The ability to control the position of the beam on the mirror, as in a cathode-ray tube, offers the potential to do local figuring not possible by conventional techniques.

D. Mirror Supports

The proper mounting of the mirrors of a telescope is crucial to its performance. The chief problem is stiction or mechanical hysteresis in the mirror-support system. For a large mirror, the unbalance forces induced into the mirror by the supports should be of the order of 0.1% of the normal load carried by the mechanism in question. Numerous support methods have been devised to meet this specification.

1. *Back Supports*

The family of back-support configurations using from three to 36 supports is shown in Fig. 10 and Table II. The simplest mechanical structure is a three-point support with each pad self-aligning against the mirror back. Variations for use on larger mirrors involve six or nine pads in groups of two or three, each group and each pad being self-aligning on each of the three ultimate attachment points. This method has been used with success on mirrors of up to 48-in. diameter.

For mirrors 60 in. in diameter and larger, one often uses individual counterweighted support pads. Each support pad is pivoted on a precision low-friction ball-bearing unit. The basic structure of a pivoted back support plus the counterwieght required with a lever-arm ratio of 6–10 means that this type of support adds a dead load of about 30–40% of the weight of the mirror. Mirrors in the range of 72–120 in. use 18 counterweight units. The 200-in. mirror uses 36 counterweighted support units.[13]

The troubles experienced with delicate counterweighted support systems have led to recent experiments in the support of mirrors by air pads. In this method, a solid mirror is supported by a rubber or plastic bag such that pressure is applied uniformly over the back of the mirror. The pressure in the bag must be varied with the zenith distance of the telescope to keep

[24] A. B. Meinel, S. Bashkin, and D. A. Loomis, *Appl. Opt.* **4**, 1674 (1965).

the force on the three collimation points constant. The 60-in. US Naval Observatory telescope successfully uses an air-bag support system with a solid, fused silica mirror. A mechanical pendulum device is used to control the pressure. The variation of weight per unit area of a large mirror with appreciable sag of the mirror surface is large enough to cause distortion to the figure of the mirror when supported by an air-pad bag. The 98-in. Newton telescope has the air support bag divided into three annuli, each to carry the proper pressure for the weight of the annular zone.

Ribbed mirrors are required for large telescope mirrors for thermal and weight reasons. A ribbed mirror offers the opportunity of carrying both axial and radial loads of the mirror at each support socket. These support systems are intricate, and much effort is being made to design and construct devices that have no cross-product forces between the radial and axial functions. While these support systems perform well, they do so only at the expense of constant care. At the present time, no satisfactory air-cushion support has been devised for mirrors with ribbed backs.

2. *Edge Supports*

While a mirror has much greater stiffness with respect to optical deflections when placed on edge, the maintenance of good figure requires special care at the edge-support interface. In mirrors larger than 36 in. one usually uses counterweighted edge supports. Edge supports are usually designed to apply push only, and, as a consequence, a relatively large number of supports are needed to distribute the load. An eight-point support system has only three to four points actually accepting loads; a 12-support system, four to five; an 18-support system, seven or eight. The 36-in. fused silica mirror for Stratoscope II, a diffraction-limited mirror, uses 12 supports that apply both push and pull. For the latter function, each support pad is cemented to the edge of the mirror.

The 36-in. Kitt Peak telescope uses edge supports in which the force-producing elements are sylphon bellows. The oil pressure in each bellows is regulated by a counterweighted unit, one for each edge support. The counterweight unit can be placed some distance from the bellows, so that this design can be used where the amount of space at the edge of the mirror cell is limited.

E. Optical Testing

Optical testing encompasses many different techniques and methods, but the two that are most frequently used in the testing of astronomical optics are (1) the Foucault knife-edge test and (2) the Hartmann screen test. Both can be done readily in a completed telescope, as well as, with certain modifications, in the optical shop; hence, one can compare the optical figure in actual use with that obtained in the optical shop. These two tests are complementary. The knife-edge test permits a qualitative evaluation of smoothness of figure, while the Hartmann test provides a quantitative evaluation of the figure itself.

The Foucault knife-edge test is performed by placing a knife-edge at the focus of the telescope. A star is brought into a position where most of the light is intercepted by the knife-edge. With good telescope tracking, the image will remain in this semiocculted condition. A small camera that is focused on the telescope mirror is placed immediately behind the knife-edge. The image in the focal plane of the camera, then, is of the telescope mirror, but the illumination is dependent upon the amounts of light from various points on the mirror passing the knife-edge. The diameter of this image is equal to the focal length of the camera lens divided by the *f*-ratio of the telescope.

If the above test is made with a bright star, the distribution of light is dominated by atmospheric seeing, but if a faint star is used, the resulting exposure of 1 or 2 min is sufficient to average out the atmospheric seeing disturbances and reveal only the mirror-figure errors. The photographs shown in Fig. 8 were prepared by the above technique.

The Hartmann test in the telescope is performed by placing a screen with holes in the incoming beam of the telescope. For ease of analysis, the screen must be made with the holes accurately spaced and the array centered on the optical axis. The holes must be large enough to keep diffraction effects small, say, 3–4 in. in diameter. Several annuli of holes, with the number of holes in each ring such that the spacing between holes is about equal, is desirable. Figure 11 shows the arrangement of holes used for tests at the 84-in. Kitt Peak telescope mirror. The Hartmann test is performed by photographing the pattern of holes with the photographic plate placed out of the best focus of the telescope. One usually goes far enough out of focus to lose the diffraction effects that can be seen between the images of the holes when the intercept is too close to the focus. As in the case of the knife-edge test, the star must be faint enough so that the exposure time for the Hartmann plate is 1 or 2 min to average out atmospheric seeing effects.

In principle, the interpretation of the test is simple, but in practice, the determination of the true figure of a mirror involves considerable numerical analysis. If the test is made exactly on-axis, then the image of the screen from a perfect mirror should have exactly the same geometry of the spots as does the screen. If the area of the mirror defined by a hole in the screen has a slope error, then the spot will be displaced from its normal position. The Hartmann plate is measured in two directions, usually in rectangular

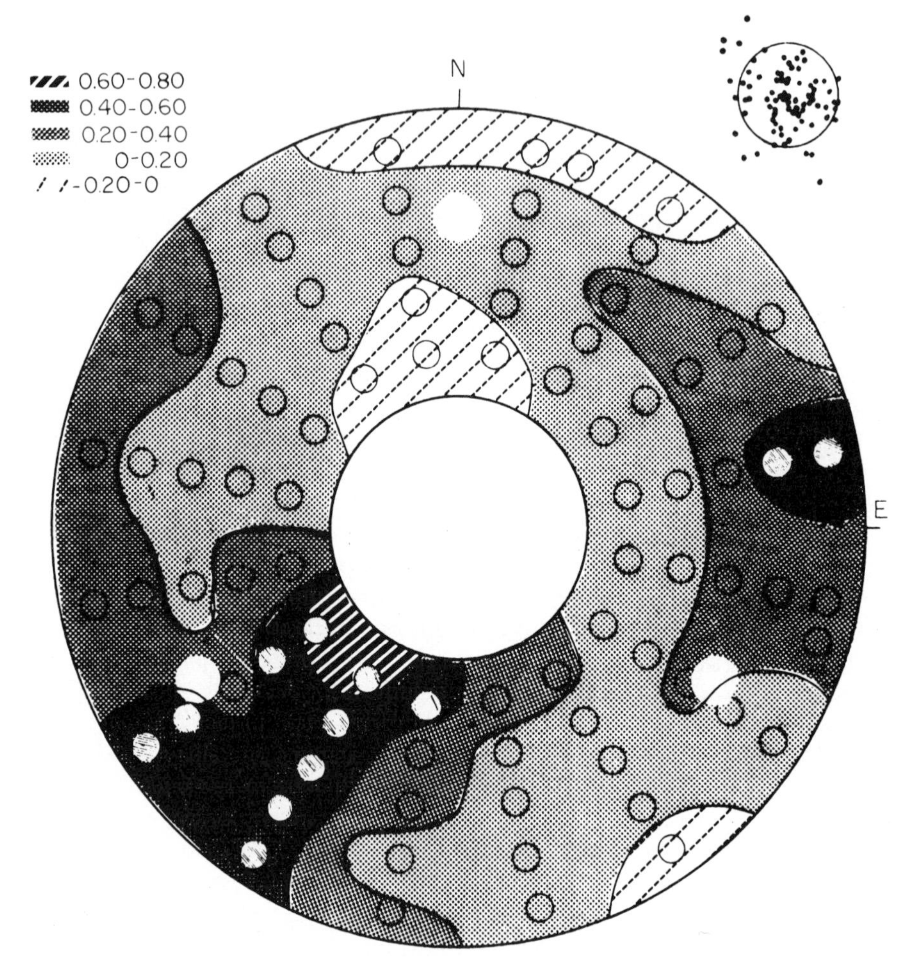

Fig. 11. Hartmann contour map of the 84-in. Kitt Peak telescope made during early tests of the mirror support system. The positions of the Hartmann screen holes are shown by the small circles. The three large circles are the positions of the collimating points. In the upper right-hand corner is the Hartmann spot diagram for a star image compared to a circle 1 arc sec in diameter. The contour interval is 0.2λ, and the Hartmann constant was 0.217.

coordinates, and the radial and azimuthal errors determined following the method given by Mayall and Vasilevskis.[25] A numerical analysis of the slope errors then yields a contour map of the mirror surface. Figure 11 shows a Hartmann contour map of the 84-in. Kitt Peak telescope primary

[25] N. U. Mayall and S. Vasilevskis, *Lick Obs. Bull.* No. 567 (1960).

mirror during an early test of the support system. The image intensity distribution computed from these Hartmann diagrams is also shown in Fig. 11.

It is claimed that the Hartmann test is not sensitive enough for the testing of mirrors that are close to diffraction-limited performance. It is a fact, however, that no astronomical mirror is yet in existence that does not have errors of sufficient magnitude to be shown readily by a Hartmann test. A useful criterion that can be applied to testing is that the Hartmann test is valid until the error pattern of the rays in the focal plane is comparable to the size of the diffraction image diameter. Only then does one need to utilize the more sophisticated and exact methods using scatter plate or other forms of interferometry.

A laboratory optical test of a telescope mirror faces the problem that one cannot test the mirror in as simple a manner as one can in actual use. If the mirror is tested at the center of curvature, the knife-edge or Hartmann test is not a null-test. The test must be corrected for a very large term due to the apparent spherical aberration; hence the resulting residual error may be quite inaccurate. One method which permits laboratory null tests involves the use of a null corrector of the type designed by Offner.[26] This corrector consists of two lenses, of which the first provides the longitudinal spherical aberration and the second the ray slope to compensate these two quantities arising when the mirror is tested at its center of curvature. The Hartmann test can also be performed in the laboratory using a null corrector of this type. The presence of auxiliary optical elements, however, causes uncertainty in the validity of the absolute figure determination. If the corrector elements are incorrectly made or incorrectly spaced, the resultant null-test will have an intrinsic error and will be useless.

IV. SYSTEM DESIGN CHOICES

The selection of a given class of mounting and of the basic *f*-ratio involves a number of choices. These choices depend not upon a simple set of factors, but upon very nebulous and often complex and conflicting factors, not the least of which is the predisposition of the customer as to the exact configuration that he wishes. The principal choices that affect the telescope design are considered below.

A. PRIMARY *f*-RATIO

The choice of the *f*-ratio of the primary mirror is basic and affects the rest of the design. The optical performance factors improve with increasing *f*-number. The smoothness of the final optical figure for a paraboloid or

[26] A. Offner, *Appl. Opt.* **2**, 153–155 (1963).

hyperboloid appears to be inversely proportional to the first power of the f-ratio. The collimation sensitivity increases as the square of the inverse f-ratio. These two factors mean that when faced with a choice of, say, an $f/2.5$ or an $f/5$ primary, the choice would be $f/5$. When the mechanical flexure of the telescope is considered, however, one finds that the flexure increases as the cube of its length. As a consequence, an $f/2.5$ optical system can be held in collimation better than an $f/5$ one.

Quite often, the most compelling factor in the basic choice of telescope design is that of cost. One of the major cost items in the construction of an observatory is the dome and building. Since the cost of the dome increases close to the 2.7th power of the diameter, and the diameter is linearly proportional to the f-ratio, it is clear why the compact $f/2.5$ to $f/3$ telescopes are now popular. These arguments leave the burden of making a good telescope upon the optician.

B. Secondary Magnification

The choice of primary f-ratio affects the magnification required by the secondary mirror in the Cassegrain and coudé configurations. The required size of the secondary mirror is solely a function of the magnification by the secondary. The diameter of the secondary, D_1, for unit diameter of the primary is given by

$$D_1 = (B + f_0)D_1 = (B + f_0)F^{-1}$$

where B is the distance behind the primary of the focus, f_0 and F_0 are, respectively, the focal length and f-ratio of the primary, and F is the f-ratio at Cassegrain or coudé focus. This diameter is for zero field of view, so the diameter of the secondary must be increased by

$$\Delta D_1 = W/M$$

where W is the width of the desired field and M is the magnification by the secondary.

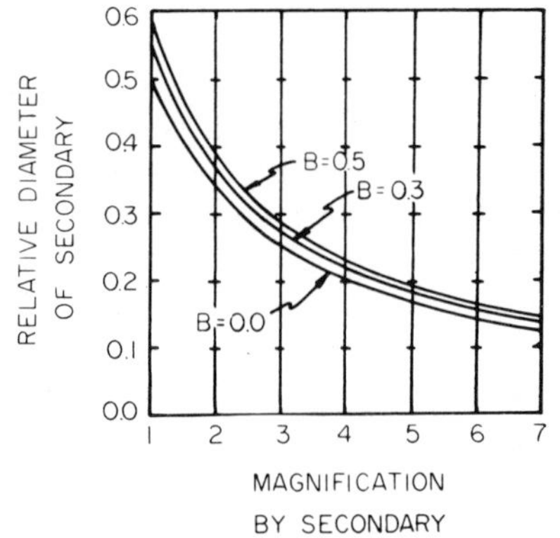

FIG. 12. Relation between the diameter of the secondary mirror of a Cassegrain configuration and the magnification of the secondary. The quantity B is the relative distance of the focus behind the primary mirror.

A graph showing the relationship between B and D_1 is shown in Fig. 12. An $f/2.5$ primary mirror and an $f/8.0$ Cassegrain focus position, e.g., requires a 26-in. diameter secondary for a 100-in. mirror at a focus 30 in. behind the primary. The size increases to 29.1 in. for a field 10 in. in diameter.

C. Secondary and Vane Obscuration

The area removed from the beam by the secondary mirror is generally of minor importance in an astronomical telescope, since the actual size of the image is much larger than the diffraction disk. In the case of a diffraction-limited telescope in space, however, the degradation of performance is appreciable. The rate of growth of the amount of energy in the diffraction rings is approximately linear for $d/D < 0.5$, so that obscurations of less than 20% do not generally cause concern.

Diffraction "rays" in the focal plane of a telescope are produced by small obstructions in the beam, such as by the support vanes for the secondary, by wires, or even by bolt heads. Each straight obstruction produces two rays 180° apart. The total amount of energy in the ray is proportional to the amount of light intercepted (for small obscurations). Wider vanes scatter the energy close to the main image, whereas narrow vanes diffract far out from the image.

D. *f*-Ratio Parameters

The choice of the *f*-ratios for the several focal positions of the telescope depends upon a number of factors. In general, one wishes to have as compact a telescope as possible. Even though the sensitivity to collimation increases as the inverse square of the primary *f*-ratio, the rigidity of the structure increases as the inverse cube of this *f*-ratio. The choice of *f*-ratio, moreover, depends upon the detector to be used, its resolution, and other limitations.

1. *Photographic Uses*

The improvement over the past few years in the speed of photographic plates with no increase in granularity has led to a major change in the optimum *f*-ratio for photography. The condition governing direct photography is that one desires to reach, within 4–5 hr., a limiting background sky fog density of approximately 0.6–0.8 due to the light of the night sky. Longer exposures are not practicable for most programs because of zenith-distance effects on refraction, extinction, and airglow. In the 1940's and 1950's, the optimum was $f/3$ to $f/4$, and prime-focus correctors for the 200-in. telescope of $f/3.3$, 3.6, and 4.5 were quite useful. Current photographic emulsions now reach limiting sky fog levels in less time, and the optimum *f*-ratios for 4–5 hr exposures are now in the range of *f*/7 to *f*/9. These mid *f*-ratios are difficult to obtain in a negative-power prime-focus corrector for an *f*/2.5 to *f*/3 primary, so the emphasis on photographic work has shifted to the Cassegrain focus, where the intrinsic field aberrations are much less severe.

Photographic work at fast *f*-ratios is important for narrowband interference filter observations. When the wavelength interval is small, the continuum due to the stars is suppressed, and emission objects and regions can be studied more effectively. In general, one wants as long an exposure as possible to integrate more flux from the emission object, and an aperture of the order of *f*/1 to *f*/2 is desired. Examples of this type of photography are given by Courtes.[14]

The use of focal reducers to change the initial Cassegrain *f*-ratio from 10–15 to an effective speed of *f*/1 to *f*/2 for narrowband photography, developed by Meinel,[13] depends upon the initial beam diameter and the final beam convergence angle. It is therefore feasible to transform the *f*-ratio of a Cassegrain focus into the equivalent of a fast, prime-focus *f*-ratio. Aside from the addition of a number of optical elements, whose reflection losses can be made very small *via* the use of nonreflecting coatings, the focal reducer does many of the functions one thought to be the sole role of the prime focus.

2. *Image-Tube Uses*

The potential development of image-tube devices for astronomy could have an influence upon telescope design, since their characteristics are somewhat different from those of photographic emulsions. In astronomy, one is dealing with such faint sources that the quantum efficiency of the detector is an important limiting quantity. The best photographic emulsion approaches a quantum efficiency of 0.5–1%. A good photoelectric surface may have a quantum efficiency in the range of 20–30%.

While efficient photomultiplier tubes have been available for astronomical uses for two decades, the achievement of equally efficient image tubes has proven an elusive goal. The history of this effort is too extensive to be included here, so only a brief review of current limitations will be given.[27]

Image tubes have a small cathode area (1-in. diameter), in contrast to the large recording surface area available with photographic plates. The

[27] W. A. Hiltner, Image Converters for Astronomical Photography, *in* "Astronomical Techniques" (W. A. Hiltner, ed.), pp. 34–62. Univ. of Chicago Press, Chicago, Illinois, 1962.

resolution of image tubes, at least in those showing improvement in information-handling capability over photographic emulsions, is of the order of 40–50 *l*/mm. As a consequence, the number of picture elements that one has available is quite limited in comparison to the case for large photographic plates. One must therefore match the focal length of the telescope to a coarser detector, but one of small diameter and limited field. Image tubes thus appear to be best adapted to Cassegrain or coudé *f*-ratio uses for the study of small areas of the sky for the recording of the faintest possible star images. An image tube could not be used at the prime focus of a telescope except for very short exposures, due to the brightness of night sky backgrounds, or if very narrow spectral pass bands are utilized.

3. *Spectrographic and Photometric Uses*

When we consider the use of a telescope for spectroscopic[19] or photometric purposes, we are generally interested in a field of good definition of only a few millimeters in diameter. One exception is important, however: when the object to be studied is very faint or invisible to the eye, the astronomer must locate the object with respect to photographs of the nearby star field, which may have been taken through a distorting field corrector. It is therefore important to have focal positions available that can be used without refractive correctors. The use of the Cassegrain focus for spectrography and photometry is therefore indicated.

In spectroscopy and photometry, we have another general problem that relates to the *f*-ratio of the focal position to be used. The adaptation of any instrument to the prime focus involves the acceptance of a wide angular cone of radiation associated with an *f*/2.5 to *f*/3 beam. In the case of a photometer, it is not possible to use a simple field lens exterior to the envelope of a 1P21 multiplier, and one must change the *f*-ratio to milder values. One definite advantage of a prime-focus spectrograph is, however, the compactness of the instrument for rather sizeable beams (4–5 in.). It is well known that the efficiency of a spectrograph on extended objects is proportional only to the beam size of the spectrograph; hence a large beam size is important for the general programs for which a prime-focus spectrograph is used. The compact size, moreover, is conducive to rigidity of mounting and freedom from flexure during long exposures.

V. TELESCOPE MOUNTINGS

A. Equatorial Class

The astronomer has traditionally used the type of mounting called the equatorial. This preferred mounting provides for motion about two axes. One axis is aligned parallel to the axis of rotation of the earth and is called the polar axis. Motion about this axis is in right ascension or hour angle. Right ascension is defined on the celestial sphere and is fixed with respect to the stars (except for precession, etc.). Hour angle is defined with respect to the meridian. The other axis is perpendicular to the polar axis and is called the declination axis.

The alignment of the telescope with the apparent axis of rotation of the earth (called the refracted pole) requires the equatorial mounting to be driven only about the polar axis, at siderial rate, in order to keep a star centered in the telescope. With such a coordinate system, the field of stars does not rotate with respect to the telescope. Other types of mounting, such as the alt-azimuth, do not have these desirable properties.

Equatorial telescope mountings generally can be classified into two groups, symmetrical and asymmetrical, with respect to the polar axis. A wide variety of mountings has evolved within these two groupings. In most cases, the aim in the design of a mounting is to provide the maximum ease of use by the astronomer, and design evaluation under this condition has moved along with the changing instrumentation. Only in the case of very large telescopes, starting at about 120 in., do the problems of mechanical design begin to take precedence over ease of use.

The most important parameter in deciding between the symmetrical and asymmetrical classes of mounting is the mass of the telescope. In the case of a large telescope, the mechanical deflections become large and the elimination of any unnecessary weight, such as is represented by the required counterweight of the asymmetrical class, is necessary.

1. *Symmetrical Equatorial Mountings*

The symmetrical class is represented in Fig. 13 by (a) the fork mount (Lick 120-in.), (b) the equatorial disk (Newton 98-in.), (c) the cradle-yoke mount (Kitt Peak 150-in.), (d) the horseshoe-yoke mount (Palomar 220-in.), and (e) the English yoke mount (Mt. Wilson 100-in.). A typical fork mount is shown in Fig. 14. These designs differ in whether the north and south bearings are on the same or opposite sides of the declination axis of the telescope. When the telescope is small and the Cassegrain focus is to be used, the north bearing must be either well north or south of the working area (cases a, b, d, or e). If the telescope is large enough for the observer to ride the telescope at the Cassegrain focus, cases (b) and (c) are desirable.

Another factor that affects the choice among the several cases shown in Fig. 13 is the type of bearing. When ball or roller bearings are to be used, the friction radius must be kept as small as possible; hence either case (a) or (e) is the usual choice. If hydrostatic oil-pad bearings are to be used, the low friction of these bearings permits the use of the larger diameters of the

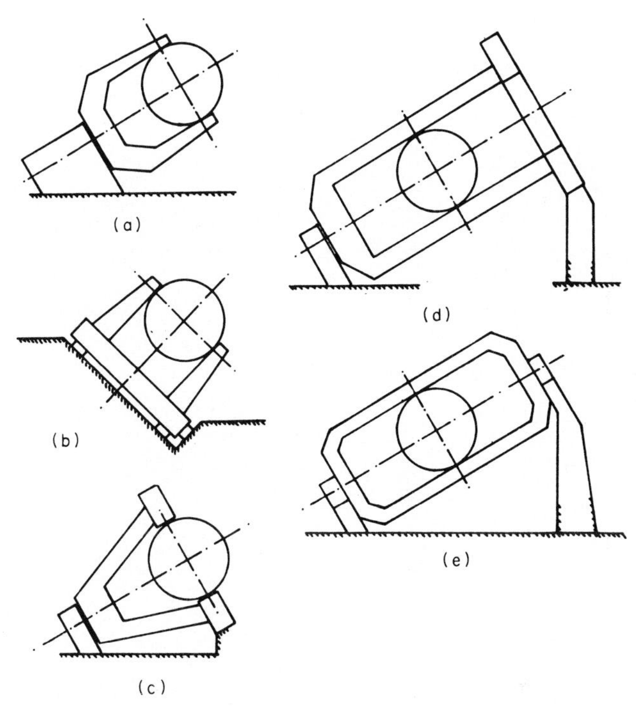

FIG. 13. Five equatorial symmetrical telescope configurations. (a) Fork; (b) polar disk; (c) yoke–cradle; (d) horseshoe yoke; (e) English yoke.

north bearing in cases (b) and (c), or case (d) can be used in order to gain the northern polar region for observation, as well as overall compactness of the right-ascension structure.

The English yoke mount has not seen much use since the 100-in. Mt. Wilson telescope; however, this design has recently been used for three telescopes of apertures from 28 to 61 in. at the Catalina Observing Station of the Lunar and Planetary Laboratory at the University of Arizona. The yoke permits a small friction radius for both north and south bearings. The full symmetry, moreover, avoids the droop problem of the fork mounting tines, and the resultant absolute pointing accuracy is even higher for a lighter-weight structure. The chief disadvantage is a limited access to the north, to about 60° declination.

A minor disadvantage is caused by a two-pier configuration such as the yoke mounting. Each time the polar axis of the telescope is moved for alignment on the north celestial pole, it is necessary to realign the worm with respect to the worm wheel. This problem also exists for the asymmetrical modified English mount (82-in. McDonald).

FIG. 14. The 60-in. US Naval Observatory astrometric telescope at Flagstaff, Arizona. Mounting by Boller and Chivens Division, Perkin-Elmer Corp.

The equatorial disk mounting first used for the 98-in. Newton telescope (Grubb-Parsons: Sisson) uses a pancake-like disk for the polar axis (Fig. 13b). Oil pad bearings on the back and sides constrain the disk and define the axis of rotation. The declination tines are erected on the face of the disk. This ingenious design is primarily suited for rather high latitudes, since the center of gravity of the entire telescope must lie well on the disk.

The 60–80 in. McMath solar telescope at Kitt Peak[28] represents an unusual design that has many uses for nonsolar astronomy. The light-gathering power of this telescope is large enough that the telescope is regularly used for siderial astronomy. This telescope, conceived by R. R. McMath and built under the scientific supervision of A. K. Pierce, uses an equatorial siderostat mounted atop a high pier. The siderostat supports a fused silica flat mirror that directs the starlight down the polar axis. The siderostat rotates about the polar axis to track the object. The drive worm wheel at the south end of the mounting is in the form of a large annulus through which the light beam passes. The declination range of the telescope is limited by foreshortening by the flat mirror to $+30°$, and by mount interference to $-40°$ declination.

The image-forming mirror of the telescope is located approximately 500 ft down the polar axis, at a point well below ground level. This 60-in. mirror, made of cast aluminum, Kanigen coated, is spherical with a focal length of 300 ft. The mirror is tilted slightly off axis and the light is intercepted by a folding flat mounted below the incoming beam, where the light is reflected to the Herschellian focus along a vertical axis in the control room.

This telescope configuration represents the ultimate in ease of operation by the astronomer. The beam of photons is focused in a large laboratory room, which can be temperature controlled, and where many elaborate arrays of instrumentation can be deployed. If a large telescope is ever built in a severe environment, such as upon the moon, a telescope configuration of this type should be seriously considered.

2. *Asymmetrical Equatorial Mountings*

The asymmetrical class is usually chosen both for convenience of observing at the Cassegrain focus and a four-fixed-mirror coudé arrangement. While such a mounting provides for all-sky access and can accommodate large Cassegrain instrumentation, the astronomer must accept a large travel of the Cassegrain focal position and an asymmetrical position

---

[28] A. K. Pierce, *Appl. Opt.* **3**, 1337–1346 (1964).

of the telescope in the dome and with respect to observations east or west of the meridian. The large focus-travel arc of an asymmetrical mounting can, however, be accommodated by use of an hydraulic floor, as, e.g., in the case of the 82-in. McDonald and 36-in. Kitt Peak telescopes. Since no fork is in the way of the floor when the telescope is at large hour-angles, one can use a simple vertical motion of the floor, as contrasted to both vertical and horizontal motion of the Cassegrain observing platform for the 84-in. Kitt Peak telescope. Several asymmetrical class mountings are shown in Fig. 15.

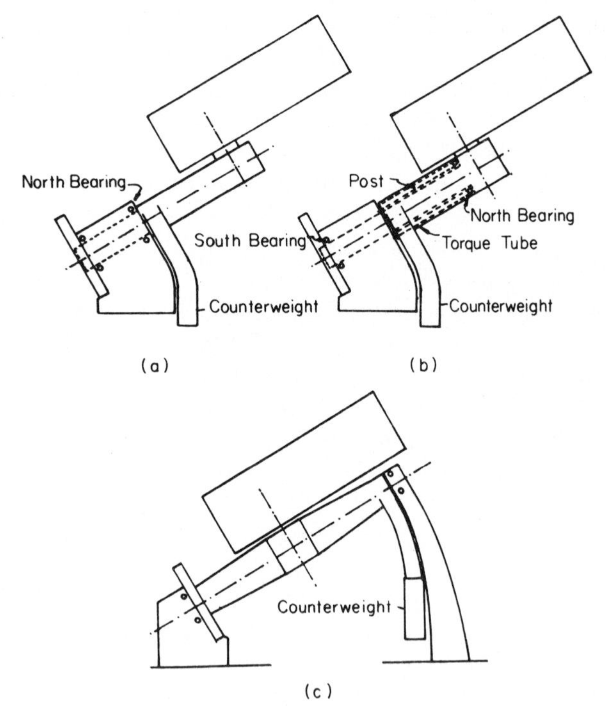

FIG. 15. Three equatorial asymmetrical telescope configurations. (a) Cantilever; (b) torque-tube cantilever; (c) modified English.

Asymmetrical class telescopes have both declination bearings on the same side of the telescope tube. The mechanical moments caused by the weight of the tube become large on small-diameter bearings, and the friction radius becomes large on plate bearings. If it is necessary to accommodate a coudé optical system, then the bearing problem becomes even more difficult, since both bearings must, in addition, be located between the telescope tube and the polar axis in order to clear the No. 4 coudé flat. Therefore the designer must increase the rigidity of both the central section of the telescope tube and the polar axis shaft.

A smaller type of asymmetrical telescope utilizing a separate counterweight torque tube is represetned by the 36-in. Kitt Peak and 40-in. Canberra telescopes. The latter has a coudé focus position, while the former is designed only for Cassegrain use.

B. Alt-Azimuth Class

In the case of very large telescopes, $D > 200$ in., even the symmetrical-class equatorial mounting is approaching a limit set by structural deflections. It is therefore necessary to consider mountings other than the equatorial. The main design approach for very large telescope structures would appear to involve the alt-azimuth mounting, in successful use for large radio telescopes for over a decade (Fig. 16). The alt-azimuth tele-

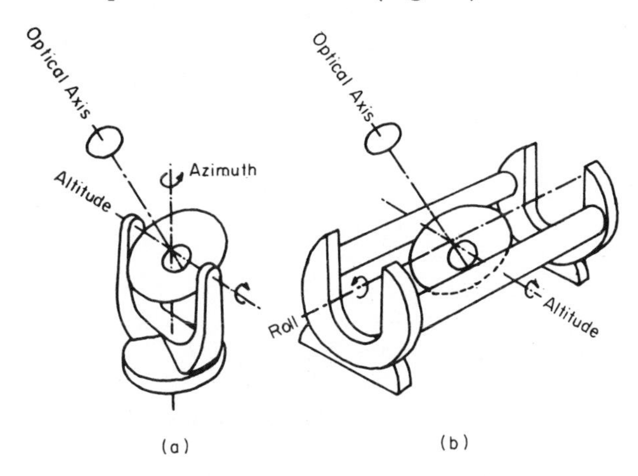

FIG. 16. (a) The alt-azimuth and (b) alt-alt telescope mounting configurations.

scope has a greater degree of symmetry than the symmetrical equatorial, since the alt-azimuth system is also symmetrical about an axis through the zenith. As a consequence, the direction of gravity always lies in one plane through the mirrors and mechanical structure.

The principal problems of the alt-azimuth mounting are: (1) angular drives are required about both the azimuth and elevation axes, (2) the angular drive rates are nonlinear, and (3) the field of view rotates. The drive rate about the azimuth axis approaches infinity at the zenith. As a consequence, an alt-azimuth telescope has a blind spot of several degrees diameter at the zenith, depending upon the maximum angular rate of drive of the telescope.

C. Alt-Alt Class

A new telescope mounting design has recently been proposed by Richardson and by Vasilevskis,[25] termed the alt-alt, shown in Fig. 16. The alt-alt mounting is identical to the alt-azimuth if the vertical axis and its rate singularity were placed on the horizon, either north–south or east–west. The most conspicuous advantage is that the drive rates do not vary as rapidly in the part of the sky of the highest useage. The design also lends itself to a horizontal coudé system with good sky coverage. The east–west orientation discussed by Vasilevskis, however, requires a fourth first-order variable angular drive rate for the No. 3 coudé flat, making four axes with rather large drive rates.

It should be noted that the alt-alt mounting does not offer a structural advantage for a big telescope, since this modification loses the important factor of complete symmetry with respect to gravity that is inherent in the alt-azimuth mounting. The structural deflection problem is quite similar to that for an equatorial; in fact, one could consider the alt-alt simply to be like the 100-in. Mt. Wilson telescope if the yoke were to be laid in a horizontal position.

D. Hybrid Classes

As for other new mountings, both Bowen[29] and Meinel have suggested some designs with advantages for very large telescopes. The Bowen concept as interpreted by Meinel and shown in Fig. 17 is essentially an inverted vertical siderostat. The image-forming mirror is mounted face down at the top of the tower and is supplied with light *via* a large flat mirror located at the foot of the tower. Since the image-forming mirror is stationary, its optical figure can be carefully adjusted and will remain in adjustment for a long interval of time.

The Meinel design (Fig. 18) combines a number of the advantages of the equatorial and alt-azimuth forms, and is probably best described as an equatorial alt-azimuth. The design is based upon a spherical primary and uses the Baker concept of reimaging correction. The primary mirror is kept in a fixed position during an increment of the observing run. The observer and auxiliary research instruments are mounted in a tubular cage that is pivoted at the center of curvature of the spherical mirror and carried on a cylindrical polar axis. This observing structure is moved in the equatorial coordinate system and accurately tracks the star as in the case of a

[29] I. S. Bowen, private communication (1965).

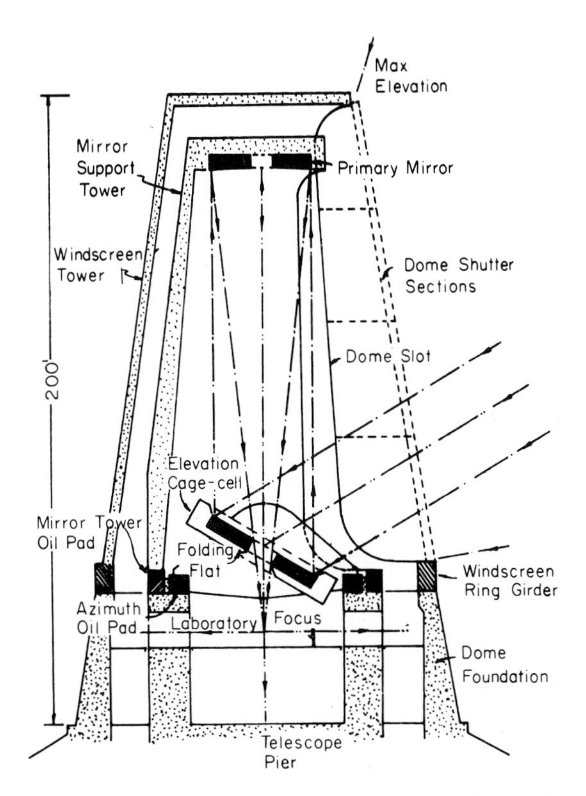

FIG. 17. Diagram of a vertical siderostatic configuration designed by Bowen and Meinel for application to a very large telescope system.

normal equatorial mounting. The equatorial–alt-azimuth mounting has a very low mass for the equatorial moving structure and largely eliminates the dome. The diameter of the moving part of the dome required for this mounting is only slightly larger than the beam diameter of the telescope.

Two telescopes of hybrid asymmetrical design were recently constructed by Century Detroit Corporation for ARPA and the University of Michigan observatory at Haleakala, Maui (Project Amos; Fig. 19). These telescopes use hydrostatic oil bearings and vacuum oil returns for all axes. The low friction and stiction of this type of bearing makes it feasible to use torque motors to reach astronomical tracking precision without the use of worm wheels. The maximum torque developed by the motors on these telescopes is approximately 3000 ft-lb. The rugged design of these mountings reflects the design specification of an absolute pointing error of $\pm 2''$ of arc.

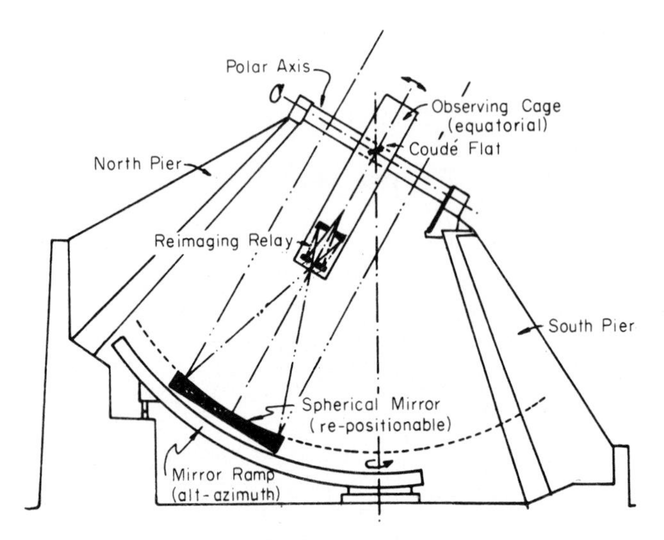

FIG. 18. A hybrid equatorial–alt-azimuth telescope design by Meinel. The primary mirror is fixed for intervals of observation.

The use of these telescopes to track satellites as well as stars requires a triaxial mounting: (1) azimuth, (2) elevation/declination, and (3) hour angle. By positioning the azimuth axis north–south, the telescope is a standard equatorial mounting for the other two axes. On the other hand, positioning the hour-angle axis horizontally, the telescope is a standard alt-azimuth for the other two axes. Since the shaft angle encoders and drives of this telescope are tied in a real-time way to an IBM 7090 computer, this telescope will permit astronomers to evaluate the feasibility of using an alt-azimuth telescope for high-precision astronomical guiding, a most valuable area of study relating to the future design of very large optical telescopes.

E. SKY COVERAGE

The choice of mounting from among the many configurations possible depends upon compromise among planned usage, the latitude of the observatory, the size and weight of the projected telescope, the cost budget, and the methods of access by the astronomer to the various focal positions. One of the most important factors in the usage category is the desired sky coverage. If one wishes all-sky coverage, horizon to horizon in all directions, then the mounting type is rather limited unless one makes concessions in the mounting structure. It must be remembered that very little useful astronomy can be accomplished within 10–15° of the horizon.

FIG. 19. The 60-in. ARPA/Michigan Haleakala telescope under construction. This Cassegrain telescope is on a triaxial mounting by Century Detroit Corporation, Detroit, Michigan.

Only for observation of comets and, occasionally, a planet is the astronomer forced to go to such large zenith distances. The resulting work is then handicapped by poor atmospheric seeing and high and variable extinction over the long tangent air path, with serious atmospheric dispersion due to refraction.

A comparison of the influence of all-sky coverage *vs* limited sky coverage for fork mountings is shown in Fig. 20. The usual fork mounting limits the south zenith distance to 75–80° at a nominal latitude of 30° in order to keep the fork tines short to avoid unnecessary flexure. The bent-fork design achieves all-sky coverage at these latitudes at the expense of a counterweight, a larger swing of the Cassegrain focal position, and inconvenience in the larger distance of the floor from the focal position to avoid interference when the fork is at ±6 h hour angle.

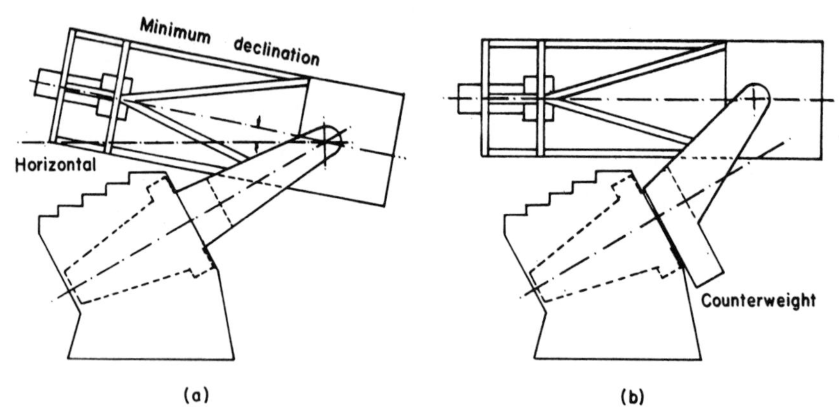

Fig. 20. Illustration of the problems of south declination limit. (a) Normal fork; (b) bent fork. The bent fork provides access to the horizon for telescopes in the 20–30° latitude range with the addition of a counterweight to the conventional fork mounting.

Limited sky coverage at high declinations is a necessary compromise of the English yoke. This limitation is sometimes elected since the area of the sky north of 65° declination is not of the highest priority for current research programs. Sky coverage under the pole is also of little use except for some long access times occasionally desirable for variable-star observations.

The alt-azimuth, alt-alt, and equatorial–alt-azimuth designs have limited sky coverage. The first two have temporary limitations due to gimbal lock—the alt-azimuth design in an angular cone of about 5–10° diameter about the zenith, and the alt-alt a similar zone at two points 180° apart on the horizon. The equatorial–alt-azimuth has limited sky coverage due to the supports for the equatorial tube and the arc length of the primary mirror ramp.

VI. EQUATORIAL TELESCOPE DESIGN

A. Telescope Tube

The telescope tube provides the important interface between the optical elements and the mechanical structure. The tube must carry the declination-axis shaft stubs and support the optics in such a way as to maintain collimation in the presence of gravitational flexure of the tube and bending moments through the declination axis. For a large telescope, the flexure of a cantilever tube would produce both translation and tilt to the optical elements. Short, continuous-skin tubes are very stiff, and the amount of decollimation can be kept within acceptable limits; however, in very large telescopes, this approach becomes unsatisfactory because of the scale-factor influence on stiffness. The truss structure introduced by Serurrier in the 200-in. telescope utilizes a simple geometry to maintain collimation in the presence of large amounts of flexure. Both the primary and secondary mirror are allowed to translate, but the parallelogram structure constrains the optical elements to move in parallel planes. The art of the truss design is to equalize the deflections of the two ends of the telescope tube.

The tube design can have a major influence on optical performance of a telescope due to internal "seeing." A telescope with a closed tube, or even a closed box adjacent to the mirror, needs special attention to the drainage of air from the tube during use, either by gravity flow or forced flow by ventilation fans.

The use of a three-mirror coudé system imposes a special problem upon the telescope tube design. The optical beam from the No. 3 mirror must emerge from the tube at a position that varies along the length of the tube, depending upon the declination of the star. The declination-box structure must therefore be cut along its length on the south side by an amount equal to the width of the coudé optical beam. This slot impairs the strength of the declination box and makes control of the telescope deflections very difficult. The solution to this problem on the 120-in. Lick telescope is to bridge the declination box by two thin webs of special steel such that the obscuration of light is not serious.

The effect of flexure of the telescope tube is to decollimate the optical system. A lateral displacement Δy of the secondary mirror from the optical axis of the primary mirror causes a growth of the image due to coma of the length ΔC given by

$$\Delta C = (3M/32F^2)(1 - eM^2)\,\Delta y$$

where M is the magnification of the secondary, F is the final f-ratio of the telescope, and e is the eccentricity ($e = -1$ for a paraboloid, $e > -1$ for a Ritchey-Chrétien).

For a rotation of the secondary through an angle α, the aberration ΔC is computed by replacing Δy by the quantity

$$\Delta y = (f - s)\alpha$$

where f is the focal length of the primary mirror and s is the separation of the secondary from the primary.

B. Drive Worm Wheels

The drive worm wheel is one of the most critical parts with respect to ultimate performance of a large telescope. In order to meet the design specifications, the wheel must be cut with the highest precision attainable, mounted with precision of centering, and lapped with the actual worm that is to be used to drive the telescope. The worm wheel can either be made massive and rigid, as in the older telescopes, or made as a flexible diaphragm constrained between guides straddling the worm, as in the newest telescopes. Both designs have yielded excellent telescope drive accuracies.

The diameter of the worm wheel must be chosen with consideration of the fact that the smoothness of drive at the focus of the telescope, of several hundred inches focal length, depends upon mechanical engagement and oil-film phenomena that occur at a radius of only tens of inches. As a rule, one should require a telescope worm wheel that is at least as big in diameter as the mirror, the preferred ratio being between 1.0 and 1.2 of the mirror diameter.

The mounting of the worm is critical to the performance of the telescope. The end-thrust bearing must be of the highest accuracy, since an error in end-thrust is directly translated into error in the sky through the radius of the worm wheel. An error of $0''.5$ on a 100-in. diameter gear is caused by only 0.0001 in. of linear motion.

The changeover of instrumentation or removal of telescope components is one particular operation that creates a potential hazard to the worm wheel and worm systems. A large telescope must be equipped with changeover bars to rigidly fix the declination tube and right-ascension structure to the base of the telescope, or, in some instances, to the floor of the dome. The worm assembly can be retracted to increase the tooth engagement clearance if large loads are anticipated during the telescope changeover.

The method of slewing a telescope in right ascension bears upon the worm problem. It has been traditional that a large telescope have a separate gear to slew the telescope. However, an analysis of the forces and wear upon a telescope worm has shown that the forces generated during slew should be negligible factors in gear performance. It is even felt that the better lubrication of the gear driving the higher worm speeds encountered during slew should lead to lower wear under these conditions than with the very slow linear contact velocities at the gear interfaces typical of tracking speed. It can therefore be concluded that any degradation of a single-gear telescope performance must be associated with accidental loadings during its lifetime and not with slew acceleration loadings.

C. Pointing Accuracy

The specification of the pointing accuracy of a telescope meets a number of problems associated with the total design of a telescope. The pointing accuracy is determined by (1) flexure of the polar structure, (2) flexure of the telescope tube, and (3) intercell decollimation of the optical elements. The total pointing error of a large telescope built according to the current state of the art is usually found to be of the order of 1–3′. Fork mountings with long tines are particularly subject to pointing errors, and careful design is necessary to keep the value less than 5′. An analysis of the pointing errors of the 120-in. Lick telescope has been made by Vasilevskis.[30]

The English yoke is perhaps the best mounting for minimum pointing error, because of the north–south symmetry about the declination axis. It should be realized that to achieve absolute pointing errors in the range of a few seconds of arc is currently beyond the state of the art of telescope-making. The 48-in. and 60-in. Haleakala (ARPA) telescopes are two in which particular care has been taken to reduce flexure down to the 2 or 3″ of arc range.

D. Telescope Drive Rates

A telescope must be moved in the two coordinate axes at several angular rates to accomplish acquisition of an object. These rates are generally divided into the following functions and approximate rates:

1. Slew: 90–120 deg/min.
2. Set: 1–4 min/sec.
3. Guide: 0.2–15 sec/sec.
4. Drift: 0–10 sec/sec.
5. Track: siderial rate 15 sec/sec.

The scaling of the ratio between slew/set and set/guide rates is usually made on the basis of accomplishing the correction of the average residual pointing error, such as would remain after slewing, in 10 sec of use of the set rate. The same rule applies for use of guide rate to correct the error remaining after use of the set rate.

---

[30] S. Vasilevskis, *Astron. J.* **67**, 464 (1962).

The above angular rates also apply when translated into units of hours, minutes, and seconds in right ascension measured at the equator. As one nears the pole, the rates in right ascension become very slow to accomplish a given correction. The 120-in. Pulkova telescope meets this problem by automatically scaling up the right ascension rates by a factor of $(1/\cos\delta)$ up to 80° declination.

The basic track rate of a telescope has other functions besides compensating for the diurnal rotation of the earth. In the first place, the apparent diurnal rate is modified by changing atmospheric refraction, so that the telescope drive is slower by a small amount. The astronomer usually determines the proper rate by empirical approximation during use. The siderial drive oscillator therefore needs to be provided with controls for a small variation of the basic rate. In the second place, the astronomer may wish to track objects with significant apparent motion, such as the moon, comets, or planets. The lunar rate is the largest variable, requiring a rate that is about 1/27 slower than siderial. The moon and comets also have significant rates of motion in declination, so that a telescope for their observation should be equipped with a tracking drive in declination.

The variation of angular rate in both right ascension and declination used for spectroscopy requires a change in rate as large as $\pm 10''$/sec about the siderial rate. This rate is used to cause the star image to drift along the slit of the spectrograph to widen its spectrum. Variable drift rates are also useful to the astronomer as a means of controlling the exposure of the spectrogram.

VII. CORONAGRAPHS

Coronagraphs are designed to permit observation of the solar corona without the assistance of the moon as an occulting disk. They represent a distinctly separate class of astronomical instruments incorporating the technical requirements imposed by the low brightness of the solar corona relative to the solar disk.

The basic limitation in observing the corona is scattered light. The light scattered by the atmosphere is a fundamental limit, with, at the best sites, a brightness at one solar radius from the limb of the sun of the order of $1\text{–}10 \times 10^{-6}$ of the sun's brightness. For this reason, coronagraphs are located at high altitudes and far from sources of atmospheric pollution. Upon occasions of a solar eclipse, the entire atmosphere is shielded by the moon, and the astronomer has a sky brightness of the order of 10^{-9} of that of the sun, and the outer corona can then be observed.

The principal problem to be overcome in a coronagraph is scattered light within the instrument. Five sources of scattered light listed by Newkirk and Bohlin[31] are as follows:

a. Diffraction at the Edge of the Objective Lens Aperture. The edge of the lens intercepts the full flux of sunlight, giving a scattered-light brightness equal to about 2×10^{-4} of that of the sun's disk. Minute irregularities in the boundary of the objective are also a source of scattering. This problem is met by the use of a field lens to image the objective lens aperture on a second aperture so that the brilliant rim of the objective is intercepted by the second aperture (Fig. 21).

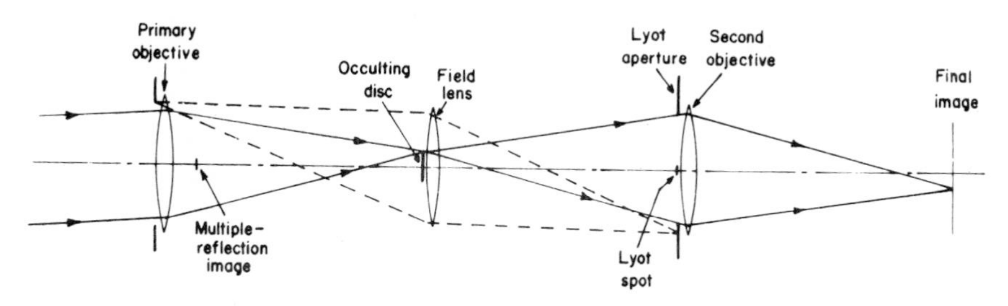

FIG. 21. Schematic optical diagram of a coronagraph.

b. Ghost Images Produced by Reflections in the Objective Lens. These have a brightness of $1\text{–}5 \times 10^{-6}$ times that of the sun. They may be trapped by a small "Lyot spot" at the point where an image of the ghost is formed by the field lens.

c. Macroscopic Inhomogeneities in the Glass. Small bubbles and seeds in the glass may give a brightness of about 10×1^{-6} of the sun's brightness.

d. Surface Imperfections on the Objective Lens. The presence of minute imperfections on the surface of the lens, such as pits, sleeks, etch marks, etc., are all sources of scattering, giving a brightness equal to about 2×10^{-6} of the sun's brightness. The optician must use the greatest care in the final polishing to obtain the smoothest surface, free from defects. The achievement of such a finish is the mark of the true artisan, the techniques used to achieve this end being not generally available.

The scattering of light by optical surfaces is one reason aluminized mirrors are not usable in coronagraphs. The surface finish of a mirror on fused quartz, e.g., can be of coronagraph quality until the aluminum is deposited, whereupon the scattering increases by several orders of magnitude. The source is probably the interfaces between microscopic areas of deposited aluminum. Designs that utilize unaluminized reflecting surfaces

[31] G. Newkirk and D. Bohlin, *Appl. Opt.* **2**, 131–140 (1963).

have been tried,[32] as have Kanigen-coated surfaces. Kanigen appeared a prospect for some coronagraphic applications, since it is an amorphous nickel phosphide material that takes a good polish and has an intrinsic reflectivity of about 0.50.

e. Body Scattering in the Glass. The material used for the objective lens has scattering within the bulk material at interfaces between the "liquid crystal" domains of the glass, giving a brightness of less than 1×10^{-6} of the sun.

The objective lens of a coronagraph usually consists of a single element to minimize both surface and body scattering. The resultant chromatic difference in focus is accommodated by moving the focal position of the occulting disk, and by achromatizing the whole system in the relay as with a Schupmann type of Mangin-mirror relay. The occulting disk is usually an inclined reflecting surface to direct the sunlight back out of the instrument with a minimum of scattered light and heating. An external occulting disk has also been used to reduce the intensity of light incident on the coronagraph objective. This arrangement can be used only with small instruments, since the occulting disk must be far from the lens, but its presence eases all of the internal scattering problems.

---

[32] J. H. Rush and G. K. Schnable, *Appl. Opt.* **3**, 1347–1352 (1964).

CHAPTER 7

Military Optical Instruments

FRANCIS B. PATRICK

US Army, Frankford Arsenal, Philadelphia, Pennsylvania

I. INTRODUCTION

A. General

1. *Functional Areas*

Military optical instruments may be regarded as a special class of optical instruments which are used by the armed forces of a nation for particular military purposes. Taken as a whole, the functional areas shown in Fig. 1 relate primarily to the control of weapon fire; hence, it is no oddity that practically all military optical instruments come within the category of, and are usually referred to as, optical fire-control instruments. Although the functional areas shown in Fig. 1 are basic to all weapon systems, a rather wide variety of military optical instruments is required to accomplish them (see Section II).

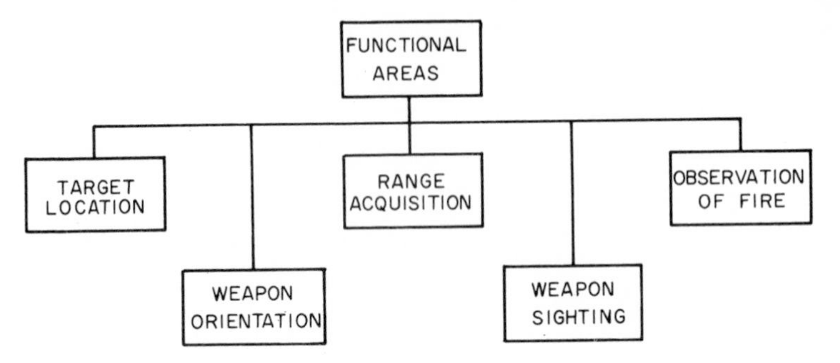

FIG. 1. Functional areas for military optical instruments.

2. *Fundamental Viewpoint*

From a fundamental viewpoint, and with but few exceptions, military optical instruments may be regarded as angular-detection or angular-measuring devices. Moreover, since these devices are largely visual in nature, they depend on the visual properties of the human eye as the final determinant. For example, the matter of detecting small, distant objects is governed mostly by the angles which they subtend and the visual acuity of the eye in perceiving them. This is also true for the process of recognition and identification of objects by the interpretation of the angular subtenses of the observed details. Further, targets can be located and weapons oriented through the measurement of appropriate directional angles by utilizing the vernier acuity of the eye. Again, the range to a distant target can be acquired through precise measurement of the angle which the base length of the range finder subtends at the target. Finally, the observation of fire is attended by the measurement of angular deflections of hits, misses, and near misses with reference to the target being fired upon.

3. *Refracting Telescopes*

The great majority of military optical instruments are in the form of refracting telescopes of moderate power, say, from $1\times$ to $20\times$ in magnification. The simple first-order theory of the telescope has been given in Vol. I, Chapter 6 of this series, and it need not be repeated here.

4. *Effectiveness of Seeing with a Telescope*

The problem is basically how much farther we can see (detect) an object with the aid of a telescope in comparison with the unaided eye. From the theory explained by König and Köhler,[1] the relative effectiveness E in

[1] A. König and H. Köhler, "Die Fernrohre und Entfernungsmesser," pp. 100–106. Springer, Berlin, 1959.

seeing with a telescope may be expressed by the equation

$$E = R_T/R_e = V_T/V_e = M[(AP/AP_0)^2(T_T)]^m \tag{1}$$

in which R_e signifies the distance from a just visible object to the unaided eye, R_T represents the distance from a just visible object to the entrance pupil of the telescope, V_e is the visual acuity of the unaided eye, V_T is the visual acuity upon viewing with the telescope, M is the magnifying power of the telescope, AP is the exit pupil diameter of the telescope, AP_0 is the diameter of the pupil of the eye, T_T is the light transmission factor of the telescope, and m is an exponent of variable power that associates visual acuity with adaptation luminance, ranging in value from about $\frac{1}{4}$ for daylight to about $\frac{1}{2}$ for night viewing conditions.

It is understandable from Eq. (1) that E becomes a maximum when the ratio AP/AP_0 is equal to unity, a condition usually realized when viewing in bright daylight. Therefore, considering T_T to be finite and taking m equal to $\frac{1}{4}$, it is plausible to say that $E_{max} = M(T_T)^{1/4}$. Now, under very dark night-viewing conditions, it is recognized that AP_0 will attain a maximum value, say, AP_M. Moreover, when AP_M exceeds AP and we take m equal to $\frac{1}{2}$, it follows that $E_{min} = (EP/AP_M)(T_T)^{1/2}$ since EP, the diameter of the entrance pupil, is equal to $M \cdot AP$. From these maximum and minimum values of E, we conclude that for daylight viewing, it is the magnifying power of the telescope which exerts the major influence as to how far we can see, whereas for night viewing, the size of the entrance pupil is the controlling factor.

Further, by assuming that the luminous level at twilight is approximately the geometric mean of the luminous levels of daylight and night, we can extend our reasoning to ascertain the efficiency of a telescope for the twilight region. We find that E_{mid}, an approximation of the twilight efficiency of a telescope, is equal to $K(M \cdot EP)^{1/2}$, where $K \approx (AP_M)^{-1/2}(T_T)^{1/3}$. It should be mentioned that in Europe, considerable credence is given to the term "twilight efficiency" in the form just given, especially in West Germany, where it has been proposed for adoption as a DIN standard, according to Köhler.[2]

Additional information concerning vision through telescopes is given by Middleton,[3] who draws attention to a number of other references on this subject.

---

[2] H. Köhler, Concerning several basic relationships in telescope observation, *Zeiss, Werkzeitschr.* **3**, 17, 65–69 (1955).

[3] W. E. K. Middleton, "Vision through the Atmosphere," pp. 128–132. Univ. of Toronto Press, Ottawa, Canada, 1952.

B. Design Considerations

1. *General*

In the design of military telescopes, it is imperative that equal consideration be given to the optical, mechanical, and maintenance-engineering aspects of the instrument. Through this triple consideration, assurance is provided that the item can be readily produced, that it will operate in military environments for the purpose intended, and, moreover, that it can be maintained without difficulty.

2. *Optical Engineering*

The optical design of a military telescope starts with the creation of an optical system which will meet all stipulated military requirements and, following execution of the lens-design phase, terminates upon completion of necessary production drawings and specifications for each of the optical components in the system. No attempt will be made here to elaborate on either the design of optical systems or lens-design procedures. Suffice it to say, however, that the design should be as simple as possible, capable of mass production, and use only standard types of optical glass.

3. *Mechanical Engineering*

The mechanical design of a military optical instrument involves consideration of a number of factors, paramount among which are ease of manufacture and maintenance, simplicity of operation, ruggedness, and reliability.

Design usually originates in the form of a study to ascertain the optimum method of mounting the various optical components. Normally, all of the lens elements are held in place by means of retaining rings, and prisms by side bonding either directly to the metal housing itself or indirectly through an intermediate metal piece. Except in very rare instances, the use of shims and scraping to attain necessary adjustment is forbidden. A further objective of this design study is to determine if "modular" construction should be attempted. In the case of large complex instruments, modular construction would be advantageous, while in the case of small instruments, the "throw-away" concept might prove more economical. Every effort must be made to employ machining tolerances which are amenable to mass production and the extensive use of jigs and fixtures in assembly operations.

The selection of materials for the instrument is important from the standpoint of weight, strength, corrosion resistance, and machinability. For this reason, aluminum is most generally specified, e.g., 24ST for tubing and retaining rings, class III for ordinary castings, and 356-T6 for investment castings. Because of its low resistance to corrosion, magnesium is seldom specified even though it is recognized as one of the lightest metals available.

Adequate sealing to prevent the ingress of moisture, dirt, or fungus spores is considered imperative. In certain instances, reliance is placed on *O*–ring gaskets, on the injection of a viscous plastic, or on metal bellows, coupled with internal pressurization with dry nitrogen, helium, or air. While static sealing is usually easily achievable, dynamic sealing of rotary shafts presents difficulties.

The strategic location of knobs and switches to facilitate operation of the instrument is considered mandatory. Also, the sizes of such knobs and switches must be adequate to permit operation with mittens under arctic conditions.

4. *Maintenance Engineering*

Maintenance engineering is performed concurrently with, and exerts an influence on, the design of new instruments. It is an important phase, and one of its main objectives is the preparation of a list of repair parts with specifications as to quantities and stockage points for issue during the life of the instrument. As many parts as possible should be standardized with respect to existing instruments, to reduce the number which must be carried in stock.

The maintenance engineer also determines the degree of maintenance to be performed at each level of maintenance. For example, at the organization level, that of the prime field user, maintenance is limited primarily to external cleaning, while at the depot level, facilities are available for complete overhaul and rebuild. In between, at the general support and direct support levels, maintenance is performed on a progressively increasing scale.

C. Common Features

1. *General*

While military optical instruments as a class exhibit a number of special features related to their intended use, some of these features such as reticles, reticle illumination, diopter adjustment, and provision for boresighting, are practically common to all instruments.

2. *Reticles*

In general, a reticle may be described as a glass disk upon which is etched a particular pattern for the purpose of indicating angular subtenses with reference to the geometric center of the reticle. Figure 2 shows a

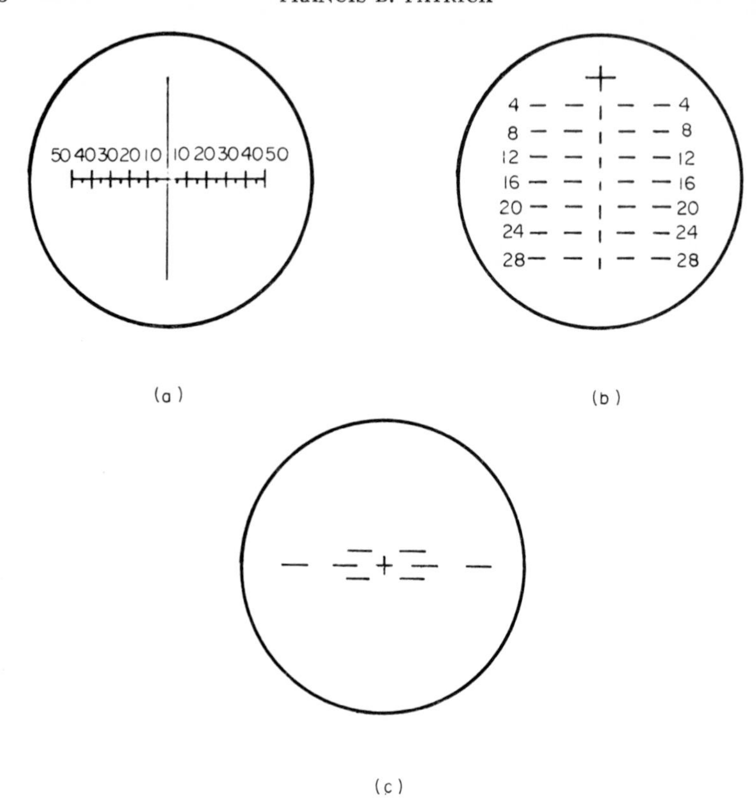

FIG. 2. Reticle patterns for military optical instruments (a) Azimuth laying type, (b) ballistic type, and (c) central laying type.

number of reticle patterns of varying degrees of complexity. When a reticle is graduated in terms of angular gun elevation vs range, it is often referred to as a "ballistic reticle."

Since reticles are placed in the focal plane of the image-forming component (usually the objective lens), computation of graduations may be performed by means of the formula

$$d = f \tan \theta + k \tan^3 \theta \tag{2}$$

where d is the linear displacement from the geometric center of the reticle, f is the focal length of the image-forming component, θ is the angular subtense with reference to the optical axis, and k is a constant which relates to the distortion of the image-forming component. If θ is small, the factor k is usually practically negligible.

Certain instruments employ a "projected reticle," i.e., the projection of a reticle pattern into the main optical system by means of a beam splitter.

Reticles for this purpose are usually in the form of an opaque metal disk with the pattern cut out, placed at the focal plane of a collimating lens and back-illuminated.

3. *Reticle Illumination*

For dusk and night use, it is necessary that it be possible to illuminate the reticle, and to be able to vary the intensity of the illumination as well. Instrument lights are normally provided for this purpose, consisting of a self-contained battery and lamp with rheostat control. The light is attached to the instrument by means of a self–locking dovetail slide arrangement in the vicinity of the reticle, and the ingress of light through a small red window serves to edge-illuminate the reticle.

4. *Diopter Movement*

On instruments having a magnifying power greater than 4×, it is normal practice to provide a longitudinal adjustment of the eyepiece to permit individual focusing of the instrument. The amount of longitudinal movement is usually indicated by a scale graduated in terms of diopters. Since one diopter is equivalent to a distance of 1m in apparent image space, the distance X at the focal plane of the eyepiece corresponding to D diopters is found from

$$X = Df_e^2 \times 10^{-3} \tag{3}$$

where f_e is the focal length of the eyepiece, and f_e and X are in millimeters.

Diopter adjustment may be accomplished internally, as by a longitudinal movement of the second erecting lens or in some cases by a prism movement. The overriding considerations as to which method to employ concern simplicity and sealing, the latter being the more important.

5. *Boresight Adjustment*

Boresighting is the process of assuring alignment between the axis of a gun tube or rocket launcher and the zero graduation on the reticle of the sighting telescope. This process requires either an angular movement of the telescope itself, or a lateral motion of the reticle with reference to the gun-tube axis.

In the former case, vertical and horizontal actuation of the telescope is accomplished by screws located in the mount, the front end of the telescope being held in a spherical seat. While this method is satisfactory for short-tube telescopes, it is not so for long-tube telescopes because of the excessive movement required of the screws. The more usual solution is movement of the reticle itself, and ingenious methods have been evolved for producing such movements.

II. REPRESENTATIVE EXAMPLES

A. Sighting Telescopes

1. *General*

Sighting telescopes may be defined as a special class of optical instrument used by the military principally in connection with the aiming of weapons in both azimuth and elevation in the direct fire mode, i.e,. when the target is within direct view and range of the weapon sighting-telescope combination.

Included within the category of the more common military sighting telescopes are such instruments as: rifle scopes, for hand-held, shoulder-fired rifles; sights, for large-, medium-, and small-caliber recoilless rifles; machine-gun sights, for tank-mounted machine guns; tank telescopes, for the main armament of armored vehicles; direct fire telescopes, for artillery-type weapons, guns, and howitzers, of all calibers; reflex sights, for air defense and aircraft (helicopter)-type weapons; and turret gun sights, for naval-gun turret-type weapons, to mention but a few.

Without exception, all military sighting telescopes are mounted directly onto the weapons which they are required to support, each is equipped with an appropriate reticle (usually of the ballistic type), and all provide the gunner with an erect, magnified image of the target being sighted on. Image erection is accomplished by either lenses or prisms, or, in some cases, by a combination of both.

2. *Rifle Scopes*

Military rifle scopes resemble commercial rifle scopes in that both types are usually constructed in the form of lens-erecting telescopes of rather short length and small diameter, say, about 10 in. long with an outside diameter of approximately 1 in. Normal military characteristics of such scopes are often quoted as: magnification, about $2.5\times$; field of view, perhaps not greater than 5° or so; exit pupil diameter, 5–8 mm; and eye distance, from 3 to 4 in. The reticle is usually installed at the front end of the telescope in the focal plane of the objective lens. Through a mechanism actuated by external knobs, the reticle can be caused to translate laterally and vertically, thus providing a means not only for boresighting but also for introducing desired azimuth and elevation corrections with respect to the bore of the weapon.

Figure 3 shows a schematic optical diagram of a typical military rifle scope, consisting of an objective lens, a reticle, an erecting lens, and an

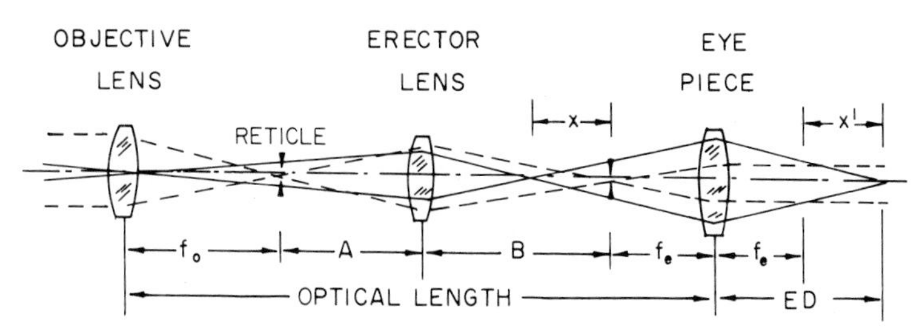

FIG. 3. Optical diagram of a rifle scope.

eyepiece. The total magnification M of the system may be expressed by the formula, $M = (f_o/f_e)(B/A)$. From the scaling of the diagram, it is apparent that the image distance B of the erecting lens is somewhat longer than its conjugate object distance A, hence, the erecting lens contributes to the magnification of the scope in the direct ratio of B to A. As to magnitudes of the ratios (f_o/f_e) and (B/A), it is a fact that the optical system design of some rifle scopes is based on attainment of the condition $(B/A) = (f_o/f_e) = M^{1/2}$, on the premise that the two ratios should be equal to each other, thus equalizing the "work" of the system.

A second point of note concerning Fig. 3 is that the principal ray, after passage through the erecting lens, intersects the optical axis at a distance x in front of the focal plane of the eyepiece. Since the distance x' on the right-hand side of the eyepiece is related to the distance x through the Newtonian formula $-xx' = f_e^2$, it follows that the shorter the distance x, the longer will be the eye distance. However, in the design of certain rifle scopes, x and x' are often made numerically equal. As an outcome of this condition, the image of the entrance pupil (objective lens) is projected into the plane of the exit pupil by the eyepiece at precisely unit magnification.

Another basis for the design of a rifle scope is to select that solution which yields the minimum value for the sum of the reciprocal focal lengths of the objective, erector, and eyepiece lenses. Naturally, such a basis leads to attainment of minimal Petzval curvature $\sum(1/n_e f)$, where n_e is the "effective" index of refraction of each of the lenses concerned. For example, suppose that it is desired to design a rifle scope of magnification $2.5\times$, eye distance 3 in., and optical length 10.5 in. with three thin lenses each possessing the same n_e. By successively assuming various focal lengths for the eyepiece, and by each time ascertaining the focal lengths of the objective and erector lenses by appropriate paraxial ray tracing, a graph like Fig. 4 could be constructed. From the graph, it is obvious (for the example cited) that $\sum(1/f)$ is a minimum when the focal length of the eyepiece is equal to 1.52 in. (actually, 1.522 in.), the focal length of the objective lens is 2.65 in., and the focal length of the erector lens is 1.53 in. From these data, it is

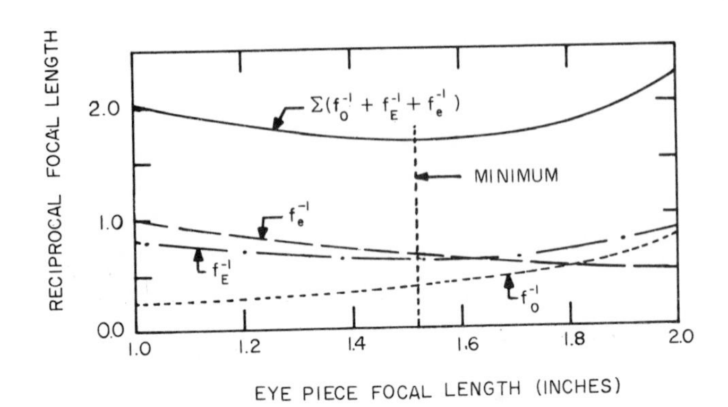

FIG. 4. Reciprocal focal length of objective lens (f_0^{-1}), erecting lens (f_E^{-1}), eyepiece (f_e^{-1}), and their sum as a function of eyepiece focal length, for a rifle scope of 2.5 × magnification, 3.0-in. eye distance, and 10.5-in. optical length.

apparent that the focal length of the eyepiece (1.52 in.) is nearly equal to one-half the desired eye distance (3 in.).

The foregoing represents but a few approaches which could be utilized in connection with design. In actual practice, however, it is not unusual to find rifle scopes in which the eyepiece is composed of two cemented doublets, and the relay system of a doublet and singlet combination, or even of two doublets, particularly in those instruments requiring extremely large exit pupils.

3. *Sights for Recoilless Rifles*

Regardless of whether the caliber of a recoilless rifle is large, medium, or small, the type of sight employed for the control of its fire is typically in the form of a rather compact prism-erecting telescope. The characteristics of such sights are usually modest, being of the order of 3 × magnification, 13° field of view, 7.5 mm exit-pupil diameter, and 1.5 in. eye distance.

Figure 5 shows a sectional view of a modern recoilless-rifle sight, the design of which reflects an application of "modular" construction. The optical system is rather conventional, consisting as it does of a cemented doublet objective lens, an inverting prism of the second type by Porro (see Vol. III, Chapter 7, Fig. 34), a ballistic reticle, and a symmetrical two-doublet eyepiece.

The objective lens is contained in a cell that threads into the front cover unit, which in turns attaches to the main body casting by means of screws. The three elements of the inverting prism are cemented together to form a single prism cluster and the whole bonded onto a plate which fits into the well of the body casting and is held in place by screws. The reticle is

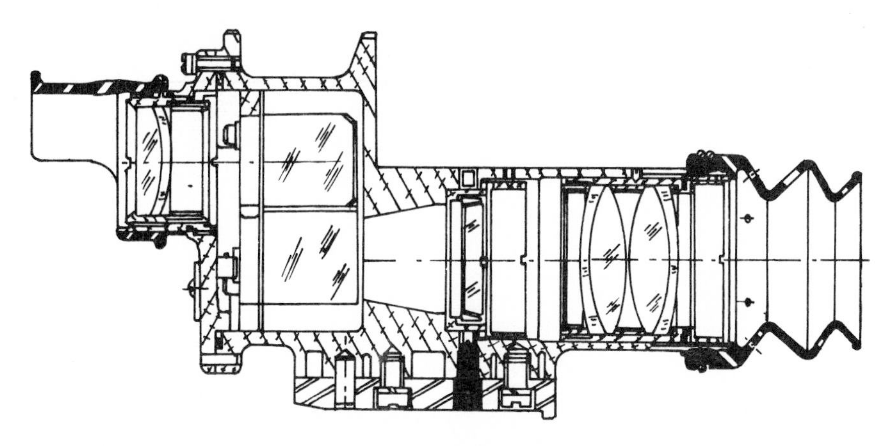

FIG. 5. Recoilless-rifle sight.

also cell-mounted, positioned as shown, and locked in place by dual retaining rings and locking screws, thus ensuring nonrotation of the reticle while the sight is in use. The symmetrical two-doublet eyepiece is housed in a cell in normal fashion such that the eyepiece can be translated along the optical axis as a unit to enable achievement of the proper factory-aligned fixed-focus adjustment. To round out the assembly, there is attached to the base of the main body casting a steel dovetail for weapon mounting purposes, a rubber eye shield clamped onto the eyepiece end of the instrument for eye protection, and a sun shade of rubber installed at the objective end of the telescope to exclude extraneous light.

4. *Machine-Gun Sights*

Machine-gun sights for the control of the fire of tank-mounted machine guns are usually installed in the cupola of an armored vehicle. Such sights are generally of the periscopic type, vertically positioned, and exhibit characteristics of low magnification, about 1.5×; wide field of view, 48°; exit pupil diameter of 4 mm; and an eye distance of 1 in.

The sight is normally connected to the machine gun through a parallelogram linkage located near the base of the instrument. By means of a mechanism within the sight, motion is transferred automatically from the linkage to the scanning prism at the top of the sight, thus permitting vertical travel of the optical line of sight in exact harmony with the line of sight of the machine gun throughout the excursion limits of the gun. Because space in the cupola of tanks is at a premium, cupola-mounted machine-gun sights must of necessity be as compact as possible. Moreover, the periscopic offset of such sights is never great and rarely exceeds 12 in. or so.

A sectional view of a machine-gun sight possessing the characteristics just described is presented in Fig. 6. The optical system comprises a slanted entrance window, a scanning head-prism, a rather complicated objective lens system, an erecting system of two cemented doublets, a mirror which deviates the optical axis 90°, a reticle, and a wide-angle eyepiece somewhat similar to the Erfle type.

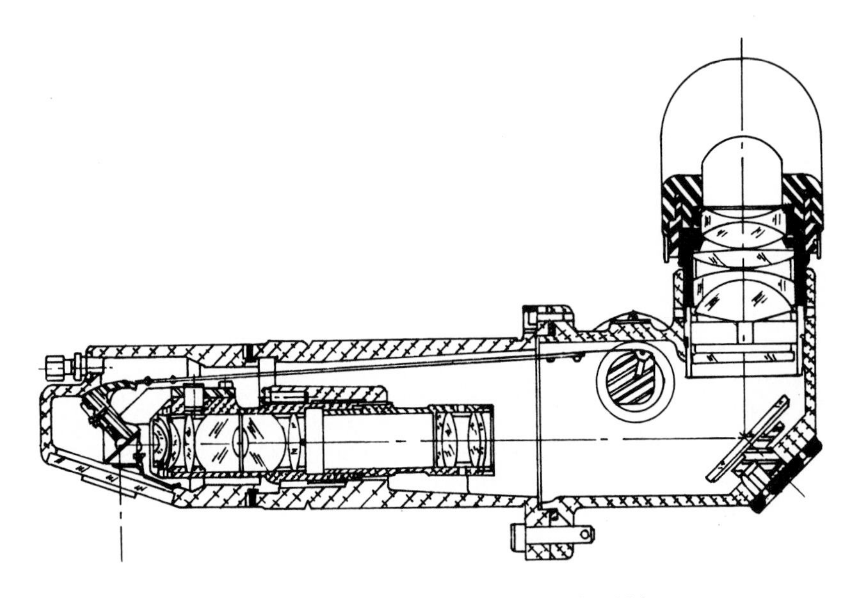

FIG. 6. Machine-gun sight (shown on its side).

Two comments are offered with regard to Fig. 6. The first concerns an explanation of the internal mechanism whereby the optical line of sight is caused to follow the elevation travel of the machine gun, while the second comment has to do with the construction of the rather unusual objective lens system.

The internal mechanism which imparts rotation to the scanning head prism is in the form of a metallic tape drive which extends from the radial arm of the head prism to the shaft which is located in the near vicinity of the reticle assembly. Although not shown on the diagram, this shaft is directly coupled to the external parallelogram linkage already referred to. Since the radius of the radial arm of the head prism is exactly equal to the diameter of the drive shaft, the scanning head-prism rotates at one-half the speed of the shaft, as required. In order to keep the tape drive tight, constant tension is applied to it through the medium of two torsion springs associated with the shaft on which the scanning head-prism is mounted.

The objective lens system of the sight consists of six separate optical elements. At first glance, it appears that this system is composed of a double Gauss objective lens preceded by two relatively thin meniscus lenses concave toward the entrance pupil located within the scanning head-prism. In actual fact, however, a real image of a distant object is formed midway between the two thick meniscus lenses, thus establishing the first focal plane of the sight. The thick meniscus and singlet lenses immediately to the right of the first focal plane function together as a single positive collective lens in forming a magnified virtual image of the real image somewhat to the left of the first focal plane and at the focus of the first erecting lens. Hence, the special arrangement of the objective–collective lens combination shown enables attainment of a low Petzval contribution and lens apertures of reduced size (0.900 in.), considering the wide field of view (48°) and the relatively short focal length (1.44 in.) of the entire combination.

5. *Tank Telescopes*

Tank telescopes are employed for the control of the fire of the main armament of armored vehicles (tanks). In general, such telescopes are in the nature of straight-tube telescopes of the lens-erecting type, and usually embody the following typical characteristics: magnification, 8×; field of view, 7.5°; exit pupil diameter, 6 mm; eye distance, 1.5 in. or so; outside diameter, some 2.5 in. (except for the eyepiece end); and overall length, about 45 in.

Figure 7 shows a sectional view of a conventional straight-tube tank

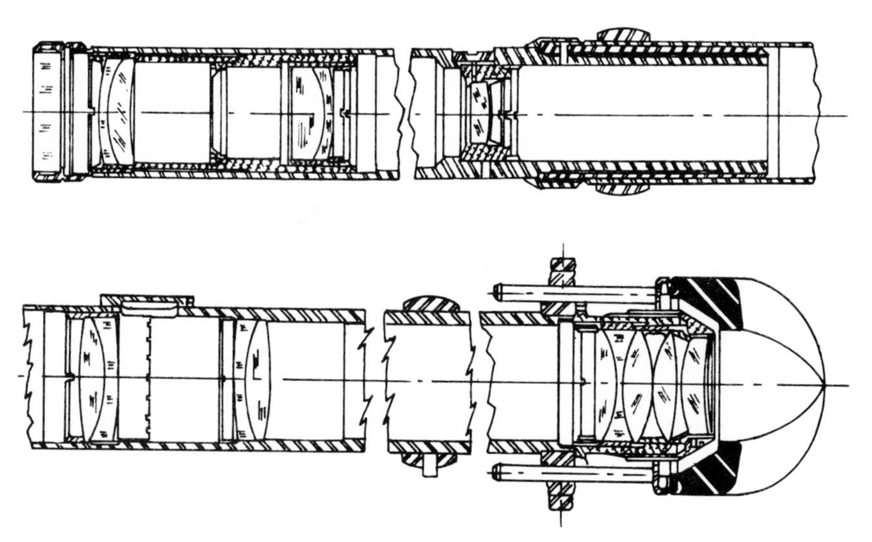

FIG. 7. Tank telescope.

telescope which depicts the details of construction and the manner of positioning and holding the various lenses. The optical system consists of an entrance window, an objective lens system of the Petzval type, a reticle lens on the plano side of which is engraved a ballistic pattern, two cemented doublet erecting lenses, and a wide-angle eyepiece. The telescope proper is supported in its mount (not shown) by means of the two narrow spherical steel bands which surround the telescope, one of which is located near the reticle assembly and the other just forward of the eyepiece. These spherical bands facilitate boresighting the telescope, since the mechanism for such is contained in the mount, and hence, the reticle lens can be locked in place.

In practically all cases, tank telescopes are designed in the form of two astronomical telescopes placed face to face with parallel light between them, since such an arrangement lends ease to the adjustment of the forward and rear portions of the entire telescope. Moreover, it will be found in most cases, that the erecting systems of tank telescopes are designed to operate at unity magnification, the advantages of which are obvious and well recognized.

Another item of importance in connection with the design of tank telescopes is the means employed for directing the path of the principal ray from the plane of the entrance pupil to the plane of the exit pupil, thereby establishing the eye distance of the telescope. In certain instances, as in Fig. 7, a reticle lens positioned at the focal plane of the objective lens system accomplishes this purpose by projecting the image of the entrance pupil into the space between the two erecting lenses, whence it is reprojected into the exit-pupil plane by the second erector and the eyepiece. However, this arrangement, simple as it may be, adds an undesirable Petzval contribution to the system, since the focal length of the reticle lens is usually rather short.

In other instances, as in the example cited by Patrick and Scidmore,[4] the rear component of a Petzval-type objective lens system can be designed to serve the dual purpose of controlling the path of the principal ray as well as contributing to control of the aberrations of the objective lens system. This arrangement works well, especially when the entrance pupil is located some distance ahead of the front component of the objective lens system.

A third possibility is to position the collective lens to the right of the focal plane of the objective lens and design it as part of the erector system. In essence, this is the scheme that was employed in the system for the machine-gun sight shown in Fig. 6. Such an arrangement is particularly advantageous when for special reasons it is desired that the objective lens system operate in the manner of a telephoto lens.

---

[4] F. B. Patrick and W. H. Scidmore, *Appl. Opt.* **3**, 427–431 (1964).

6. *Direct-Fire Telescopes*

Direct-fire telescopes for the control of the fire of artillery weapons in the direct-fire mode are usually constructed in the form of elbow telescopes, and installed on weapons in mounts with the line-of-sight axis parallel with the bore of the gun, and the eyepiece axis at right angles to it. For towed artillery weapons, direct-fire telescopes are normally rather compact, prism-erecting, elbow telescopes. However, for self-propelled artillery applications, a considerable longitudinal extension of the line-of-sight axis is often required; hence, direct-fire telescopes for such applications are usually of an extended lens-erecting type with inclusion of a penta prism at the eyepiece end to provide the desired elbow feature.

The nominal characteristics of current direct-fire elbow telescopes are of the order of: magnification, 8×; field of view, 8°; exit-pupil diameter, 7 mm; and eye distance about 1 in., not too different from the characteristics of direct-fire straight-tube, tank telescopes.

A sectional view of a modern direct-fire elbow telescope having the characteristics just mentioned is shown in Fig. 8. Perhaps the more

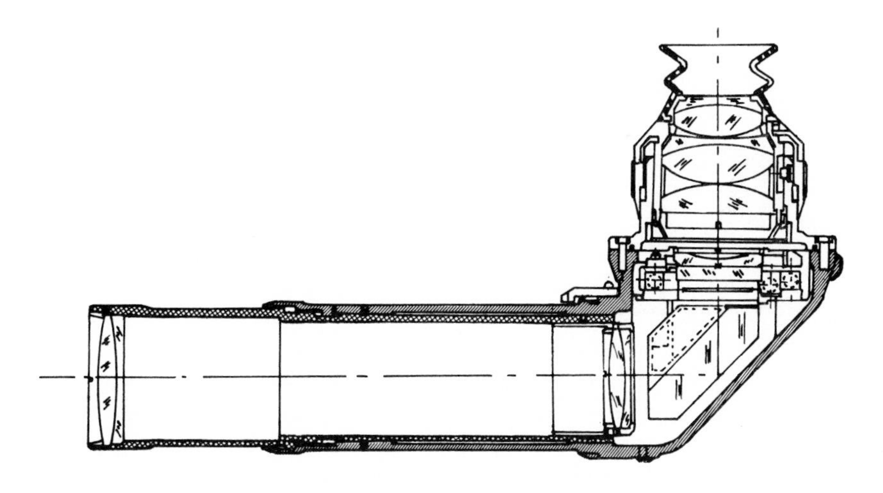

FIG. 8. Direct-fire elbow telescope.

important features associated with this instrument relate to the fact that the objective lens system is of a modified Petzval type, the Amici prism is held in place by bonding (rather than by straps and screws), a dual reticle system is employed (one fixed and one capable of vertical movement) and a rubber bellows is incorporated at the eyepiece end of the telescope to prevent the ingress of moisture.

7. *Reflex Sights*

A reflex sight is perhaps one of the simplest optical devices ever used by the military in connection with lead-computing sights for rapid-fire air defense and early fighter-aircraft weapons. More recently, reflex sights have been used to direct the fire of armed helicopters. While reflex sights are not generally regarded as sighting telescopes *per se*, but rather as collimated sighting devices, they are included here by reason of their widespread usage and close functional similarity with unit-power sighting telescopes.

In general, a reflex sight consists of an illuminated reticle, an objective lens, and a beam-splitter plate as indicated in Fig. 9. The function of the

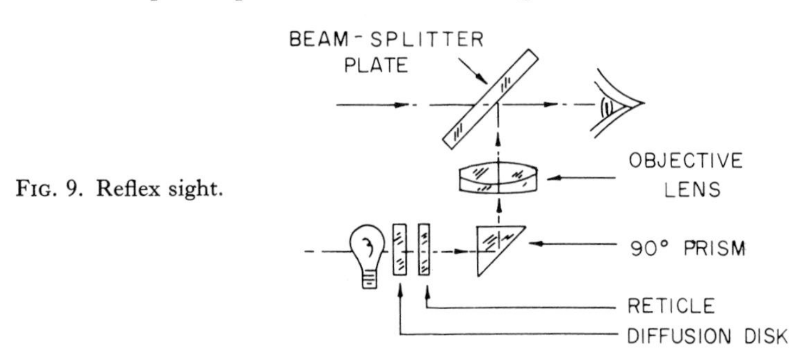

FIG. 9. Reflex sight.

beam-splitter plate is to deviate the path of the collimated beam into the eye of the observer, so that the image of the reticle appears to be coincident with and superimposed on the target being observed through the beam-splitter plate.

By mounting the sight in the manner shown and rotating it about a horizontal axis, the reticle will appear to move in elevation with respect to the target. Also, by rotating the sight about a vertical axis, the reticle will appear to move in azimuth with respect to the target. In both instances, the apparent angular movement of the reticle in relation to the angular rotation of the sight will be in the ratio of 1 : 1, which is clearly advantageous, because it enables gross simplification of the mechanism which imparts rotation.

8. *Turret Gun Sights*

Turret gun sights are used on board ship in naval gun turrets for control of the fire of naval weapons. Figure 10 shows the optical diagram of a typical naval turret sight (US Navy Turret Periscope Mark 20) the characteristics of which are: magnification, 8×; field of view, 5°; exit-pupil diameter 5 mm; and eye distance, 33 mm. While the optical diagram of

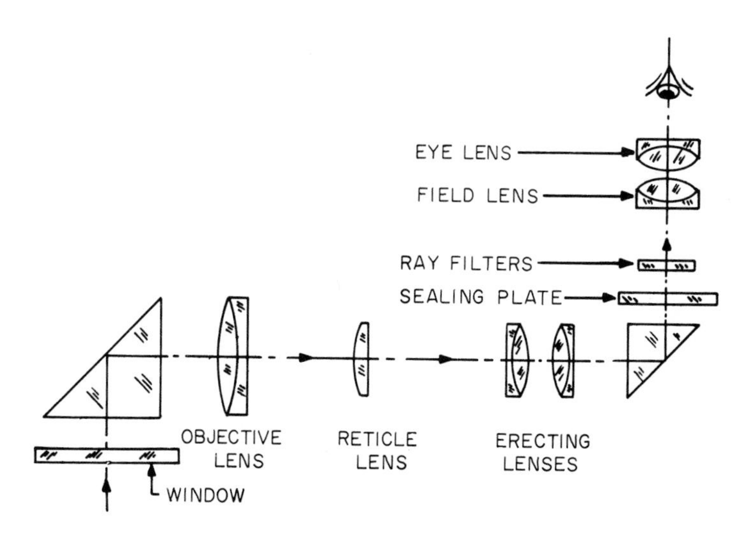

FIG. 10. Naval turret gun sight (shown in its side).

Fig. 10 is depicted in a horizontal position, in actual use, the instrument is positioned vertically. For this reason, the offset distance between the line-of-sight axis and the eyepiece axis indicated on the diagram should be regarded as the periscopic height of the instrument.

The optical system of the sight consists of a rectangular entrance window, a scanning head-prism (capable of rotation in elevation $\pm 15°$), a cemented doublet objective lens, a reticle lens with an etched cross line on the plano side, a symmetrical lens-erecting system composed of two cemented doublets, a right-angle eyepiece prism, a window for sealing the eyepiece portion of the instrument, selective ray filters (red, green, variable-density polarizing, and clear), and a symmetrical two-doublet eyepiece. Rotation of the scanning-head prism in elevation is accomplished through a hand wheel conveniently located at the base of the instrument. The hand wheel is connected to a long drive shaft, at the upper end of which is a lead screw engaging a gear segment attached to the head prism.

Further information concerning mechanical details of the instrument just described may be found in a Navy publication.[5] Also included in that publication is much useful information on Navy-type range finders, submarine periscopes, a ship-mounted binocular, and a tilting-prism telescope gunsight.

[5] Opticalman 1 and C, pp. 138–150. Bur. of Navy Personnel, U.S. Government Printing Off., Washington, D.C. (1966).

B. Prism Binoculars

1. *General*

Prism binoculars are used by the military as an optical aid for observation in the detection, recognition, and identification of hostile and friendly targets, and for determining their angular locations by reference to known landmarks, compass headings, or triangulation. Binoculars are also used to observe the effects of fire, whether impacts are on the target, to the right or left of it, beyond it (overs), or not quite to it (shorts).

A further use of binoculars is in the conduct of air-search rescue operations over land and sea. In such instances, because of the vibration of the aircraft and its movements in pitch, yaw, and roll, it is usually necessary to stabilize the viewing axis of the binocular in some manner, either internally or externally. A similar situation arises in connection with land vehicles (tanks) moving over uneven terrain, and naval vessels ploughing through rough waters, that would require stabilization of the line of sight of their sighting telescopes. Various means such as pendulums, dash pots, gyros, and stabilized platforms have been utilized to achieve stabilization. However, because of its complexity this matter will not be discussed further here, and the reader is advised to consult texts on the subject, such as those by James *et al.*[6] and Evans.[7]

2. *Construction*

A military binocular usually consists of two prism-erecting telescopes (called barrels) joined together in close alignment by a pivoted hinge pin. The hinge permits controlled separation of the two barrels for accommodation to the interpupillary distance of the user. Since the interpupillary distance of most individuals varies from about 58 to 72 mm, there is attached to the hinge a scale with spaced graduations between those limits for reference purposes.

It is imperative that the optical axes of both barrels be initially adjusted and maintained parallel with the hinge pin and with each other within certain limits in two directions in order to avoid eye strain through prolonged use of the binoculars. Lack of parallelism in the vertical plane is termed dipvergence, and the normal tolerance for it is about 8′ of arc Lack of parallelism in the horizontal plane is called divergence or convergence, depending on whether the two optical axes diverge or converge

---

[6] H. James, N. Nichols, and R. Philips, "Theory of Servomechanisms." McGraw-Hill, New York, 1947.

[7] W. R. Evans, "Control System Dynamics." McGraw-Hill, New York, 1954.

toward the eyes. The normal tolerance for divergence is 90′ of arc, and for convergence, 0′ of arc.

Because a difference in magnification between the two barrels would result in the formation of unequal-sized retinal images, it is normal to specify that the magnification difference between the two barrels must not exceed 5% of the magnification of either one.

3. *Features*

One of the most important features of a binocular is that it enhances stereoscopic vision (depth perception). Just how much farther we can perceive in depth with a binocular is governed by the magnification M of the binocular, the distance between the centers of the entrance pupils of the binocular B, the interpupillary distance IP of the user, and the maximum distance R_s at which an object can be perceived in depth by the unaided eye. These quantities are related by

$$E_s = (R_s'/R_s) = M(B/IP) \tag{4}$$

where E_s is a measure of the stereoscopic efficiency of a binocular, and R_s' represents the maximum distance at which an object can be seen in depth with the binocular.

Now, R_s depends on the interpupillary distance of an observer and his stereoscopic acuity. According to Martin,[8] the lower limits of stereoscopic acuity (binocular parallax) for individuals have been experimentally determined to be about 5″ of arc or less for highly experienced observers under favorable circumstances and conditions of steady-state observation, about 10″ of arc for experienced observers for instantaneous observations under favorable conditions, and up to whole minutes of arc for observers without experience. Martin further points out that there are some individuals who appear to be lacking in the stereoscopic sense altogether.

The increased distance at which objects can be seen with a binocular as compared with the unaided eye has already been discussed in the introductory portion of this chapter, and simplified equations were derived for the cases of daylight, twilight, and night-viewing. We must also consider the relative efficiencies of hand held and statically-mounted binoculars to ascertain the effect of hand tremor on the efficiency of viewing with a binocular. Information on this subject may be found in Table I,[9] which is based on data given by Köhler.[2] We see by the last column that as the magnification is increased, the efficiency of seeing with hand-held binoculars is reduced, which is in agreement with experience.

---

[8] L. C. Martin, "Technical Optics," Vol. II, 2nd ed., p. 161. Pitman, London, 1960.

[9] K. Brunnckow, E. Reeger, and H. Siedentopf, *Z. Instrumentenk.* **64**, 86 (1944).

TABLE I

TELESCOPE EFFICIENCIES MEASURED WITH DAY-VISION, UNCOATED ZEISS FIELD GLASSES[a]

$M.EP$	Telescope efficiency in day vision, measured supported $L_{supported}$	$\frac{L_{supported}}{M}$	Telescope efficiency in day vision, measured freehand, $L_{freehand}$	$\frac{L_{freehand}}{M}$	$\frac{L_{free}}{L_{supp}}$
6×30	5.02	0.84	3.95	0.66	0.79
8×30	6.40	0.80	5.00	0.62	0.78
7×50	6.48	0.93	4.55	0.65	0.70
10×50	9.12	0.91	5.62	0.65	0.62
15×60	11.75	0.78	6.48	0.43	0.55

[a] See Brunnckow *et al.*[9]

We can extract another piece of information from Table I by reference to our derived equation for daylight viewing, namely, $E_{max} = M(T_T)^{1/4}$. Since M is given in the first column of the table, and E_{max} appears in the second column, we deduce that the third column yields values of $(T_T)^{1/4}$. Upon raising each value in the third column to the fourth power, we arrive at the light transmission factor of each binocular. The values thus obtained appear reasonably close to the expected light-transmission factors of uncoated binoculars.

4. *Hand-Held Binoculars*

Figure 11 shows the optical system for a typical 7×50 hand-held military binocular. The system consists of a Porro-prism erecting system, a cemented doublet objective lens, a two-piece Kellner eyepiece, and a reticle mounted in the left barrel. The weight of the entire binocular is about 53 oz and the F-ratio of its optical system is about 3.9.

By way of contrast, Fig. 12 shows the optical system of a new 7×50 military binocular which is intended to supplant the binocular just referred to. This new binocular is described in a paper by Yoder,[10] in which he points out that the new optical system permits a weight reduction of about one-half, a size reduction of about one-third, and a reduction in overall length of about one-fourth in comparison with the former binocular. These gains were obtained by increasing the relative aperture of the optical system to $f/3$. The objective lens was converted into a three-element telephoto to provide a shorter distance from front vertex to image plane and to permit

[10] P. R. Yoder, *J. Opt. Soc. Am.* **50**, 491–493 (1960).

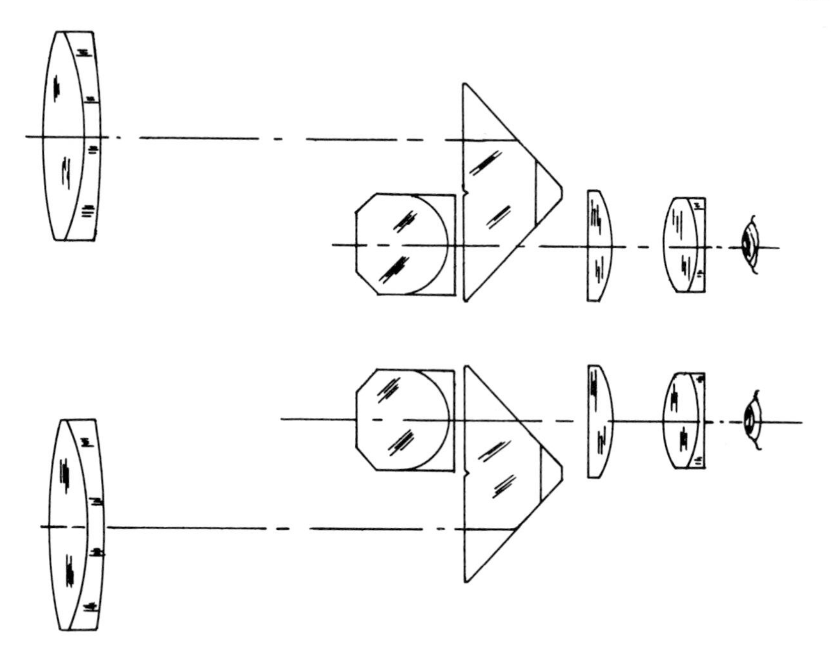

FIG. 11. Prism binocular, former type.

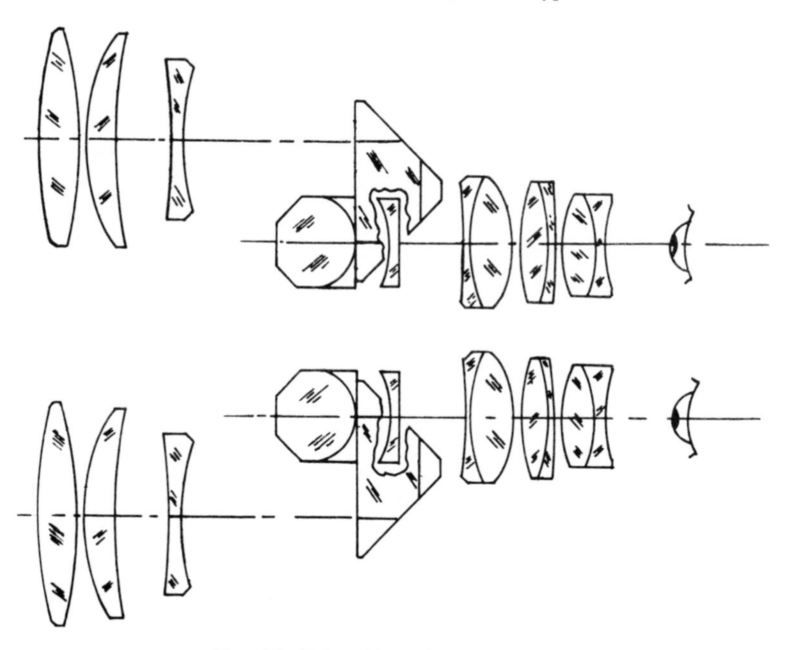

FIG. 12. Prism binocular, newer type.

greater flexibility for control of aberrations. The eyepiece was converted from the Kellner type to the three-cemented-doublet type shown, mainly for better control of the off-axis aberrations. In order to approximate the original eye distance with the new reduced focal length and much thicker eyepiece, a negative lens was introduced to the left of the reticle plane to cause the principal ray to diverge prior to entering the field lens of the eyepiece.

5. *Tripod-Mounted Binoculars*

A schematic diagram of a tripod-mounted 10 × 45 binocular is shown in Fig. 13. This type is used by commanders of artillery batteries and is

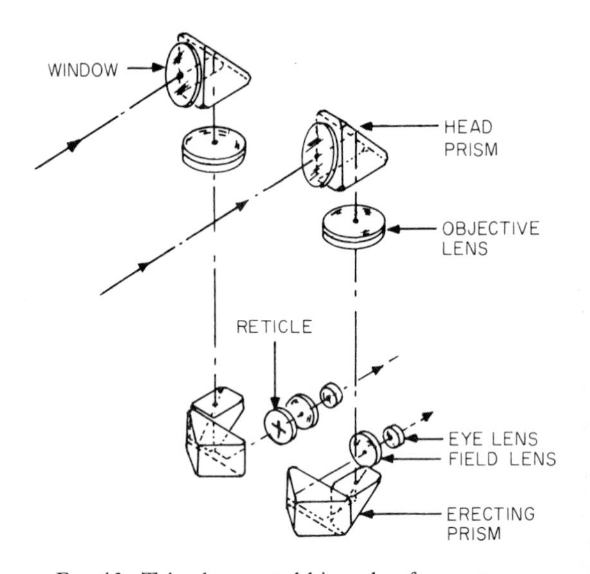

FIG. 13. Tripod-mounted binocular, former type.

often referred to as a Battery Commander's Telescope or a Spotting Telescope.

The optical system of this binocular makes use of a separated Porro prism, thus providing the necessary vertical offset (periscope height). The remainder of the optical system is conventional and consists of a top window, a cemented doublet objective lens, a reticle in the right barrel, and a wide-angle eyepiece.

The instrument is capable of 360° rotation about its vertical axis, and at the base of the binocular is an external horizontal reading circle graduated in angular mils for the measurement of azimuth angles. In addition, there is a mechanism for tilting the telescope forward and backward so as to elevate or depress the line of sight in a vertical plane. Interpupillary adjustment is achieved by pivoting one barrel of the instrument about a hinge pin located near the top of the binocular in the plane of the objective lenses.

Figure 14 is a schematic diagram of a new tripod-mounted binocular which is intended to replace the instrument previously described. This modern instrument is worth study, as it exhibits a number of features and innovations not previously discussed.

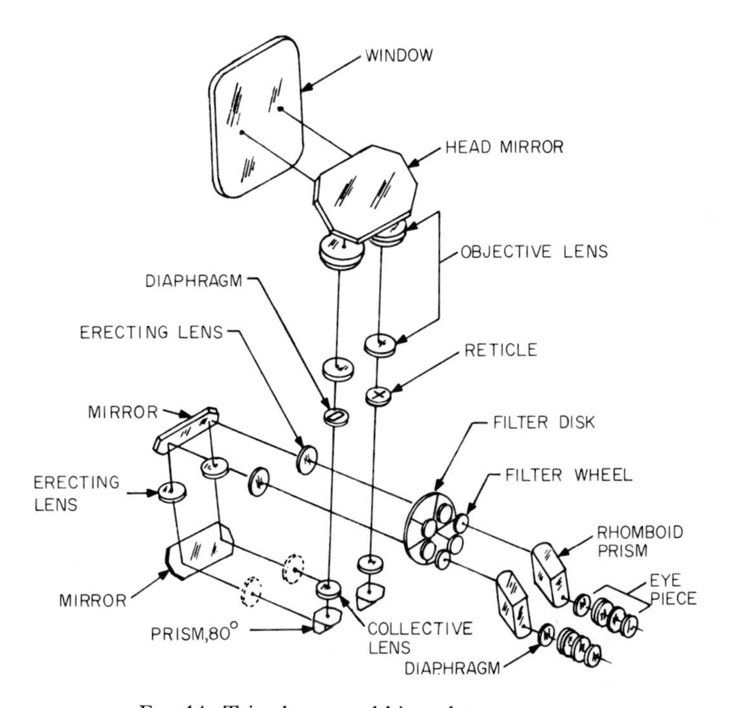

FIG. 14. Tripod-mounted binocular, newer type.

The binocular is a selective, dual-power instrument of 10× and 20× with a 50-mm entrance-pupil diameter, and its periscopic height corresponds to that of the previous instrument. It is capable of 360° rotation about its vertical axis, and its line of sight can be elevated 45° or depressed 28°. An interpupillary adjustment from 58 to 72 mm is provided through rotation of the left rhomboid prism and eyepiece combination.

The optical system in each barrel is basically that of a lens-erecting telescope, inasmuch as the three mirrors and the prism in the optical path constitute two rhomboids and thus do not affect image erection. In tracing through the optical system of one barrel, we find, in the following order:

an entrance window and tilting head mirror for elevation scanning (common to both barrels), a two-doublet objective lens system, a reticle, a collective lens, an 80° prism, the first right-angle mirror, the first erecting lens, the second right-angle mirror, the second erecting lens, a filter disk and wheel combination, a rhomboid prism, a diaphragm (field stop), and an Erfle eyepiece.

A selective magnification change is accomplished by 180° rotation of the combination of the two mirrors and the four erecting lenses shown on the lower left-hand side of the diagram. This rotation interposes the second erecting lenses into the dotted positions shown between the 80° prism and the first mirror.

Another feature of this system is the projection of elevation and azimuth scales into the field of view of the right eyepiece by means of two projection systems not shown on the diagram. Through attachment of the scales directly to the rotatable parts, any backlash in the gearing does not contribute to scale-reading error. A further feature of this binocular is the inclusion of a filtering device which permits 12 possible filter combinations from density zero to density seven to be obtained in progressive steps.

C. Panoramic Telescopes

1. *General*

A panoramic telescope is a military optical instrument which is employed to aim a gun or howitzer in azimuth. The telescope mount travels in azimuth with the weapon, carrying the line of sight of the telescope. The telescope provides the mechanism for setting the azimuth angle, while the mount or other associated equipment supplies the elevation angle and cross-leveling mechanism.

The characteristic feature of a panoramic telescope is that it maintains an erect image regardless of whether the line of sight is directed forward, to the side, or to the rear of the observer. The upper or rotating head of the instrument may be rotated through any desired angle up to 360° in the horizontal plane without requiring the observer to change his position.

In discussing the subject of panoramic telescopes, we will first examine the optical properties of an older type, discuss a study of its error structure, and describe how the findings of that study have been incorporated into the design of a new panoramic telescope.

2. *Former Type*

Figure 15 shows an optical schematic of a former type of panoramic telescope. The optical system is rather simple, and, in sequence, consists

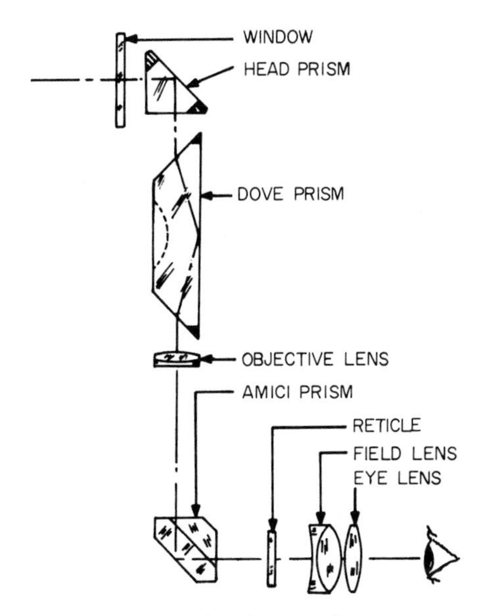

FIG. 15. Panoramic telescope, former type.

of an entrance window, a tilting head prism, a rotatable Dove prism, an objective lens, an Amici prism, a reticle, and an eyepiece. As mentioned previously, the head prism can be rotated through 360° in azimuth, and in order to keep the image erect, the Dove prism must be made to rotate at $\frac{1}{2}$ the speed of the head prism, by suitable gearing. Because the end faces of a Dove prism are at an angle of 45° rather than normal to the optical axis, the Dove prism cannot be used in converging light; hence, it is positioned in parallel light in front of the objective lens. The Amici prism could, of course, be replaced by a pentaprism if preferred.

3. *Error Structure*

The error structure of a panoramic telescope of the type shown in Fig. 15 has been analyzed by Hollis and Scidmore,[11] and the results of their analysis show that there are three basic errors which affect performance: a circle error, a plumb travel error, and a horizontal-plane error.

The authors define circle error as the difference in azimuth angle subtended by two points—a reference point and a target point—as read on the azimuth scale of the panoramic telescope, and the true azimuth angle.

---

[11] W. W. Hollis and W. H. Scidmore, Analysis of error structure of M12 panoramic telescope. Memorandum Rept. M60-12-1, U.S. Army, Frankford Arsenal. Armed Services Tech. Information Agency, Arlington, Virginia (1959).

Plumb travel error they define as the azimuth error introduced as a result of an elevation or depression of the line of sight and evidenced by the failure of the line of sight to track a true vertical line. They define horizontal-plane error as a deviation of the line of sight from a true horizontal plane as the instrument is rotated in azimuth.

Of these three errors, the authors aver that the first two are the more serious. However, they go on to say that the contributing causes of the circle error are the inclination of the Dove-prism axis with the vertical axis of the head prism, the pyramidal and collimation errors of the Dove prism, and the mechanical tolerances of the components. With regard to the causes of plumb travel error, the authors enumerate the deviation of the projected line of sight from the reticle from perpendicularity with the elevation axis, nonperpendicularity of the headprism normal with respect to the elevation axis, deviation of the head-prism axis from a horizontal plane, plus mechanical tolerances on components. In fairness to the authors, it should be mentioned that they went further with their analysis and developed explicit equations expressive of the errors mentioned, and validated their equations through computation and experimental tests.

From this explanation of the error structure, we can draw some rather definite conclusions as to how the system can be improved. With respect to circle error, it appears that the primary cause is the fact that the Dove prism is in front of the reticle. We also find that the plumb travel error is due principally to the head prism and its relation to the reticle. Hence, we conclude that in a new instrument, the reticle should be combined with the head prism and rotate with it in azimuth, and that the derotating prism should follow the reticle. The plumb travel error can be reduced by designing the mount such that when the gun is elevated, the panoramic telescope remains stationary. These features have been incorporated into the new instrument to be described.

4. *New Type*

The optical system for a new panoramic telescope is shown in Fig. 16. As may be seen from the diagram, the system incorporates a lens-erecting telescope, a head prism, a Schmidt (Pechan) type of derotating prism, a pentaprism, a reticle, and a front window. The window, head prism, objective lens, and reticle are mounted in a single tube and rotate together in azimuth as a complete unit. Since the reticle is in front of the derotating prism, the latter does not contribute to the circle error. The erecting-lens assembly is of the thick-meniscus (low-Petzval-curvature) type which was referred to earlier in Section II,A,4. Angle-reading errors have been eliminated by the incorporation of a digital readout counter in place of the usual divided circle.

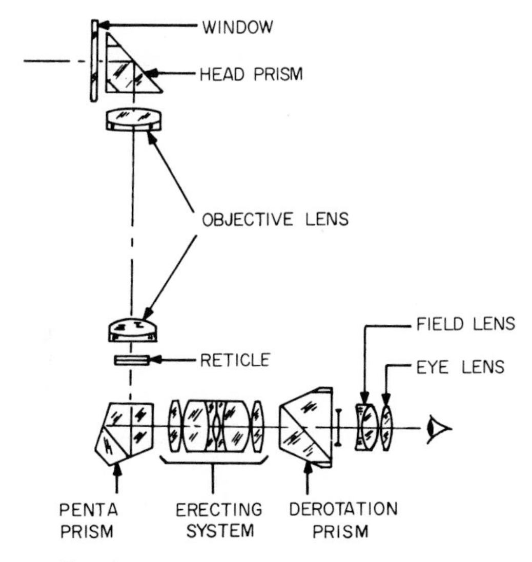

FIG. 16. Panoramic telescope, newer type.

D. PERISCOPES

1. *General*

A periscope may be defined as a type of optical instrument which exhibits a decided offset between the line of sight and viewing (eyepiece) axis. In most cases, the offset is in a vertical direction. By reason of this offset, observation and weapon pointing can be conducted from confined locations, e.g., from entrenchments, or the interiors of armored vehicles (tanks), naval craft (submarines) or aircraft (helicopters). In covering the subject of periscopes, three items will be discussed: tank-driver's periscopes tank-gunner's periscopes, and submarine periscopes.

2. *Tank-Driver's Periscopes*

A tank-driver's periscope is essentially of boxlike construction consisting primarily of two mirrors set at 45° and offset by about 9 in. vertically (Fig. 17). There are differences between models, which relate mainly to physical size and manner of construction. For example, in some models, the entire central portion of the periscope is made of a solid plastic material to which are cemented at the top and bottom, back-surfaced glass mirrors and glass entrance and exit windows, the whole being encased in a metal housing. Another variation is in the form of a single-piece, rather large, rhomboid prism of glass, also surrounded by metal. In other forms, the

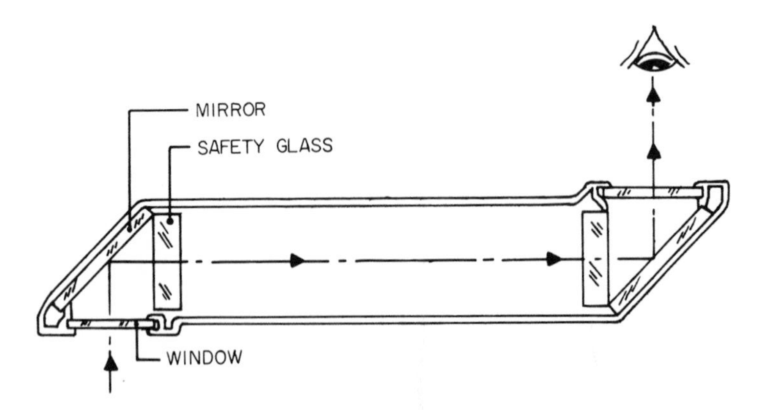

FIG. 17. Tank-driver's periscope (shown on its side).

central body is divided into two or more sections separated by air spaces to avoid glass splatter into the user's eyes if the exposed frontal portion of the periscope is forcibly shattered.

As can be seen in Fig. 17, the entrance and exit windows of the periscope are rectangular in shape, the width of each window being considerably greater than its height. In normal periscopes of this type, the ratio of width to height is about 4 : 1. In various periscope sizes and constructions, the vertical field of view varies from about 20° to 40°, and the horizontal field of view from about 80° to 160°.

3. *Tank-Gunner's Periscopes*

An optical schematic of a typical tank-gunner's periscope is shown in Fig. 18. From this diagram, it is apparent that the entire instrument consists of two periscopes, one of unity power (for observation) and one of high power (for weapon pointing). The principal optical components of the unit-power periscope are the common entrance window, the common rotatable head mirror, a lower reflecting mirror, and an exit window. The principal optical components of the high-power periscope, in addition to the common entrance window and rotatable head mirror, are an objective lens, a roof pentaprism, a reticle, and an eyepiece.

Concerning the high-power system of the periscope, we note that the combination of the rotatable head mirror and the roof pentaprism constitute a modified version of a Hensoldt inverting prism, and we see also that the lower portion, consisting of the objective lens, roof pentaprism, reticle, and eyepiece, could be mechanically mounted into a small housing to form what may be termed a separate, lower, elbow telescope.

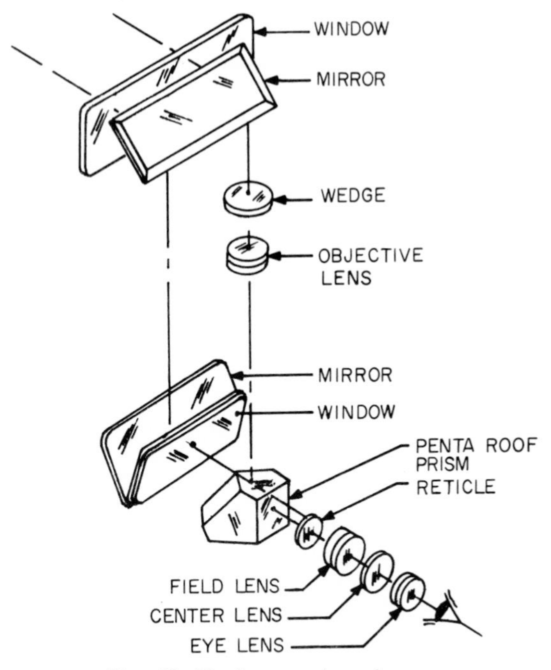

FIG. 18. Tank-gunner's periscope.

As to integration of the low- and high-power optical systems into a unified periscope, reference is made to Fig. 19, which is an exterior view of the instrument. For ease of description, we will consider it to consist of three units, designated *A*, *B*, and *C*. Unit *A* is the main body of the periscope, into which is mounted the common entrance window, the common rotatable mirror, and the optical components of the unit-power optical system. Also mounted in unit *A* is a mechanism which is connected to the shaft output of unit *B* and the rotating head mirror. This mechanism

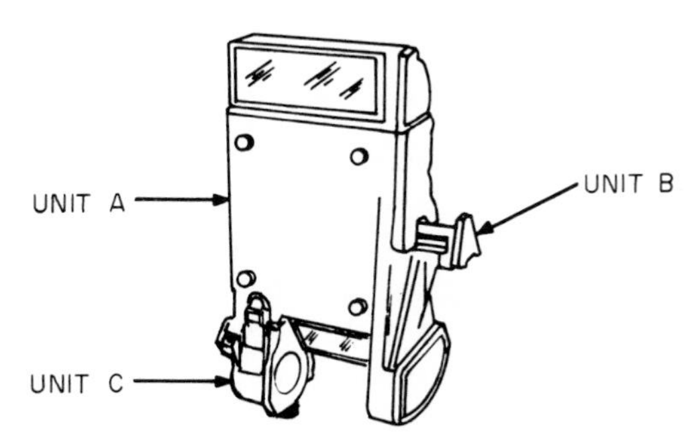

FIG. 19. Exterior view, tank-gunner's periscope.

provides a half-speed rotation to the mirror with respect to the shaft output of unit *B*. Unit *B* is a connector between the periscope and the elevation mechanism of the weapon, through a device known as a differential ballistic drive. Unit *C* is the lower elbow telescope assembly previously mentioned, and it is attached to unit *A* by trunk latches.

By means of the apparatus just described, as the weapon itself is elevated or depressed, the line of sight of the periscope will faithfully follow the angular position of the axis of the bore of the weapon. Moreover, by introducing a known angular correction into the differential ballistic drive (for very practical reasons), there can be produced a constant angular difference between the line of sight of the periscope and the axis of the bore of the weapon.

Returning to the periscope itself, numerous variations of the optical system shown in Fig. 18 are possible. The first variation concerns the elimination of a portion of the left side of the unit-power system, and providing, in its place, a duplicate of the high-power system on the right-hand side of the periscope, thus converting the periscope into a binocular high- and low-power periscope. A second variation is to detach the lower elbow telescope and substitute for it an infrared-viewing elbow telescope, to provide an active IR night-viewing capability. A third variation is to modify the construction of the lower elbow telescope by withdrawing the opaque-lined reticle, converting the roof pentaprism into a beam-splitting prism, and providing an entrance window just behind the prism in prolongation of the eyepiece axis. In front of this entrance window, a reticle projection system would be placed to project an image of a reticle into the focal plane of the eyepiece. Conceivably, the reticle could be actuated vertically and horizontally by suitable electromechanical mechanisms to introduce desired vertical and horizontal angular lead information with respect to the target into the main sighting system of the periscope. While the three variations just discussed represent but a few of the possibilities that could be achieved, the reader is reminded that the stark simplicity of Fig. 18 represents but a beginning.

4. *Submarine Periscopes*

Periscopes for attack-type submarines are in reality extremely long lens-erecting telescopes varying in overall length (vertical offset distance) from some 20 ft to about 40 ft and beyond. A greatly simplified optical diagram of a submarine periscope, not to scale, is shown in Fig. 20.

The first optical component of the periscope (aside from the fixed entrance window) is a rotatable head-prism, to provide elevation scan of the line of sight in a vertical plane. The azimuth scan of the periscope is accomplished through physical rotation of the entire instrument from the

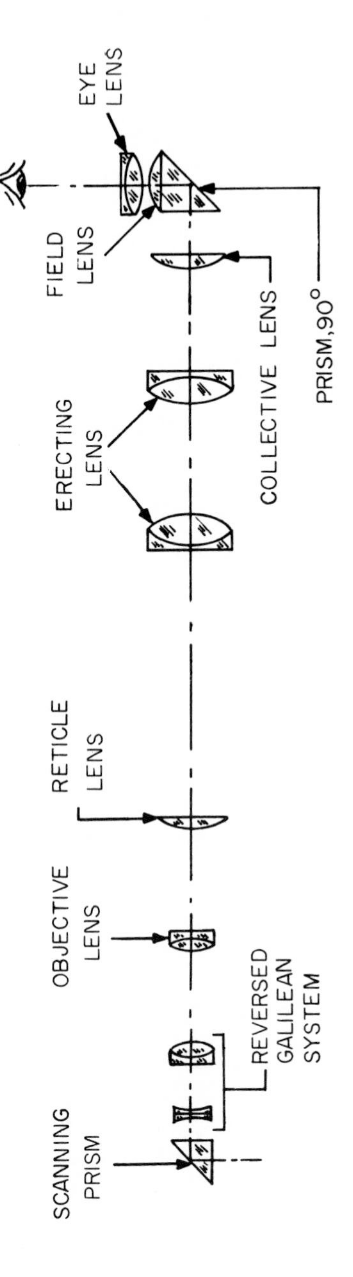

FIG. 20. Submarine periscope (shown on its side).

user's position at the lower (eyepiece) end by means of handles provided for that purpose, with assistance perhaps from an electric motor driving unit. In the diagram, the rotatable head prism is shown in the form of a normal right-angle prism. However, in practice, it is usually of the "double Dove," or sky-searching, type, which provides more extensive coverage in the vertical plane, reaching almost to the zenith.

Following the head prism is a reversed Galilean telescope, the function of which is to provide the convenience of selective, dual-power (lower magnification with wider field of view) through "flip" in and out of the two components of the Galilean telescope.

Immediately following the reversed Galilean telescope, is the main body of the periscope. We can visualize this as consisting of two astronomical telescopes with a rather wide separation between them. We could also consider the periscope as a system in which the focal length of the objective lens is equal to the magnification of the periscope multiplied by the focal length of the eyepiece, the magnification of the erecting system being unity; hence, the two erecting lenses could be of the same focal length and identical with each other.

As to the eyepiece, it could be of a modified Kellner type with its field lens cemented directly to the lower right-angle prism, as shown. To round out the optical system, there could be installed collective field lenses at or near the front and rear image planes to influence the path of the principal ray as described earlier.

The first-order layout of an optical system for an actual submarine periscope has been considered briefly in Vol. I, Chapter 6, Section VI,4. Moreover, the Navy publication[5] previously referred to contains much useful information concerning the optical design, characteristics, maintenance, and repair of modern submarine periscopes.

E. Articulated Telescopes

1. *General*

In recent times, straight-tube, lens-erecting telescopes have lost their appeal as weapon sights for control of direct fire of the main armament of tanks. Advances in modern weaponry require that straight-tube telescopes be much longer, to the extent that undue vertical head movement on the part of the user is required. Not only does this impose an inconvenience and discomfort on the user, but it even detracts from the usefulness of the telescope at or near the extremities of excessive head movement. These objections can be overcome through conversion of what was formerly a straight-tube, lens-erecting telescope into an articulated telescope. In the discussion which follows, three methods will be explained for accomplishing this conversion.

2. *Discussion*

One method of converting a lens-erecting telescope into an articulated telescope is that described by Martin[12] and attributed to Heyde. In essence, it requires the introduction of a vertically offset rhomboid system between the objective lens and the erecting system. Articulation is accomplished by rotating the entire objective lens system and the top reflector of the rhomboid as a complete unit about the midpoint of the lower reflector of the rhomboid. Simultaneously, the lower rhomboid reflector must be rotated about its horizontal axis at $\frac{1}{2}$ the speed of rotation of the upper system.

The second method is to utilize the principle of a panoramic telescope by placing it, not in a vertical plane, as commonly used, but with its main axis horizontal. Thus, what were formerly the azimuth and elevation scanning motions now become the elevation and azimuth travel motions. With this arrangement, it can be seen that no eye movement would be required while the head prism of the panoramic telescope is rotated in elevation.

The third method of achieving articulation is to use the rotational properties of the second type of Porro prism system. This method may be explained by reference to Fig. 21, which is a top view of the optical system of an articulated telescope. The prism system consists of two small, right-angle prisms and a larger, Porro prism, the latter being actually two right-angle prisms combined in one solid piece of glass. Note also that the relative positions of the three prisms constitute two rhomboids, hence they have no effect on the inversion or reversion of the imagery. Since this prism is positioned behind the objective–reticle combination, any errors associated with it do not contribute to sighting errors.

Articulation is accomplished through vertical rotation of the forward member of the telescope about the horizontal axis of the first right-angle prism. In order to correct the image lean caused by this rotation, the rear member of the telescope is simultaneously rotated about the horizontal axis of the second right-angle prism by an equal amount but in opposite direction to the front member. Therefore, the front and rear members of the telescope operate on a one-to-one ratio, and can be connected by suitable gearing to accommodate that ratio. The Porro prism is constrained to maintain the offset distances shown in the diagram, and it is usually desired that the rear member of the telescope be reasonably horizontal. Thus, when the front member of the telescope is elevated through a given angle, the Porro prism must incline at $\frac{1}{2}$ that angle from the horizontal, imparting to the rear portion of the telescope very slight longitudinal and vertical movements.

---

[12] L. C. Martin,[8] p. 71.

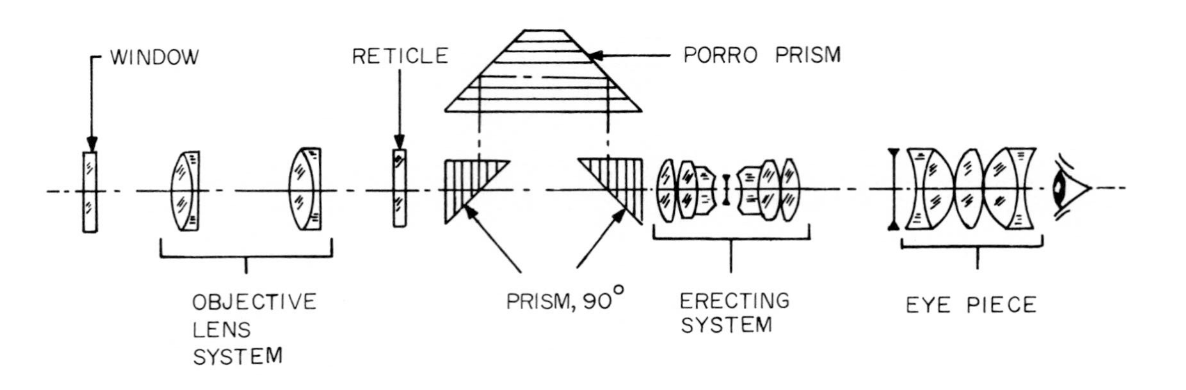

Fig. 21. Optical system of an articulated telescope.

While the explanation just given may seem somewhat complicated, this latter method of achieving articulation is exceedingly simple to apply, and, in view of its other many advantages and merits, exhibits what may be termed attraction and appeal.

F. Bomb Sights

The basic elements of bombing from aircraft may be explained by reference to Fig. 22, which represents a particular case of bombing, that of horizontal flight, either upwind or downwind, with the target stationary on the ground. We suppose that the aircraft is flying on a straight course (zero drift) and from left to right in the direction *ABC* at a constant speed u_0 and

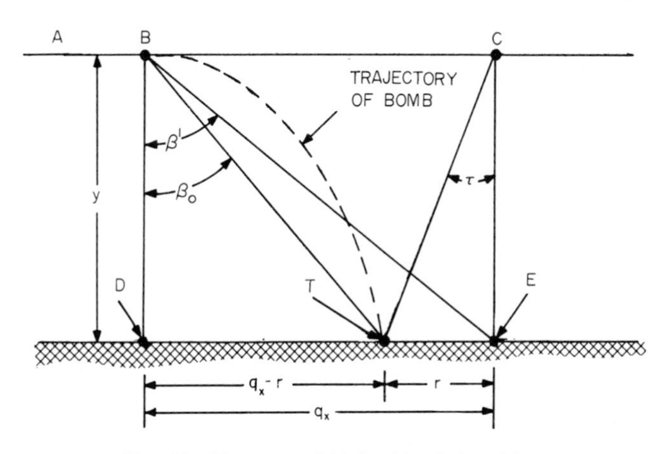

Fig. 22. Elements of high-altitude bombing.

constant altitude y with the intent of bombing the target T. At the moment the aircraft reaches the point B, the bombs are released. The bombs then travel in an approximately parabolic curve from B to T under the influence of gravity, the forward velocity of the aircraft, and air resistance (drag). The function of the bombsight is to locate the point B at which the bombs are to be released.

It is noted from Fig. 22 that the position of B is uniquely defined by the line from B to T and the angle β_0. Now, the line from B to T constitutes the line of sight from the bombsight to the target, and is governed by the angle β_0 with respect to the true vertical BD. Hence, all that need be done by the bombadier is to preset the angle β_0 into his sight before reaching the point B, and observe through his bombsight until the target T intersects the range cross-line of the reticle. When this occurs, the point B has been reached and the bombs are released.

The magnitude of the angle β_0 can be ascertained by an equation based on the geometry of Fig. 22, information contained in bombing tables relating to t_w (time of flight of the bomb) and the angle τ (angle of trail of the bomb due to air resistance), and from known (or assumed) values of aircraft velocity u_0, altitude y, type of bomb being dropped, and the structure of the atmosphere. From Fig. 22, it is apparent that $\tan \beta_0 = (q_x - r)/y$, where q_x is the product of u_0 by t_w, and the linear trail r of the bomb is equal to y times $\tan \tau$. If we replace q_x/y by $\tan \beta'$ and substitute $\tan \tau$ for r/y, we obtain

$$\tan \beta_0 = \tan \beta' - \tan \tau \tag{5}$$

as the equation which must be satisfied for the special case referred to.

The problem before us is: What type of optical system should be utilized for the bombsight, and what precautions should be observed in connection with it?

One solution is to utilize the principle of a panoramic telescope, but mounted for downward viewing through an opening in the floor of the aircraft, and oriented such that the 360° scanning motion would operate in a vertical plane for target tracking and measurement of β_0 and β'. The elevation motion of the panoramic could be used to set off desired lateral deflection angles. The magnification of the intended optical system should be reasonably high to assure positive identification of a comparatively small target at T, with a reminder that the slant range BT would probably be relatively long.

It is imperative that the bombsight system should be capable of orientation about a true vertical and, further, that the true vertical be maintained throughout the entire bomb run, otherwise the predicted angle β_0 will unknowingly shift, resulting in a bombing error. This feature requires continuous cross-leveling of the bombsight, preferably by automatic means, coupled with complete and continuous stabilization of the desired line of sight.

For more complete information on the subject of bombing from aircraft, the reader is referred to the work by Hayes,[13] which served as a basis for the material presented here.

G. Night-Vision Devices

The advent of high-performance, infrared-sensitive, image-converter tubes and their successors, infrared- and visible-light-sensitive image-intensifier tubes, permits the design of special military optical instruments

[13] T. J. Hayes, "Elements of Ordnance," pp. 488–512. Wiley, New York, 1938.

for performance of the functions denoted in Fig. 1 at extremely low levels of ambient illumination, extending well beyond the region of twilight into that region typified as deep night darkness.

In discussing the subject of night-vision devices, emphasis will be placed on the optical systems involved, utilizing as needed only the general characteristics of the tubes mentioned. These characteristics are discussed in Vol. II, Chapter 6, and in brochures available from the various manufacturers.

Reduced to its simplest terms, a night-vision device consists of an objective lens system, the electron-optical element, and an eyepiece. The image of an object formed by the objective system on the photocathode surface of the electron-optical element is transferred to the screen of that element, where a visible image is formed which is viewed with the aid of an eyepiece acting as a magnifier. Hence, the electron-optical element acts in the manner of a lens-erecting system in a telescope. Perhaps the simplest of the night-vision devices is the sniperscope, which is represented in schematic form in Fig. 23.

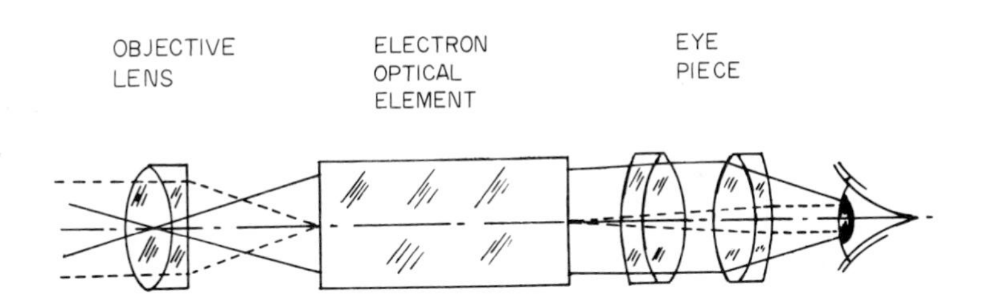

FIG. 23. Schematic diagram of a sniperscope.

The overall requirements for a specific night-vision device are usually stated in terms of the linear dimensions of the target, the distance or range at which it is desired to recognize the target, the desired field of view, and the magnification. Other parameters such as minimum length or weight, and maximum diameter, may or may not be specified. The principal characteristics of the electron-optical element are its photocathode diameter, resolution, contrast transfer function, distortion, and its magnification or demagnification.

From this multiplicity of constraints, the problem is to design an optical system which is an acceptable compromise between conflicting requirements of range, field of view, magnification, and perhaps weight. As an example, let us assume that the resolution of the electron optical device is R_T line pairs per millimeter. Standard practice usually calls for the target image to cover at least 7.5 line pairs for recognition to be possible. If the angular size of the target requiring recognition is θ, then

$$\tan \theta = 7.5/R_T f_0 \tag{6}$$

or

$$f_0 = 7.5/(R_T \tan \theta) \tag{7}$$

where f_0 is the focal length of the objective lens.

Once f_0 has been determined, the field of view is known, since the photocathode diameter is fixed. If the field of view is too small for the intended application, some tradeoff between maximum recognition range and field of view may be required.

For good low-light-level performance, a lens system of the highest possible aperture is required. However, it will of course be obvious that for many combinations of R_T and θ, the resulting size of an $f/1$ lens would be impractical, and there must again be a tradeoff, this time between maximum recognition range and entrance-pupil diameter.

Upon establishment of a value for the focal length of the objective lens, the desired magnifying power of the night-vision device can be obtained upon assignment of an appropriate focal length for the "eyepiece" based on a knowledge of the demagnification of the electron-optical element, in the same manner as for a lens-erecting telescope.

H. Pancratic Telescopes

1. *General*

A telescope with a built-in feature which permits the user a choice of different magnifications is called a pancratic telescope. Although application of the pancratic principle to military optical instruments is not too widespread, in its selective, dual-magnification form, it has been applied to submarine periscopes, tank range-finders, tank telescopes, observation telescopes, etc., and in its continuously variable form, to stadiametric range finders, tank telescopes, and air-defense sighting telescopes, among others.

2. *Dual Magnification*

Selective, dual magnification may be attained in various ways. One method is through use of different focal-length eyepieces mounted in a swivel turret at the eye end of the instrument. A second method (see Fig. 20) is by means of a reversed Galilean telescope located in front of the objective lens of a high-power telescope. Through installation of a suitable mechanism, the two components of the Galilean can be caused to "flip" in or out of the system as desired, thus altering the magnification. This method is rather appealing, for it is simple to apply and easy to operate. For these reasons, this method is used extensively in connection with submarine periscopes.

Still another way of attaining selective, dual magnification is much like the method just described, except that actuation is applied to the erecting lenses of the system, either by longitudinal translation along the optical axis, or by "flipping" them in or out (see Fig. 14) of the optical path. Quite naturally, this method can be applied effectively only if the erector lenses are of comparatively small diameter.

3. *Continuously Variable Magnification*

The feature of continuously variable magnification can be accomplished by application of the zoom principle to either the objective lens system or the erecting lens system of a telescope. Since details of the former are given in Vol. III, Chapter 3, and details concerning the latter are covered by Martin,[14] further explanation in this chapter is not deemed necessary.

III. RANGE FINDERS

The primary purpose of a military range finder is to determine the range (or distance) from a point of observation to a distant target. In general, military optical range finders may be divided into three broad classes: stadia range finders, self-contained-base range finders, and time-based (laser) range finders. Each of these classes will be examined in some detail by reference to a number of representative examples and by discussion of the theory, error structure, advantages, limitations, and applications of each.

A. STADIA RANGE FINDERS

a. Theory. All stadia range finders are based on the principle that a target of width W at a range R subtends an angle θ, and through knowledge or estimation of W and the measurement of θ, the range R can be ascertained from

$$\theta = W/R \qquad \text{or} \qquad R = W/\theta \tag{8}$$

We can conceive of a stadia range finder as being in the form of a small, hand-held device such as that shown schematically in Fig. 24. It consists of a low-power Galilean telescope, a beam-splitter prism, a rotatable mirror,

[14] L. C. Martin,[8] pp. 48–51.

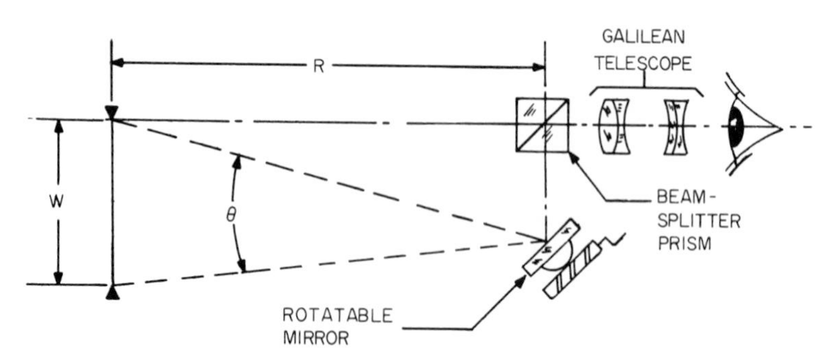

FIG. 24. Stadia range finder (schematic).

a mechanism for rotating the mirror (a worm and worm wheel is indicated on the diagram), and a reading scale, the latter displaying the range R in terms of selected widths W, as governed by the angle θ. Operation of the device is easy. All that one needs to do is to select a target, estimate its width, rotate the mirror until one end of the target seen in one beam coincides with the other end of the target seen in the other beam, and then apply a measured rotation of the mirror to superimpose the two images on one another. The range R is then read from the range scale opposite the value of W on the width scale.

b. Error Structure. The error structure of a stadia range finder can be studied by differentiating Eq. (8) and expressing the result in the form

$$dR/R = dW/W - (R/W)\,d\theta \tag{9}$$

It is apparent from Eq. (9) that the percentage error in range depends on the algebraic sum of two quantities, the percentage error in estimating the width W of the target, and the product of (R/W) times the angle error $d\theta$.

B. SELF-CONTAINED-BASE RANGE FINDERS

a. Theory. Unlike a stadia range finder, a self-contained-base range finder involves a measurement of the angle θ subtended by the base B of the range finder at the target distance R in accordance with

$$\theta = B/R \tag{10}$$

No estimate of target size is now required, but the angle θ is much more difficult to measure than in the stadia range finder.

Self-contained-base range finders may, in general, be divided into three categories: the stereoscopic type, the coincidence type, and the two-magnification (or azimuth) type. Further, the coincidence type is usually subdivided into the superposed-image type and the split-field (erect or invert) type, depending on the mode of presentation of the image at the eye. Since the fundamental distinctions and differences which apply to these various range finders are excellently covered by Cheshire[15] and Gardner,[16] no attempt will be made here to distinguish among them. Instead, attention will be devoted to a discussion of specific devices employed for measuring the angle θ, a brief description of two modern military tank range-finders, and an explanation of the error structure applicable to these two range finders.

b. Range Compensators. Literally, the "heart" of a self-contained-base range finder is the device employed for the measurement of the angle θ. In modern terminology, such a device is usually called a "range compensator" or just a "compensator." Under the heading "deflecting devices for coincidence range finders," Cheshire[15] lists seven different types of compensator. These are: a longitudinally traveling prism in a convergent beam, a tilting prism in a parallel beam, oppositely rotating prisms in a parallel beam acting as a variable-power prism, rotary end reflectors, laterally displaced object glasses, a tilting plane-parallel plate in a convergent beam, and differently magnified images brought into coincidence by rotation of the range finder about a vertical axis. At one time or other, each of these seven different types of compensator has been used in military range finders.

There is yet another type of compensator not mentioned by Cheshire which is employed in modern military range finders. This is known as the "sliding lens" type. Basically, it consists of a zero-power lens combination composed of two equally weak lenses, one positive and one negative, and functions as a variable-range wedge through the lateral translation of one lens with respect to the other. In addition, a graduated range scale is usually coupled directly to the positive lens and moves with it, thus eliminating one instrumental error. Figure 25 shows a diagram of a typical sliding-lens compensator. The principle of its operation is based on satisfying the equation

$$\theta = B/R = D/f \tag{11}$$

where f is the focal length of the positive lens, D is its lateral displacement for measurement of θ, and θ, B, and R have the same meaning as before. [Note: Since the range scale is attached directly to the sliding lens, D is also the linear distance on the range scale from the infinity mark to the particular graduation mark of a range R; hence Eq. (11) is also utilized to

---

[15] F. J. Cheshire, Range finder, short-base, *in* "Dictionary of Applied Physics" (Sir R. Glazebrook, ed.), Vol. IV, pp. 633-654. Macmillan, London, 1923.

[16] I. C. Gardner, Elementary optics and applications to fire control instruments, pp. 85–96. U.S. Government Printing Off., Washington, D.C. (1924).

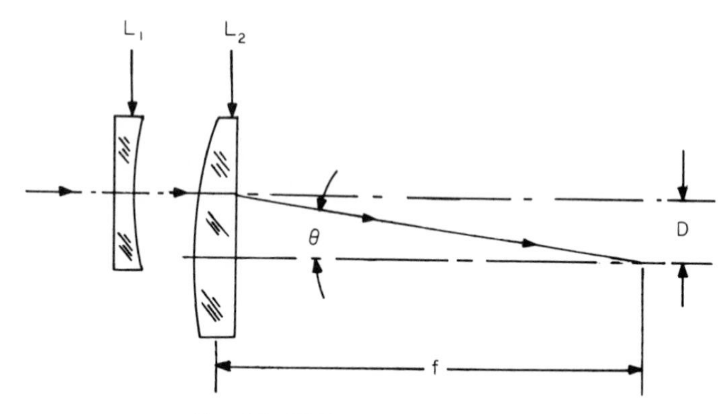

FIG. 25. Sliding-lens compensator.

compute range-scale graduations.] Upon differentiating Eq. (11), we gain an insight into the error structure of a sliding-lens range compensator:

$$dR/R = df/f - dD/D \tag{12}$$

which will be referred to later on.

c. Range Finder No. 1. Figure 26 shows the optical system for a modern military optical range finder of the stereoscopic type for use in an army tank. For the purpose of analysis, the optical system may be conveniently divided into four major parts: a right-hand sighting system, a left-hand sighting system, a collimator system, and a sliding-lens compensator. The right-hand sighting system is essentially a lens-erecting panoramic telescope, as is evident from the list of the various optical components starting from the end penta reflectors, to the eye lens of the eyepiece. The left-hand sighting system is exactly symmetrical with the right-hand system except for the sliding lens compensator system which is on the right-hand side. The collimator system serves two purposes; first, it projects the right and left stereo reticles into the two main sighting systems and, secondly, it projects the range scale, the gun-laying reticle, and the ammunition scale into either the right or left main sighting system, depending on the selected position of the scale transfer prism.

The pertinent features of this range finder are: stereoscopic measurement of range; elevation of the line of sight with no movement of the eyepieces; location of the eyepieces to suit the restricted position of the gunner; use of the left-hand sighting system as an offset sighting telescope in the event that the right penta reflectors are destroyed by enemy gunfire; azimuth and elevation boresight adjustment of the gun-laying reticle; halving adjustment of the stereo reticles; internal wedge adjustment (with a graduated correction scale) for observer calibration of the range finder

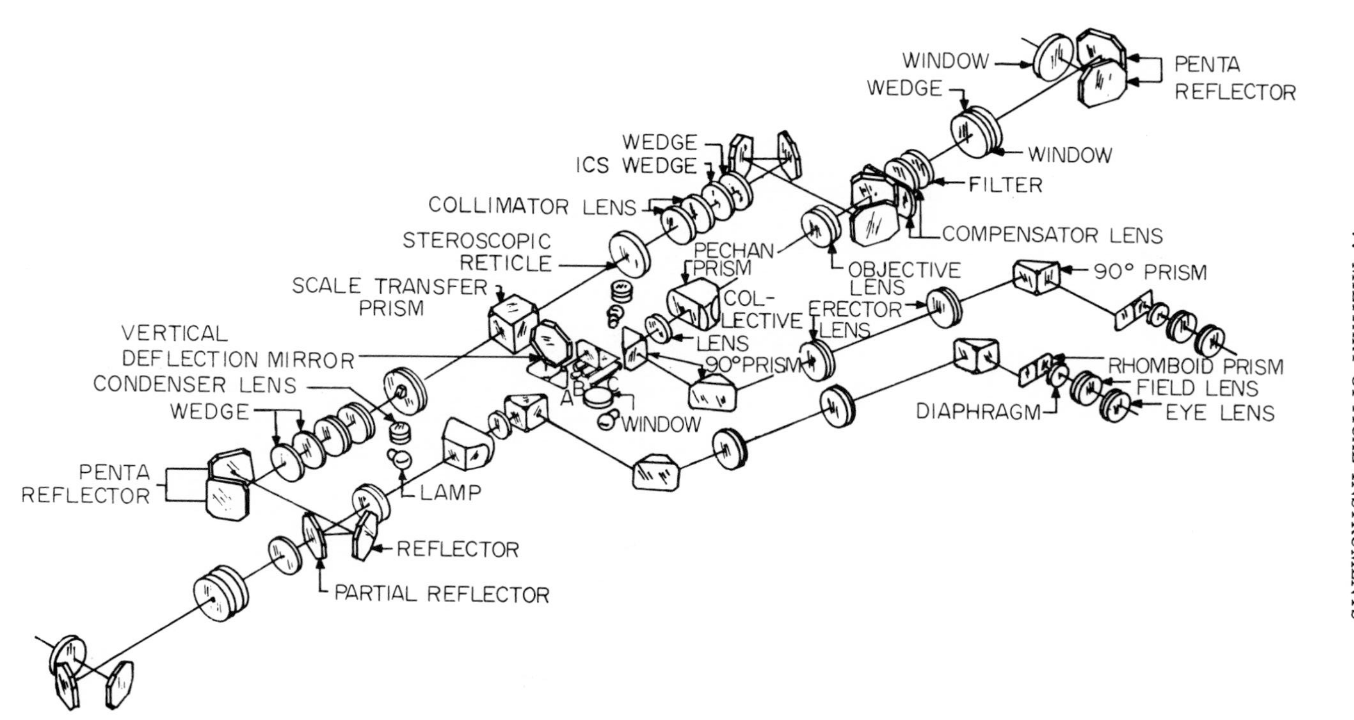

FIG. 26. Stereoscopic range finder No. 1. Scales: A, range scale; B, gun laying reticle; C, ammunition scale.

on a target of known range; replaceable, factory-aligned right and left end boxes (window, penta reflectors, and wedge); and incorporation of a ballistic system which converts range data obtained by the operator into superelevation data, and transmits it to the gun *via* a superelevation transmitter, thus automatically positioning the gun in elevation. The description of this range finder has of necessity been brief, but further and more complete details concerning methods of installation, operation of the ballistic mechanisms, and other features of the optical system may be found in the patent issued to Deal and Patrick.[17]

d. Range Finder No. 2. The optical system of a second stereoscopic range finder which was also conceived by this author is shown in Fig. 27. Basically, the optical system consists of a left viewing system, a right viewing system, a collimator system, and a sliding-lens compensator system. Insofar as the two viewing systems are concerned, the range finder may be regarded as an unbalanced system, since the left viewing system is in the nature of a lens-erecting telescope of some considerable length, whereas the right viewing system is a rather short, prism-erecting telescope. The reason for this imbalance is the necessity for positioning of the two eyepieces at a considerable distance to the right of center of the baseline of the instrument, in order to accommodate the restricted position of the operator.

Despite the imbalance exhibited by this range finder, there is an almost perfect match of the tangential and sagittal curvatures and other aberrations between both viewing systems. This was achieved through exceedingly careful design of the lens-erecting telescope system.

The collimator system for this range finder is not as complicated as the one shown in Fig. 26. This is due to the elimination of various unneeded optical components in the central portion of the collimator system, plus a different method of beam projection, from double penta reflector systems on each side as shown in Fig. 26, to a single Porro mirror system on each side as noted in Fig. 27.

An important feature of this particular range finder, which is not apparent from the diagram, is the fact that the longitudinal axis of rotation of the range finder is about a line through the center of rotation of an average individual's head, which is located at, or near, the position of his ears. This feature is very convenient from a human-engineering viewpoint, for the operator experiences no neck strain as the range finder is elevated or depressed. A further feature of the optical system of this range finder is that it can be converted from a binocular, stereoscopic type into a monocular, full-field, coincidence type through the simple expedient of introducing a rhomboid beam-splitter prism combination to direct the optical path from the left viewing system into the path of the right viewing system, just in front of the right eyepiece.

---

[17] C. T. Deal and F. B. Patrick, U.S. Patent 2,857,816 (filed July 23, 1956).

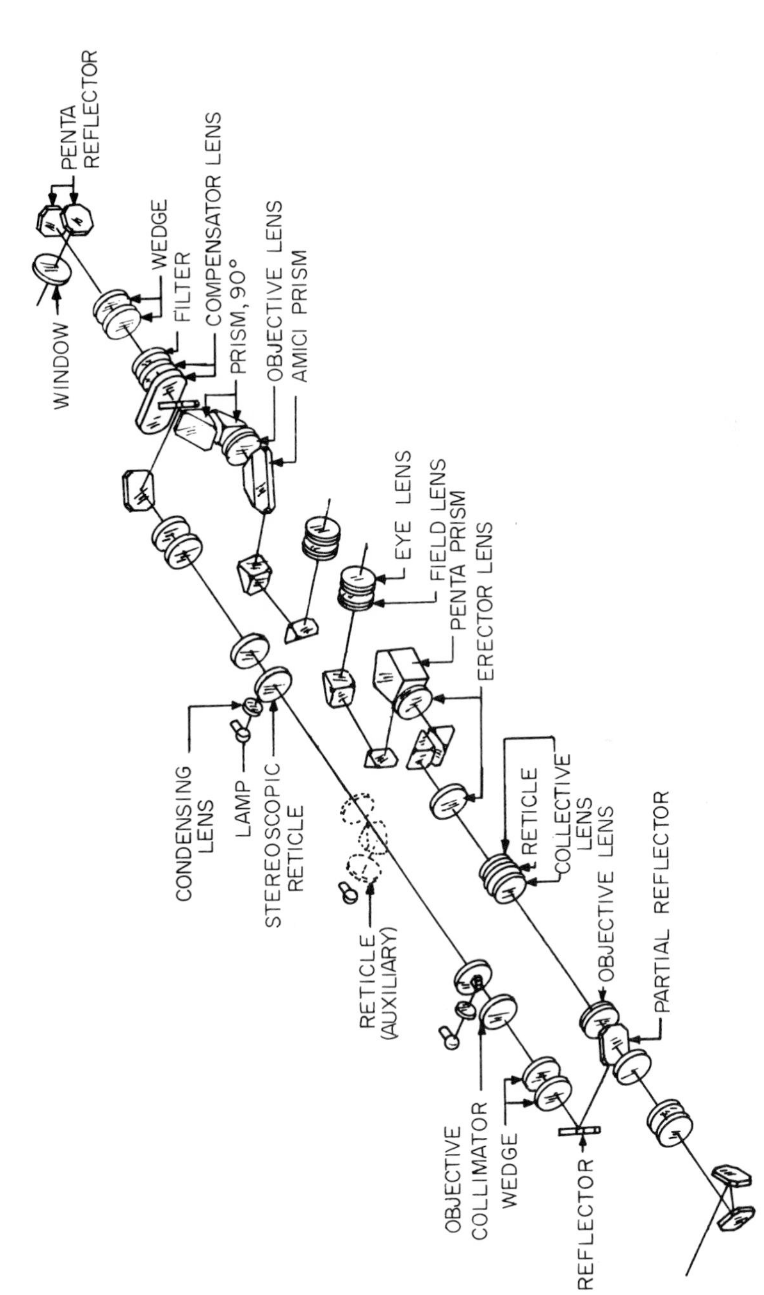

FIG. 27. Stereoscopic range finder No. 2.

e. Error Structure. The error structure of any particular range finder can be studied by differentiation of the range equation in the usual way. The quantities involved are the base length, the acuity of the observer's eye. and the accuracy of the compensating mechanism used in the instrument. The effects of flexure, stratification of the air inside the range finder, and variation of temperature from one part of the apparatus to another must all be taken into account. Some error studies on actual range finders have been reported by the author.[18, 19]

C. Time-Based Optical Range Finders

a. General. Time-based optical range finders offer a desirable means for obtaining more accurate measurement of range than self-contained-base range finders. Recent advances in electronic circuitry permit highly accurate measurement of the time of transit of light pulses from the observer to the target and back again. Also, application of the discovery of coherent laser radiation has made possible a great extension in range coverage as compared to previous incoherent optical radars.

In covering the subject of time-based optical range finders, discussion will center on a particular military version, a review of its optical system and prediction of maximum operating range, plus a brief remark on safety.

b. Discussion. Information in the way of diagrams, photographs, and explanations concerning the intended use, principles of operation, pertinent features, and problems encountered during the course of development of the military laser range finder XM23 are contained in an article by Abell,[20] to which the reader is referred. In this connection, it should be pointed out that similar efforts either have been or are being performed by other laboratories, according to Lengyel.[21]

An optical schematic of the XM23 is shown in Fig. 28. As noted on the diagram, the optical system is rather compact. The objective lens of the transmitter is in close proximity to the common objective lens assembly, which, through the medium of the beam-splitter prism, serves as the objective lens for the visual telescope and the receiver lens for the returned laser-beam signals. By this arrangement, boresight alignment between the receiver and viewing axis is constantly maintained.

The maximum operating range of a laser range finder like the XM23 depends on a number of factors, but in the final analysis, requires that the

---

[18] F. B. Patrick, *J. Opt. Soc. Am.* **52**, 1318(A), (1962).

[19] F. B. Patrick, *Appl. Opt.* **2**, 277–281 (1963).

[20] W. T. Abell, *Ordnance* **50**, 201–204 (1965).

[21] B. A. Lengyel, "Introduction to Laser Physics," p. 286. Wiley, New York, 1966.

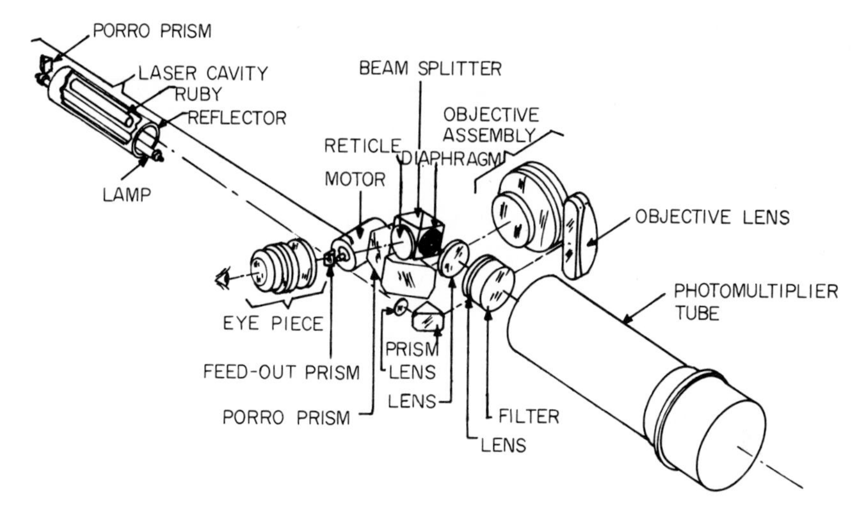

FIG. 28. Optical schematic XM23 laser range finder. (Courtesy of the US Army.)

power of the laser signal received at the detector (photomultiplier tube) be equal to or above a certain level. This is normally that associated with background (sunlight) noise or, in the absence of background noise, that associated with dark current and/or quantum noise.

From elementary radiometric considerations, it can be shown that when a target is situated normal to the range finder, the angular subtense of the target is the same as that of the laser beam, and the field of view of the receiver is equal to the angular width of the laser beam, the maximum operating range can be predicted from the equation

$$P_s = P_{\mathrm{T}} A_{\mathrm{r}} K / \pi R^2{}_{\max} \geq P_{\mathrm{b}} \geq P_{\mathrm{t}} \tag{13}$$

where P_s represents the peak power of the received signal from the target at maximum range $R_{\max}$; P_{T} is the peak power of the transmitted laser beam; A_{r} is the area of the receiver lens; K is an overall factor that includes atmospheric transmittance, reflectance of the target, transmittance of optics, and angular orientation of the target; P_{b} is the rms power due to background noise; and P_{t} is the rms power for threshold detection in the absence of background noise.

For further information on laser range finders, including elaboration of the equation just given and techniques for assessing accuracy of ranging, false-alarm rates, probability of detection, and values with regard to P_{b} and P_{t}, the reader is referred to a publication[22] by the Radio Corporation of America.

[22] Electro-optics handbook. RCA, Burlington, Massachusetts (1968).

It is recognized that the energy emitted by lasers is powerful, and for that reason definite safety precautions should be observed in order to avoid injury, particularly to the retina of the eye. A rather extensive listing of general and specific precautions to be observed in connection with lasers is to be found in a technical report[23] published in *Laser Focus*. Another report, that by Mac Keen *et al.*,[24] provides information on hazards associated with various materials used in laser systems and in irradiation studies. Both reports are replete with references to the considerable work that has been accomplished in this field.

GENERAL REFERENCES

F. Auerbach, "The Zeiss Works" (transl. by R. Kanthack). Foyle, London, 1927.

L. Bell, "The Telescope." McGraw-Hill, New York, 1922.

A. E. Conrady, "Applied Optics and Optical Design" (2 vols.). Dover Pub., New York, 1957, 1960.

A. Gleichen, "The Theory of Modern Optical Instruments" (transl. by H. H. Emsley and W. Swaine). Stationery Office, London, 1921..

N. Günther, "Fernoptische Beobachtungs und Messinstrumente." Wissenschaftliche Verlagsgesellschaft MBH., Stuttgart, 1959.

D. H. Jacobs, "Fundamentals of Optical Engineering." McGraw-Hill, New York, 1943.

O. K. Kaspereit, Designing of optical systems for telescopes. Ordnance Tech. Notes No. 14. U.S. Army Ordnance Dept. (1933).

H. C. King, "The History of the Telescope." Griffin, London, and Sky Publ. Cambridge, Massachusetts, 1955.

G. S. Monk and W. H. McCorkle, "Optical Instrumentation." McGraw-Hill, New York, 1954.

Climatic extremes for military equipment, Military Standard, MIL-STD-210A. U.S. Government Printing Off., Washington, D.C., (1957).

Design of fire control optics. Ordnance Corps Manual ORDM 2-1 (2 vols.). U.S. Government Printing Off., Washington, D.C. (1952).

Elementary optics and applications to fire-control instruments. War Dept. Tech. Manual TM9-2601. U.S. Government Printing Off., Washington, D.C. (1948).

Fire control material: General specification covering the manufacture and inspection of. Military Specification, MIL-F-13926A(MU). Defense Supply Agency, Washington, D.C. (1962).

Opticalman 1 and C, and Opticalman 3 and 2. Bur. of Naval Personnel, U.S. Government Printing Off., Washington, D.C. (1966).

Optical components for fire control instruments; General specification governing the manufacture, assembly and inspection of. Military Specification, MIL-0-13830A. Defense Supply Agency, Washington, D.C. (1963).

Optical Design. Military Handbook, MIL-HdBK-14-1. Defense Supply Agency, Washington, D.C., 1962.

Optical terms and definitions. Military Standard, MIL-STD-1241. U.S. Government Printing Off., Washington, D.C. (1960).

---

[23] The American conference of governmental industrial hygienists, *Laser Focus* **4**, 19, 50–57 (1968).

[24] D. Mac Keen, S. Fine, and E. Klein, *Laser Focus* **4**, 19, 47–49 (1968).

CHAPTER 8

Surveying and Tracking Instruments

M. S. DICKSON and D. HARKNESS
W. and L. E. Gurley, Troy, New York

I. DEFINITIONS

It is well to establish certain definitions in surveying so that distinctions can be made in the mind of the engineer. These topics have been discussed in great detail by Davis.[1]

A. Plane Surveying

Plane surveying is surveying which neglects the shape of the earth, and confines itself to considering the earth's surface as a plane. It has several forms:

a. Route Surveying. This is the surveying necessary for the location and construction of lines of transportation or communication, such as highways, telephone and power lines, railroads, canals, and pipelines.

b. Construction Surveying. This is the surveying necessary, e.g., for locating foundations for buildings, establishing level lines, plumbing, steel erection, etc.

---

[1] R. E. Davis, Surveying, *in* "Civil Engineering Handbook" (L. C. Urquhart, ed.), 4th ed., Section I. McGraw-Hill, New York, 1959.

c. Cadastral Surveying. This is the surveying necessary to establish property lines, in connection with the value, ownership, and transfer of land.

d. Topographic Surveying. This is the surveying necessary to obtain the location of points on the earth's surface in both horizontal and vertical planes so that maps may adequately portray the configuration of the earth's surface.

e. Hydrographic Surveying. This is the surveying of bodies of water so that adequate information may be available for navigation, as well as concerning water supply and subaqueous streams.

f. Mine Surveying. This is a combination of route and topographic surveying both on the surface and below the surface of the earth. This type of surveying is needed not only to direct tunneling operations, but also to establish boundary claims for mineral rights beneath the surface.

g. Land and Boundary Surveying. This is the surveying necessary to establish property lines for deed purposes, and for generating city plot maps.

B. Geodetic Surveying

Geodetic surveying is surveying which takes into consideration the contour and shape of the earth. Route, construction, topographic, and control-coordinate surveying are the same here as in plane surveying, except that measurements are corrected from a plane to a spheroidal surface.

C. Photogrammetric Surveying

Photogrammetric surveying employs the use of either terrestrial or aerial cameras to produce precise photographs from which measurements for mapping can be made. Many different types of plotting devices for photogrammetric mapping have been developed, such as the Multiplex Projectors, the Kelsh Plotter, and the Autograph, to name a few.

D. Astronomical Surveying

Astronomical surveying is the surveying necessary (1) to obtain the altitude and azimuth of celestial bodies at specified times so that the exact position of the station on the earth can be computed as to latitude and longitude, and (2) to determine azimuths from true north, primarily generated from sights on Polaris, the north star.

II. MEASUREMENTS

1. *Purpose*

All measurements made in surveying are performed to locate points precisely on the surface of the earth with respect to each other, both in a horizontal plane and in vertical planes.

In Fig. 1, if the sloping distance AC is measured by instrumentation

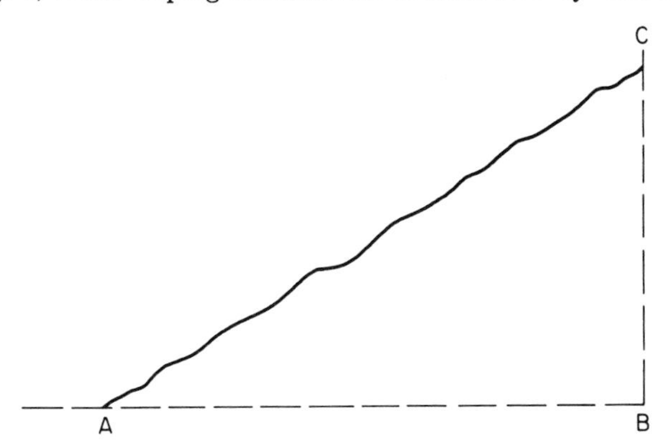

FIG. 1. The quantities to be measured in a survey.

described in the following paragraphs, and the vertical angle is measured, then the vertical distance BC, or the height of C above B, is easily obtained. Similarly, the horizontal distance AB can be obtained trigonometrically. All linear measurements made on the surface of the earth are taken as horizontal distances, or reduced to horizontal distances. All vertical measurements are made to obtain the heights of points on the surface of the earth above some datum, generally mean sea level.

2. *Angular Measurements*

a. Horizontal Angular Measurements. These are made to obtain the direction of imaginary lines with respect to each other, or from an established north–south line, either true or magnetic.

b. Vertical Angular Measurements. These are made to obtain sufficient data so that vertical measurements can be obtained trigonometrically.

III. INSTRUMENTS FOR LINEAR MEASUREMENT

A. HORIZONTAL

1. *Measuring Tapes*

Tapes of various types and lengths are used for linear measurements. The most common are:

a. Metallic Tapes. These are made of woven fabric, usually with metal wires or threads woven longitudinally to prevent excessive stretching. These tapes are mainly used for rough measurements.

b. Steel Tapes. These are bands of high-carbon steel of various widths, thicknesses, and lengths, The most common tapes are graduated each foot, with end feet graduated to tenths or hundredths of a foot. Metric tapes are also available.

c. Invar Tapes. These are used for the most precise measurements because of their low coefficient of expansion with temperature. Tapes used for precise measurements are standardized by comparison with bench standards. Standard conditions of comparison are: temperature 68°F, or 20°C, tension 10 lb, supported throughout. When conditions differ from standard, corrections to the length must be made for temperature, tension, or support. Since all distances must be reduced to horizontal, slope corrections must also be made. In geodetic measurements, the distances are also reduced to sea level.

2. *Optical Methods of Measurement*

a. Stadia Method. In surveying, the method used most often is probably the stadia method. A graduated rod is observed through a tachymeter, which is essentially a telescope equipped with a pair of parallel crosswires.[2] The distance of the rod from the telescope is then a known multiple of the intercept observed on the rod between the two crosswires. If a simple telescope is used, the computed distance is from the anterior focus of the objective lens, and a correction must be applied as indicated in Figs. 2 and 3. The internal focusing, or anallatic telescope, first described by I. Porro in 1850, employs an internal negative focusing lens to bring the image of the object into the plane of the reticle lines. The anallatic point from which the distance of the rod is measured can now be made to coinicde very closely with the central pivot of the instrument; in this case, the instrument constant $(f + c)$ is for all practical purposes zero.[3] Almost all surveying-instrument telescopes are now of the internal focusing type.

b. Range Finder. The range finder is another optical device for measuring distances. In simple form, it consists of two prisms at a fixed base distance from each other (Fig. 4). Prism A is fixed, and prism B is rotated by a mechanism which is calibrated to indicate distance. In use, a direct sight is taken through A to an object. Prism B is then rotated to bring the

[2] A. König and H. Köhler, "Die Fernrohre und Entfernungsmesser," pp. 362–378. Springer, Berlin, 1959.

[3] E. W. Taylor, A new perfectly anallatic internal-focusing telescope, *Trans. Opt. Soc.* **25**, 200–208 (1924).

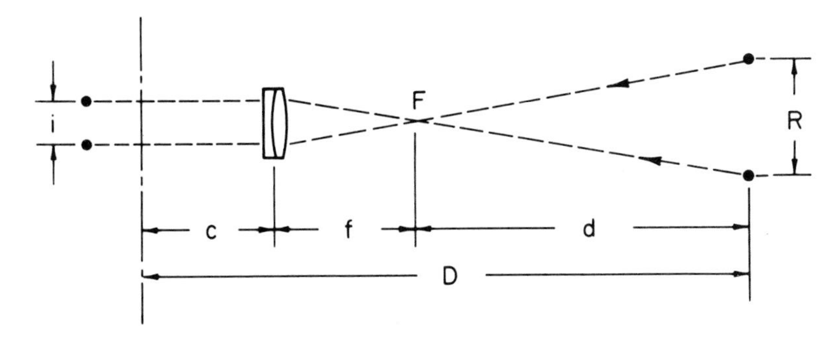

FIG. 2. Principle of stadia surveying. Here R is the rod intercept and F is the anterior focus of the telescope objective. The range D is given by $(f/i)R + (f + c)$. The factor (f/i) and the correction $(f + c)$ are both known constants. The factor (f/i) is usually made equal to 100 by proper spacing of the stadia lines. The correction $(f + c)$ is generally about 1 ft.

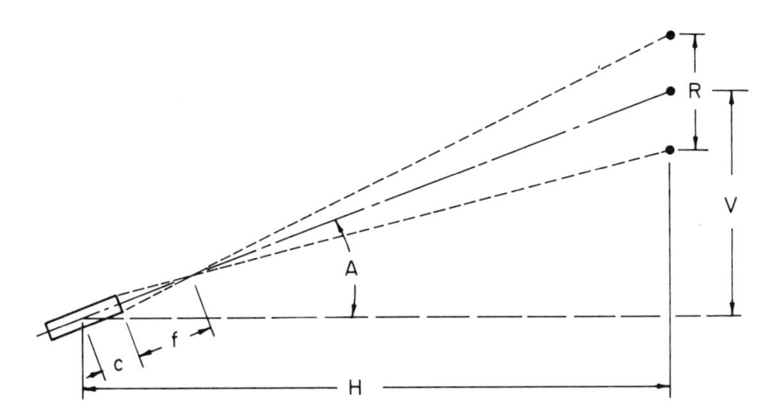

FIG. 3. Stadia surveying with an inclined rod intercept. Here, $H = (f/i)R \cos^2 A + (f + c) \cos A$, and $V = \frac{1}{2}(f/i)R \sin 2A + (f + c) \sin A$.

image of the object into coincidence with the direct sight. The calibrated movement of prism B indicates the distance. Range finders, because of their relatively low accuracy, are little used in surveying. They are discussed in Chapter 7.

c. Subtense Bar. The subtense bar consists of two targets fixed at a given distance apart. Usually, these are at the ends of an Invar tape under tension. The subtense bar is used in conjunction with a theodolite capable of precise angular measurements (Fig. 5). In use, the subtense bar is set up at the end of the line to be measured, and is oriented perpendicular to the line by a small telescope. A theodolite is placed at the other end of the line,

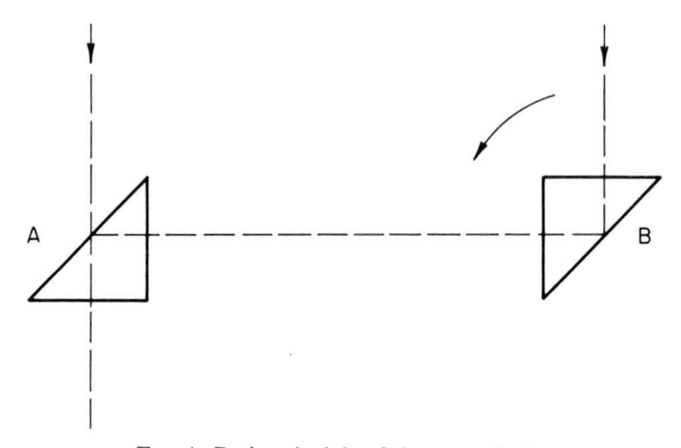

FIG. 4. Basic principle of the range finder.

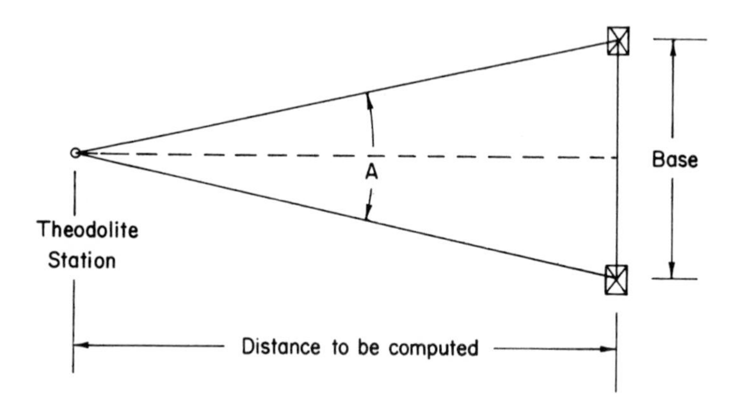

FIG. 5. Principle of the subtense bar.

and the angle subtended by the two targets is precisely measured. The distance from the theodolite to the bar can then be computed.

3. *Electronic Measurements*[4,5]

Electronic distance-measuring devices are fairly new. The most popular are the tellurometer[6] and the electrotape. Both of these devices are based on microwave transmission to a target. The electrotape uses a phase

[4] Electronic Surveying (1960 developments). Publ. 767, Natl. Acad. of Sci., Washington, D.C. (1960).

[5] *Can. Surveyor* **15**, Nos. 7 and 8 (1961); also **16**, No. 2 (1962).

[6] M. M. Thompson, ed., "Manual of Photogrammetry," 3rd ed., Vol. I, p. 364. Am. Soc. of Photogrammetry, Falls Church, Virginia, 1966.

comparison technique by measuring the phase delay of a radio microwave between two units, either of which can perform the measurement.

4. *Time-Based Range Finders*

Distance can also be measured by a light beam through a device known as a geodimeter.[6] A beam of light is sent to a reflecting mirror target and returned. The time of transmission of the beam to the target and return is measured, and is converted into distance by the known velocity of light. Lasers have recently been introduced as a light source for such devices, and this has increased their range as well as their accuracy.

B. VERTICAL

As it is not usually possible to measure vertical distances with a tape, such measurements are generally made by rod and level. A typical level is shown in Fig. 6. When this instrument is set up on its tripod and leveled, the telescope can be rotated about its spindle in the vertical axis. The optical

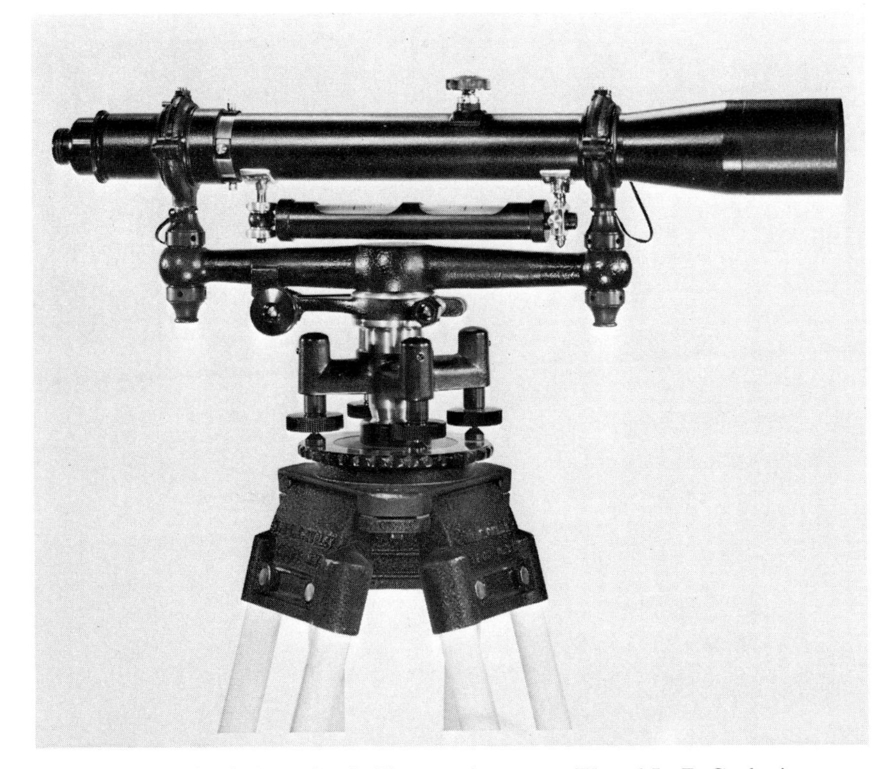

FIG. 6. Typical wye level. (Photograph courtesy W. and L. E. Gurley.)

axis of the telescope then generates a horizontal plane; hence, any particular sight taken is a horizontal line of sight. A graduated rod can then be set on any point of known elevation and a horizontal sight taken on the rod. This, then, would give the so-called "height of the instrument," or the elevation of the line of sight above the point of known elevation. The rod can then be moved to a point of unknown elevation and a similar sight taken. If the rod reading is then subtracted from the "height of the instrument," the resulting figure would be the elevation of the point on which the rod is held. For the most precise measurements, corrections for the curvature of the earth, for refraction, for temperature, and for other factors must be made. A modern "autoset" level is described in Volume IV, p. 237, of this series.

Referring back to Fig. 1, it can be shown that vertical measurements can be computed if the slope distance is measured by the stadia method or with a subtense bar, and the vertical angle is also measured.

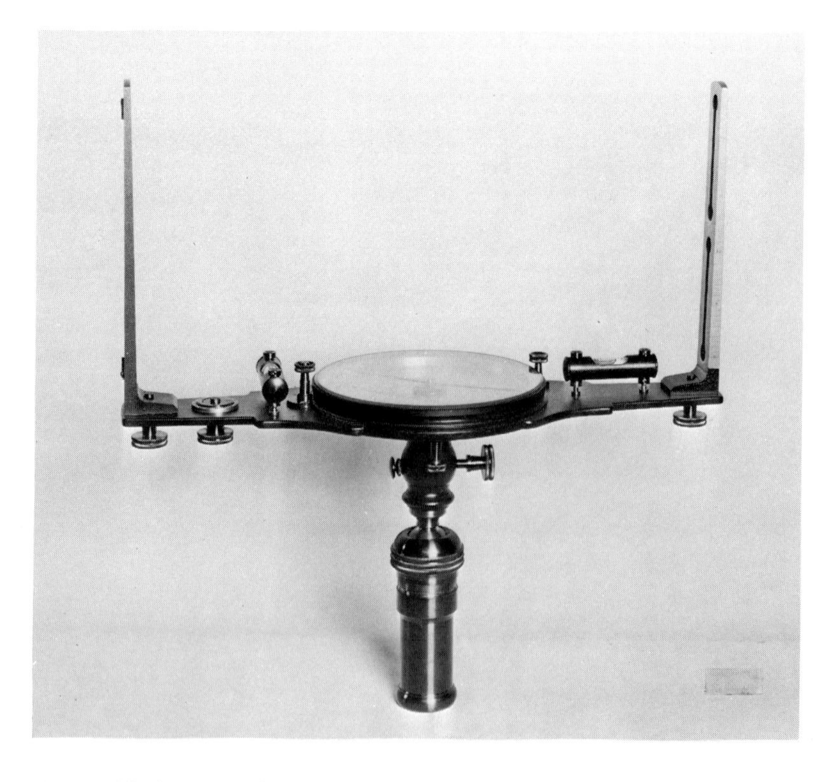

FIG. 7. Typical surveying compass. (Photograph courtesy W. and L. E. Gurley.)

IV. INSTRUMENTS FOR ANGULAR MEASUREMENTS

A. Horizontal Angles

Horizontal angles can be measured by several different instruments of varying accuracy.

1. *The Compass*

The least accurate instrument for this purpose is the compass, which measures angles from magnetic north. Other angles can also be measured by sighting at two different points and taking the difference in readings from the compass needle. Since compasses are generally graduated to the nearest 30′ of arc, greater accuracy than this cannot be obtained. A typical compass is illustrated in Fig. 7.

2. *Alidade and Plane Table*

Horizontal angles can be plotted and subsequently measured through the use of an alidade and plane table (Figs. 8 and 9). The telescopic alidade is merely a small telescope mounted to a straight-edged blade, the straight edge being parallel to the optical line of sight. Using stadia methods of measurement, rays can be drawn on plotting paper to many outlying points from one's assumed position, and these points plotted, thus making a map "as you go." Angles between points can be measured from the plot using a protractor. These measurements are accurate to approximately $\frac{1}{4}°$.

B. Horizontal and Vertical Angles

1. *The Transit*

The most widely used instrument for the measurement of both horizontal and vertical angles is the transit.[7,8] This instrument is illustrated in Fig. 10. The term "transit" indicates that the telescope can be fully rotated about a horizontal axis (or "transited"). There are many instruments on the market that are called transits, but since the telescope cannot be fully rotated, this is a misnomer. The term "tilting level" would be more appropriate for this type of instrument.

---

[7] L. C. Martin, "Optical Measuring Instruments," pp. 61–74. Blackie, London, 1924.

[8] A. König and H. Köhler,[2] pp. 234–243.

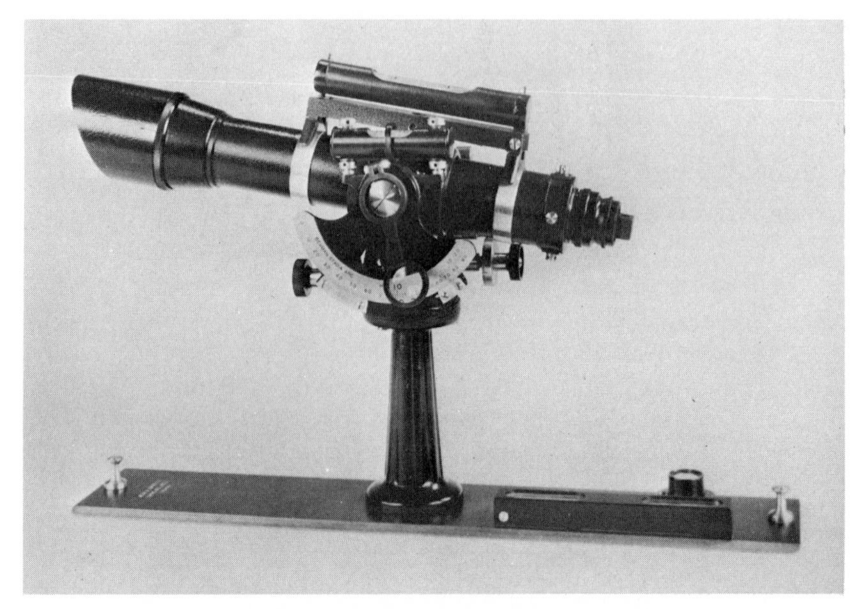

FIG. 8. Typical high-standard alidade. (Photograph courtesy W. and L. E. Gurley.)

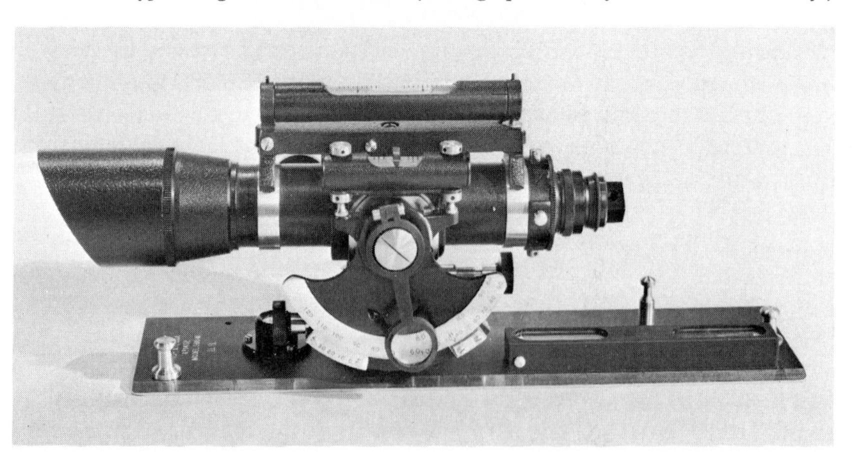

FIG. 9. Typical low-standard or mountain alidade. (Photograph courtesy W. and L. E. Gurley.)

The transit consists of an upper, or vernier plate to which are attached A-shaped standards supporting the telescope, and a lower plate to which is fixed a horizontal, graduated circle. The upper and lower plates are fastened to the inner and outer spindles, respectively, the two axes of rotation being coincident with and at the geometric center of the graduated circle. The outer spindle is seated in the tapered socket of the leveling head.

FIG. 10. Cross section of a typical American transit. (Photograph courtesy W. and L. E. Gurley.)

Near the bottom of the leveling head is a ball-and-socket joint which secures the instrument to the foot plate, yet permits rotation of the instrument about the joint as a center.

When the lower plate is rotated, the outer spindle revolves in its socket in the leveling head. The outer spindle carrying the lower plate may be clamped in any position by means of the lower clamp-screw. Similarly, the inner spindle carrying the upper plate may be clamped to the outer spindle by means of the upper clamp-screw. After either clamp has been tightened, small movements of the spindle may be made by turning the corresponding tangent-screw. The axis about which the spindles revolve is called the vertical axis of the instrument.

Level tubes called plate levels are mounted at right angles to each other, one on the upper plate and one on one of the standards. They are provided for leveling the instrument so that the plane of the horizontal circle will be truly horizontal when observations are made. Four leveling screws, or foot screws are threaded into the leveling head and bear against the foot plate; when the screws are turned, the instrument is tilted about the ball-and-socket joint. When all four screws are loosened, pressure between the sliding plate and the foot plate is relieved, and the transit may then be shifted laterally with respect to the foot plate. From the end of the spindle and at the center of curvature of the ball-and-socket joint is suspended a chain with hook for the plumb line. The instrument is mounted on a tripod by screwing the foot plate onto the tripod head.

The telescope is fixed to a transverse horizontal axis which rests in bearings at the upper extremity of the standards. The telescope may be rotated about this horizontal axis and may be fixed in any position in a vertical plane by means of the telescope clamp-screw. Small angular movements about the horizontal axis may then be made by turning the telescope tangent-screw. Fixed to the horizontal axis is the vertical circle, and attached to one of the standards is the vertical vernier. Beneath the telescope is the telescope level tube.

On the upper plate is the compass box (see Fig. 10). If the compass circle is fixed, its N and S points are in the same vertical plane with the line of sight of the telescope. The compass boxes of some transits are so designed that the compass circle may be rotated with respect to the upper plate, so that the magnetic declination may be laid off and true bearings read. At the side of the compass box is a screw, or needle lifter, by means of which the magnetic needle may be lifted from its pivot and clamped.

Summing up the several features: (1) the center of the transit can be brought over a given point by loosening the leveling screws and shifting the transit laterally; (2) the instrument can be leveled by means of the plate levels and the leveling screws; (3) the telescope can be rotated about either the horizontal or vertical axis; (4) when the upper clamp-screw is tightened, and the telescope is rotated about the vertical axis, there is no relative movement between the verniers and the horizontal circle; (5) when the lower clamp-screw is tightened, and the upper one is loose, a rotation of the telescope about the vertical axis causes the vernier plate to revolve, but leaves the horizontal circle fixed in position; (6) when both upper and lower clamps are tightened, the telescope cannot be rotated about the vertical axis; (7) the telescope can be rotated about the horizontal axis and can be fixed in any direction in a vertical plane by means of the telescope clamp and tangent-screws; (8) the telescope can be leveled by means of the telescope level tube, and hence the transit can be employed as an instrument for direct leveling; (9) by means of the vertical circle and vernier, vertical angles can be determined, and hence the transit is suitable for trigonometric leveling; (10) by means of the compass, magnetic bearings can be determined; and (11) by means of the horizontal circle and vernier, horizontal angles can be measured.

An optical plummet (a small telescope sighting vertically downward) has been recently added to the American transit, thus eliminating the plumb bob string. This is also illustrated in Fig. 10.

2. *The Theodolite*

The term "theodolite" was originally used to designate any instrument which measured both horizontal and vertical angles. With the advent of the American transit, the term "theodolite" was dropped, except in Europe. Later, when more accurate and more precise measurements were demanded by Federal agencies for establishing triangulation networks, the transit was increased in size so that circles could be graduated with finer increments to permit readings to a few seconds of arc. To distinguish this instrument from the smaller transit, the term "theodolite" was used. Thus, "theodolite" came to mean a more precise and more accurate surveying instrument than the transit (see Vol. IV, p. 228).

During World War II, the European type of theodolite[8] was demanded by the US Army because these instruments were of high accuracy, precisely made, permitted angles to be read rapidly. These theodolites contain horizontal and vertical circles which are read optically through a microscope placed alongside the main telescope eyepiece. Each reading of the horizontal or vertical circle is obtained automatically as the mean of two readings at opposite points of the circle, and is therefore free from errors due to eccentricity. The accuracy of the instrument with a 4-in. horizontal circle is comparable to that of an 8-in. theodolite of ordinary design. This type of theodolite, illustrated in Figs. 11 and 12, has gained popularity in the United States. Today the term "theodolite" is applied quite generally to an optically read instrument, regardless of whether it reads to seconds of

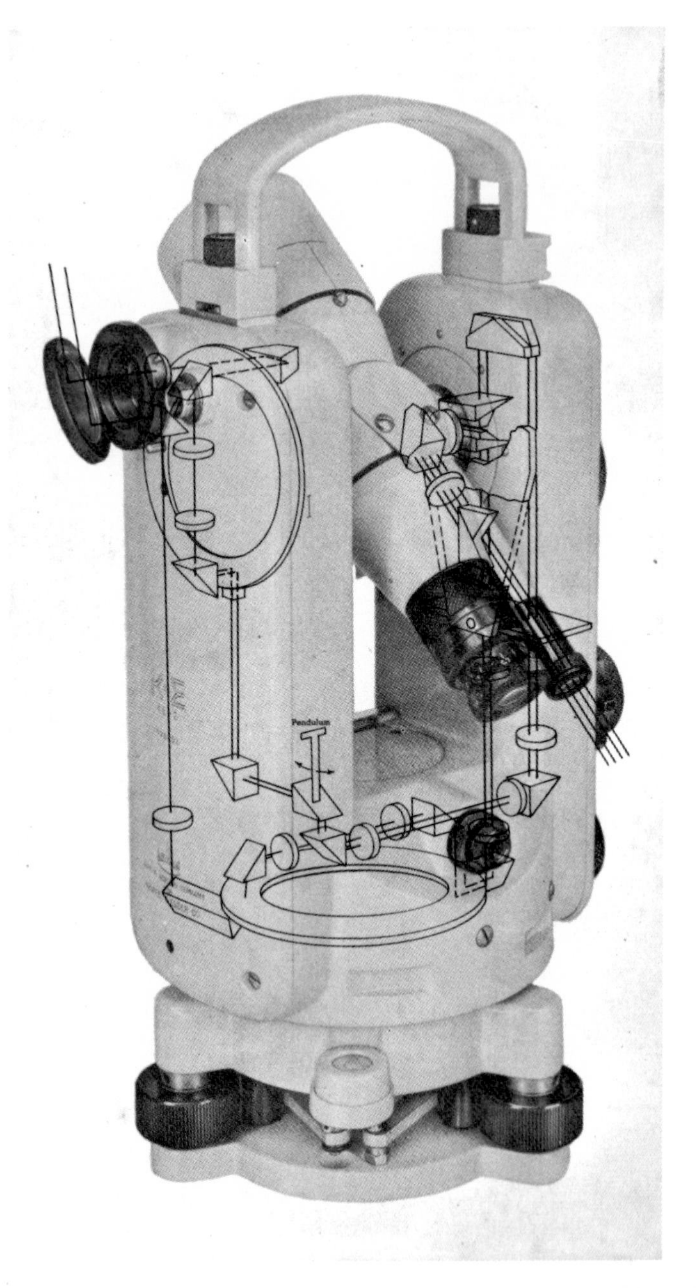

FIG. 11. The Model KE-2 1″ theodolite. Made by Askania for Keuffel and Esser, Inc.

FIG. 12. The Kern DKM-3 theodolite. (Photograph courtesy Kern and Co., Aarau, Switzerland.)

arc, or minutes, or even to degrees only. Thus, today, the term "transit" is applied to a vernier-read instrument and the term "theodolite" to an optically-read instrument.

3. *Sextant and Astrolabe*

Two other surveying instruments, used primarily for the measurement of vertical angles, are worthy of mention, the sextant and astrolabe.

The sextant is, in general, a hand-held instrument used on shipboard or in aircraft to measure the vertical angle from the horizon (real or artificial) to the sun or to navigational stars at a given instant of time. The angles so obtained can be used in computations to obtain a "fix," i.e., the latitude and longitude of the ship's position on the earth's surface. The sextant is an ancient instrument and its construction is well known.[9] Aircraft sextants have a built-in artificial horizon, and various types have been described. One of these is the bubble sextant dating from World War I.[10] This instrument contains a circular spirit level, the radius of curvature of the upper glass surface being equal to the focal length of a collimating lens which serves to project the image of the bubble to infinity. The sun and the bubble can then be brought into coincidence by tilting a beam-splitting mirror in the usual way (Fig. 13).

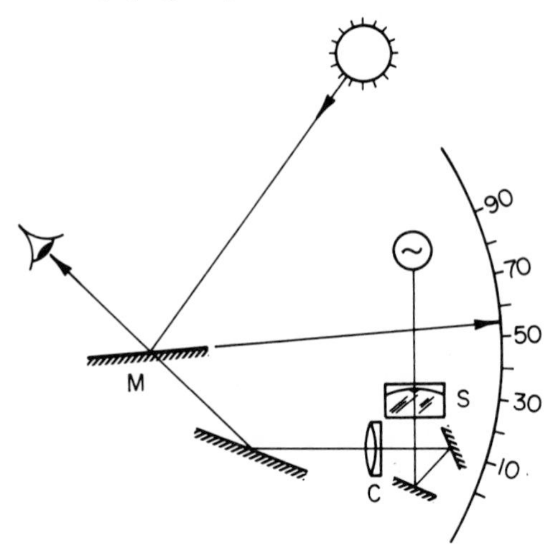

FIG. 13. Principle of the bubble sextant. Sunlight is reflected to the observer's eye by the unsilvered tilting mirror M. The bubble of a circular spirit level S, which can be illuminated by a small lamp, is seen through M after reflection in three fixed mirrors, the bubble being at the focus of a collimator C. The scale indicates the altitude of the sun above the horizon.

The prismatic astrolabe, introduced by Claude and Driencourt about 1900, consists of a small telescope in front of which is a prism giving a fixed angle of elevation (60°) relative to an artificial horizon formed by a mercury trough.[11] The instrument can be rotated about a vertical axis, and it permits the observer to determine the exact instant of time at which a navigational star is at precisely 60° above the horizon. Many stars can be read

[9] L. C. Martin,[7] pp. 76–84.

[10] L. B. Booth, The aerial sextants designed by the Royal Aircraft Establishment. The Optical Convention of 1926, pp. 720–728.

[11] T. Y. Baker, Exhibit of a prismatic astrolabe, *Trans. Opt. Soc.* **24**, 110 (1923).

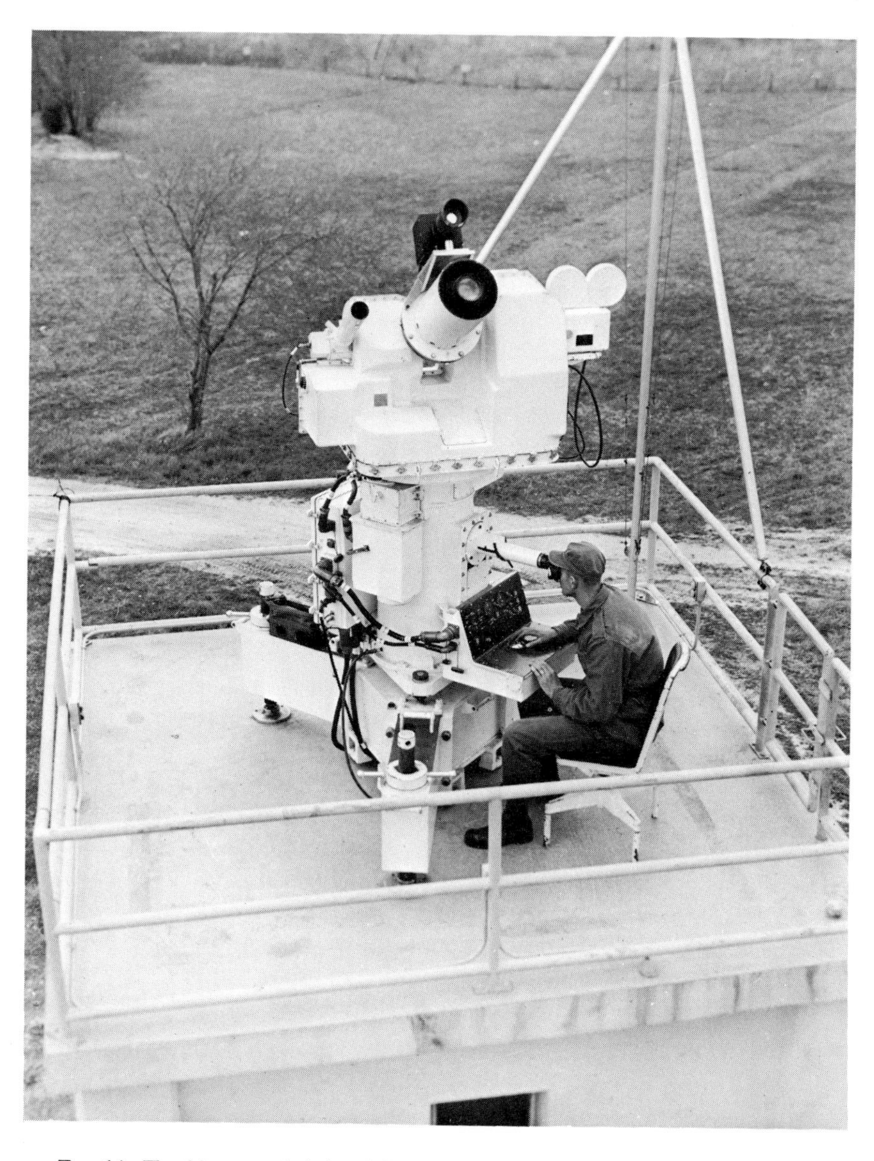

FIG. 14. Tracking encoded theodolite. (Photograph courtesy American Machine and Foundry Company, Stamford, Connecticut.)

in a short period of time, and a simple computation gives the latitude and longitude of the point of observation.

There are many configurations and modifications of these various surveying instruments, but they all solve the surveying problem in basically the same way.

4. *Surveying Cameras*

The latest modifications in surveying are the application of cameras to surveying instruments and the encoding of horizontal and vertical-angle readout. More specifically, cameras have been applied to theodolites[12] so that terrestrial photographs can be made on which are superimposed the angular readings. This is done to increase the ease and rapidity of taking readings, the actual measurements being obtained in an office, from the photographs, rather than from recordings of the information in the field.

Photoelectric encoders have been built into theodolites so that angular measurements can be transmitted to computers, which in turn immediately tabulate the solution of the surveying problem. A large theodolite of this type is shown in Fig. 14. Fixed encoded theodolites are used primarily for tracking high-velocity planes, missiles, satellites, and space vehicles.

V. TRACKING SYSTEMS

Instruments used for tracking rockets and other high-velocity objects in the sky are used for two principal purposes: (1) to observe the object itself to detect any changes in its attitude, shape of exhaust flame, etc., and (2) to determine its trajectory in space. The first requirement calls for the use of a long-focus lens with high resolution; good optical contrast is also required because the object itself is often of vanishingly low contrast against the sky. The second requirement is met by the use of special tracking instruments. The earliest of these was the simple balloon theodolite for the visual tracking of sounding balloons.[13] Today, when rockets and ballistic missiles must be tracked, much more sophisticated instruments are used. Excellent summaries of the tracking problem and of appropriate instrumentation have been given by Schendel[14] and Reuyl and Carrion.[15]

---

[12] M. B. Strain and J. B. Case, Terrestrial Photogrammetry, *in* "Manual of Photogrammetry" (M. M. Thompson, ed.), 3rd ed., Vol. II, pp. 919–959. Am. Soc. of Photogrammetry, Falls Church, Virginia, 1966.

[13] T. F. Connolly, A new form of balloon theodolite, *Trans. Opt. Soc.* **24**, 326–328 (1923).

[14] A. H. Schendel, Optical tracking instrumentation. *J. Soc. Motion Picture Television Engrs.* **67**, 237–241 (1958).

[15] D. Reuyl and W. Carrion, Optical tracking methods and instruments at the Ballistics Research Laboratories, *J. Soc. Motion Picture Television Engrs.* **71**, 505–508 (1962).

1. *Cine-Theodolites*

During and since World War II, many types of cine-theodolite[16,17] have been constructed for tracking a rapidly moving object in the sky, notably by Akeley in the US, Askania in Germany, and Contraves in Switzerland. This instrument is basically a motion-picture camera using 35 mm perforated film, coupled to and aligned with a pair of tracking telescopes which can be used by two observers to track the target, one observer tracking in elevation and the other in azimuth. Each telescope has an X-wire which is accurately boresighted to a set of fiducial marks in the film plane of the camera. Smoother tracking is obtained by continuous motor drives, the rates of which are controlled by the observer.

On each frame of film is recorded, in addition to the target photograph, the frame number, the time, and the elevation and azimuth angles of the focal-plane fiducial marks, so that by measurement of the distance of the target image from the fiducial marks on the film, the exact angular position of the target can be determined for each frame. Of course, several such records must be obtained simultaneously from different observing stations in order to locate the target in space.

2. *Improved Tracking Instruments*

As rockets became more powerful, the requirements of a tracking instrument became more demanding, and during the past 15 years, a series of elaborate and very precise tracking systems has been developed. One of the first of these was the M-45 tracking camera (nicknamed the "Gooney Bird")[18,19] This was based on a 50 caliber machine-gun mount, with a single observer, and was equipped with a Mitchell 35 mm camera with 24- or 48-in. lenses, and also a high-speed camera if necessary.

By the early 1950's, much larger tracking telescopes were being developed, such as the Navy "Intercept Ground-based Optical Recorder" (IGOR),[20] with a reflecting telescope of 18-in. aperture and 96-in. focal length. The image could be further magnified five times by a supplementary lens if desired. Either 35 or 70 mm film could be used, the usual frame rate being 60 per second.

---

[16] L. Goldberg, Phototheodolites *in* "Optical Instruments"; Vol. I of Summary Technical Report of Division 16 N.D.R.C., pp. 528–550 (1946). A TI 32670, Armed Services Tech. Information Agency, Arlington 12, Virginia.

[17] S. M. Lipton and K. R. Saffer, *J. Soc. Motion Picture Television Engrs.* **61**, 33–44 (1953).

[18] M. A. Bondelid, *J. Soc. Motion Picture Television Engrs.* **61**, 175–182 (1953).

[19] S. E. Dorsey, *J. Soc. Motion Picture Television Engrs.* **65**, 631–635 (1956).

[20] S. M. Lipton, Multipurpose optical tracking and recording instrument (IGOR), *J. Soc. Motion Picture Television Engrs.* **62**, 450–459 (1954).

About 1957, a much more sophisticated tracking telescope was built by the Perkin-Elmer Corporation known as the Recording Optical Tracking Instrument (ROTI).[21] As originally planned, this instrument (MARK I), comprised a pair of reflecting telescopes, the upper being a 16-in. Newtonian with a focal length of 100 in., which could be increased by 2×, 3×, 4×, or 5× by a turret of relay lenses. The lower telescope was a 16-in. Schmidt of focal length 50 in., which could be increased optically to 75 or 100 in. A later version called ROTI Mark II contained only one telescope, a Newtonian of 24-in. aperture and the same focal lengths as before, namely, 100–500 in. This system has been described in detail by Economu.[21, 22] In use, it is mounted in a small dome of the type familiar to all astronomers. With these long focal lengths, it is necessary to adjust the focus as the rocket moves away from the instrument, data for the focus adjustment being derived from a radar unit nearby. A very high resolution is claimed for this instrument, limited generally by the turbulence or diffusion of the atmosphere.

GENERAL REFERENCES

In addition to the references cited, there is an extensive literature on the subject of surveying and surveying instruments. A few titles are listed here for reference.

R. M. Abraham, "Surveying Instruments." Casella, London, 1931.

C. B. Breed and G. L. Hosmer, "Higher Surveying." Wiley, New York, 1962.

R. C. Brinker and W. C. Taylor, "Elementary Surveying," 4th ed. Intern. Textbook. Co., Scranton, Pennsylvania, 1961.

R. E. Davis and F. S. Foote, "Surveying Theory and Practice," 4th ed. McGraw-Hill, New York, 1963.

R. E. Kiely, "Surveying Instruments." Columbia Univ. Press, New York, 1947.

P. Kissam, "Surveying for Civil Engineers." McGraw-Hill, New York, 1956.

J. C. Tracy, "Surveying Theory and Practice." Wiley, New York, 1947.

---

[21] G. A. Economu, A. telescope system for missile and satellite photography, *Phot. Sci. and Engrs.* **3**, 35–40 (1959).

[22] G. A. Economu, V. Luban, and M. H. Mehr, *J. Soc. Motion Picture Television Engrs.* **67**, 249–251 (1958).

CHAPTER 9

Medical Optical Instruments

JOHN H. HETT

Consultant, American Cystoscope Makers, Inc., Pelham Manor, New York

INTRODUCTION

Optical instruments for use in the visual examination and photography of the body cavities and internal organs, through natural or artificial openings, are generally known as "endoscopes." There are many types, of course, each suited to a particular application, and some of the best known will be described here.

Endoscopes are basically very thin periscopes, and they often carry a small lamp at the far (distal) end, although in some types, the lamp may be at the eyepiece (proximal) end, the light being conducted down the tube by fiber optics or other means. Some endoscopes are also equipped with small operating tools, snares and the like, for use by the surgeon.

More recently, fiber-optic bundles have been incorporated into endoscopes in place of a series of lenses. The advantages are a much more flexible tube and a brighter image.

I. HISTORICAL NOTES

The discussion of medical optical instruments given here is by no means exhaustive, but is planned to emphasize the design of the chief instruments now in use. Similarly, the medical references represent a selection, always subject to debate, from a vast literature.

The simplest optical instrument for the examination of body cavities is undoubtedly the open tube. These were known in India at least five centuries before Christ, and, with rather minor modifications, are still in use, particularly in bronchoscopy, esophagoscopy and proctoscopy. Endoscopy slumbered for over two millenia, awaiting improvement in illumination.

Not until 1795 did Bozzini improve the illumination through an open tube by using at the proximal end a 45° mirror with a hole in it and a Desormeaux lamp burning alcohol and turpentine.[1,2] Using the same lamp, Kussmaul,[3] in 1868, attempted gastroscopy with a rigid tube and a flexible obturator. In 1879, Nitze[4] placed lenses in an open tube to build the first cystoscope. Illumination was provided by a hot platinum wire loop which was in a water-cooled area. At about this time, Edison perfected the incandescent bulb, which opened a new era in endoscopy, since it allowed the use of wider visual fields and sometimes higher magnification.

In 1881, Mikulicz,[5] now using the Edison lamp and lenses in a rigid tube with a 30° distal offset, built the first usable gastroscope. Then, in 1932, a great improvement was achieved by R. Schindler and G. Wolf with the development of the flexible lens gastroscope.[6]

II. OPTICAL THEORY OF MEDICAL INSTRUMENTS

Since medical instruments are neither telescopes nor strictly microscopes in design, we shall consider here some of their fundamental optical properties.

The optical diagram in Fig. 1 represents a simple medical instrument consisting of (1) an objective and (2) an eyepiece. An object Y is in space

---

[1] P. Bozzini, *J. Prakt. Heilkunde* **24**, 107–124 (1806).

[2] A. J. Desormeaux, Endoscope and its application to the diagnosis and treatment of the genito-urinary passages, *Chicago Med. J.*, 1867. R. Fergus Sons, Printers.

[3] A. Kussmaul, *Deut. Z. Chir. Leip.* **58**, 500 (1900–1901).

[4] M. Nitze, *Wiener Med. Wochschr.* **29**, 649, 688, 713–715 (1879).

[5] J. Mikulicz, Ueber gastroskopie und oesophagoskopie, *Wiener Med. Presse* **12**, 1405, 1437, 1473, 1505, 1537 (1881).

[6] R. Schindler, "Gastroscopy," 2nd ed. Chicago Univ. Press, Chicago, Illinois, 1967.

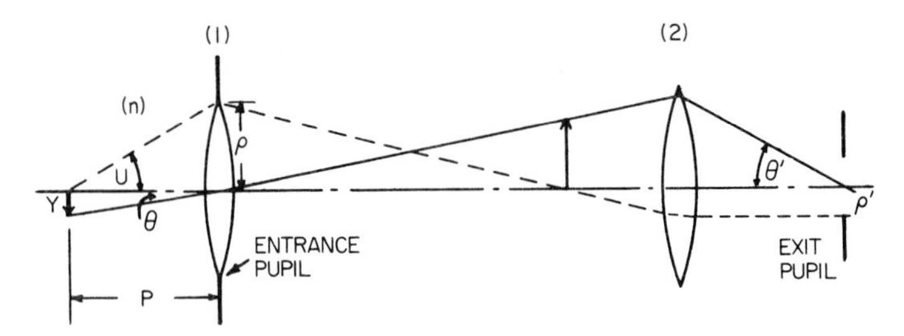

FIG. 1. Optical diagram of a simple medical instrument. Lens (1) represents the objective system and (2) the eyepiece system. Shown are the spatial relationships of lenses, object, image, and pupils.

of refractive index n. The space at the eyepiece is assumed to contain air with index unity. The radius of the entrance pupil is designated by ρ and the radius of the exit pupil (where the eye is located) by ρ'.

The Lagrange invariant (hnu)[7] can be applied to the entrance and exit pupils, giving

$$\rho n\theta = \rho'\theta' \tag{1}$$

Hence the angular magnification of the instrument is

$$M_{\mathrm{i}} = \theta'/\theta = n\rho/\rho' \tag{2}$$

The angular size of the object as presented to the instrument is Y/P, where P is the distance from the object to the entrance pupil. The instrument thus presents an image angle at the eye of

$$\theta' = \frac{Y}{P} M_{\mathrm{i}} = \frac{nY}{P} \frac{\rho}{\rho'} \tag{3}$$

One may define the visual magnification ("magnifying power") M_{v} of the instrument as the ratio of the angle subtended at the eye by the image to the angle subtended by the object at a conventional viewing distance D, usually taken as 10 in. Then,

$$M_{\mathrm{v}} = \frac{\theta'}{Y/D} = \frac{nD}{P} \frac{\rho}{\rho'} \tag{4}$$

In using Eq. (4), it should be noted that the magnification M_{v} is inversely proportional to the working distance P, the remaining factors being constant. Hence a plot of M_{v} *vs* P, as shown in Fig. 2 for an actual instrument such as a culdosdope, is a hyperbola. Graphs like this are characteristic of many medical instruments.

---

[7] "Applied Optics and Optical Engineering," Vol. 1, p. 210, Eq. (8). Academic Press, New York, 1965.

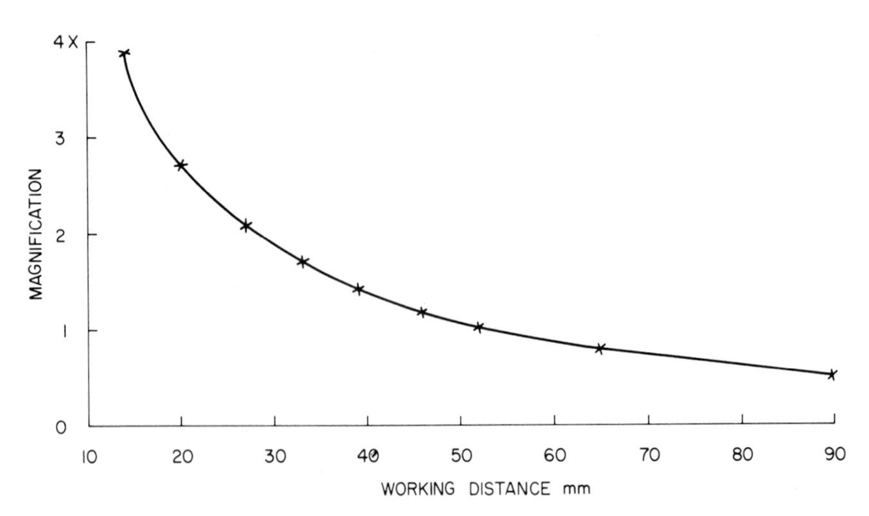

FIG. 2. Graph of magnification *vs* working distance for a culdoscope. The general form is typical for endoscopes.

The conventional viewing distance D varies with the observer. The observer, when viewing at close range, as in medical work, tries to produce as large an image on his retina as his accommodation or spectacle correction will allow. This distance is usually in the range 8–14 in., or less for uncorrected myopes.

The numerical aperture of the instrument is

$$nU = n\rho/P = M_v\rho'/D \tag{5}$$

A. Apparent Luminance of Images

It is well known that, if one neglects instrument losses, the apparent luminance of an image is equal to the physical luminance of the object. With medical instruments, however, even the axial losses may be considerable, because of the large number of optical elements used. Also, the illuminance in the image space falls off as a function of field angle because of vignetting and other factors.

If one compares the axial brightness of an object viewed through an instrument with that seen by the unaided eye, we have

$$\text{Relative luminance} = t\rho'^2/\rho_e^2 \tag{6}$$

where ρ' is the radius of the exit pupil of the instrument, ρ_e is the radius of the pupil of the eye, and t is the transmittance factor.

This equation is true provided that the exit pupil of the instrument is smaller than the pupil of the eye. It is clear that a medical instrument will need strong illumination at its distal end just to provide the observer's eye with adequate illumination. For example, if the observer's eye has a pupil diameter of 4 mm and the instrument $1\frac{1}{2}$ mm, the luminance of the object must be seven times greater than with the unaided eye to appear equally bright. Assuming an instrument loss of 25%, the luminance of the object must be 8.75 times greater.

In the medical case, the actual object luminance is determined not only by the source brightness, but also by the reflection coefficient of the mucosa. This coefficient varies in different parts of the body, but an average may be taken as 17%. This value was obtained with a spot photometer at the mucosa of the inner cheek of an adult.

B. Sources of Illumination

A very small incandescent lamp has long been the method of illuminating the object field of an endoscope. It was seldom satisfactory for photography because of the relatively low light levels obtained through the instrument.

To increase the illumination, many attempts have been made to use external lamps of high intensity and to convey the light down the endoscope. The use of a glass rod as a light-transmitting medium was described by Smith as early as 1899.[8] In 1914, Roccavilla of Italy designed a peritoneoscope which permitted the light source to remain outside the abdomen.[9] Thompson[10] described the use of a fused quartz rod with an external lamp in 1934. This use of the quartz rod was then largely neglected until 1951, when cystoscopes using these rods were described by Fourestier *et al.*[11] Meanwhile, in 1940, Brubaker and Holinger,[12] using a strong external light reflected down an esophagoscope, were able to take Kodachrome pictures.

The high intensity available from xenon flash lamps led to their use at the distal end of a gastroscope by Debray and Housset in 1956.[13] Several types of endoscope are now made in France and Germany using distal

---

[8] D. D. Smith, U.S. Patent 624,392 (1899).

[9] A. Roccavilla, *Riform. Med. Napoli* **30**, 991 (1914).

[10] J. L. Thompson, U.S. Patent 1,965,865 (1934).

[11] M. Fourestier, A. J. Gladu, and J. Vulmière, French Patent 1,036,992 (1951); also U.S. Patent 2,699,770 (1955).

[12] J. D. Brubaker and P. Holinger, *J. Biol. Phot. Assoc.* **10**, 83 (1941).

[13] C. Debray and P. Housset, Photographie en couleur à travers le gastroscope flexible: utilization d'un flash electronique, *Semaine Hop.* **32** (39/5), 2238–2243 (1956).

xenon lamps. The xenon lamp is also used extensively as an external source with quartz-rod light transmission.

Recently, fiber-optic light-transmission systems have beeh described by Wallace.[14] This system consists of a 150-W DLS Sylvania lamp with elliptical reflector which images the filament on the end of a fiber-optic light carrier. The carrier, in turn, is attached to another group of fibers at the proximal end of the instrument. Analogous systems were later developed in Germany. The advantages of the fiber-optic illuminating system compared to the quartz rod are as follows:

1. Fiber-optic systems do not show diffraction effects in the object field, such as occur with quartz-rod systems.
2. At the distal end of the instrument, the fibers may be conveniently arranged around the entrance pupil to direct the light most effectively in the object field.
3. Fiber optics may be used in instruments of very small diameter, since the fibers may be distributed efficiently, as, e.g., in annular form.
4. With fiber optics, the light source may be conveniently removed some distance from the instrument. Thus, the operator avoids contact with hot lamps.
5. Finally, a single, high-intensity light source may be used with many different types of endoscope.

With the incandescent lamp, the illumination is approximately inversely proportional to the square of the working distance, a relation which also holds for fiber-optic illumination for long working distances. In the fiber-optic case, directional effects are often noted. For short working distances, the output end of the fiber bundle should be considered as a luminous surface. If round, the fiber bundle may be considered as a circular disk of luminance B. Then on the central axis, the object illuminance E is given by

$$E = \pi B \sin^2 u \tag{7}$$

where u is the angle subtended by the radius of the disk at the object.

III. THE FLEXIBLE-LENS GASTROSCOPE

This instrument is normally about 1 m long and about 13 mm in diameter. The lower distal half is flexible and the upper part rigid. In addition to a lamp at the distal end, an air-channel provides for the inflation of the stomach.

A typical optical system for this instrument is shown in Fig. 3. The objective system is a reversed Galilean telescope with an Amici prism

[14] F. J. Wallace, *J. Urol.* **90**, 324 (1963).

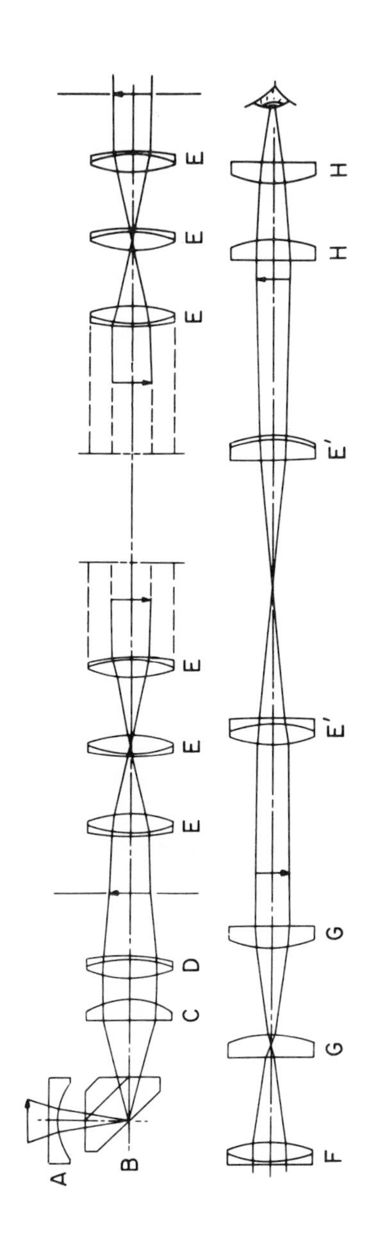

FIG. 3. The flexible-lens gastroscope. A reversed Galilean objective (*A–D*) is followed by six sets of three lenses each (*EEE*) to provide a flexible section. This is followed by a rigid telescope (*F–H*).

indicated at B. The first focal plane is chosen conjugate to a working distance of about 50 mm from the first surface of the lens A. The diameters of lenses C and D may be taken as 6 mm. The figure shows the paths of the upper and lower chief rays. The field of view is about 55°. The entrance pupil lies in the Amici prism and is 0.4 mm in diameter. The exit pupil is 1.67 mm in diameter. The working distance is from 100 mm to 5 mm.

To determine the visual magnification, we use Eq. (4), taking $D = \frac{1}{4}$ m (25 cm); $n = 1.0$; $P = \frac{1}{20}$ m (5 cm); $\rho/\rho' = 1/4.17$; whence $M_V \approx 1.2$.

The gastroscope relay system is designed to perform the usual dual functions. First, it must relay the primary focal plane back to the focal plane of the eyepiece, but it must also relay the entrance pupil back near the last relay lens so that the eyepiece may form the exit pupil at the desired eye-point. In the flexible gastroscope, in order to allow bending, the focal lengths of the relay lenses E are chosen as short as practical. The allowable bend is 35°, which is achieved by using 18 lenses in the flexible section in groups of three. In practice, it is difficult to distribute the flexure evenly among the relay lenses. The best approach has been to use a pair of concentric, but width-tapered, phosphor bronze springs. The taper avoids excessive bending at the junction with the rigid portion and at the same time permits greater flexibility near the distal end, thus allowing the tip of the instrument to follow the greater curvature of the stomach.

One can see from Fig. 3 that the last focal plane of the relay system is imaged at infinity by lens F. This, in turn, is viewed by the telescopic system designated by lenses G through H.

Though the lens gastroscope has been used with success for many years, it has several limitations, including: (1) The rigid portion is difficult for the patient to accept, so that the observation time is limited. Some patients, such as those with cervical immobilization, cannot be gastroscoped. (2) The illumination was not sufficient to allow the use of color motion picture photography.

Many attempts have been made to improve photography through the gastroscope, notably with the electronic flash system of Debray and Housset[13] and the swallowed camera system of Uji.[15]

Debray's system involves, at the distal end, in addition to an incandescent bulb for viewing, a gas-discharge lamp through which as much as 45 J could be discharged in 1 msec. A camera is mounted at the proximal end and its shutter synchronized with the lamp flash. Still color pictures of very high quality have been obtained with it. However, this system has not been widely adopted in this country because of the risks involved in discharging from condensers several hundred volts through the flash lamp.

[15] S. Tasaka and S. Ashizawa, *Bull. Am. Gastros. Soc.* **5**, 12–15 (1958).

IV. THE ADVENT OF FIBER-OPTIC SYSTEMS

The need for a truly flexible gastroscope motivated most of the original research work on fiber-optic systems. Thus, as early as 1930, Dr. R. Schindler, the distinguished gastroscopist, and Lamm experimented with drawn glass fibers in an attempt to transmit optical images.[16] Plastic fibers were also tried by many experimenters. The early attempts failed, usually because of light loss through the walls of the fibers. van Heel[17] succeeded in coating glass fibers with plastic. Properties of fibers and winding techniques were studied in London by Hopkins and Kapany.[18] The whole subject of fiber optics is discussed in Vol. IV, Chapter 1 of this series.

Satisfactory image-carrying fibers were not available until Curtiss perfected the glass-on-glass drawing technique.[19] Working with Hirschowitz, he produced a prototype gastroscope, flexible for its entire length.[20] This instrument has been further developed and also clinically evaluated by several authors.[21]

A. FIBER GASTROSCOPE

The arrangement of the fiber flexible gastroscope, with motion picture camera mounted, is shown in Fig. 4. At the distal end, under two separate quartz windows, are placed a 10-V lamp and the Amici prism. The prism is followed by a pair of doublets mounted on a slider which allows these lenses, controlled by a rack and pinion at the proximal end, to focus the gastric mucosa directly onto the distal end of the fiber bundle at plane A. Each achromat is 6 mm in diameter, the pair having a focal length of 7.9 mm. The object field may be focused from approximate contact with the cover plate to 15 cm working distance. The field of view is approximately 45° at 2.5 cm working distance from the cover glass. The fiber-bundle diameter is approximately 5 mm. At the proximal end (plane B) a $10\times$ fully corrected eyepiece is used for viewing the end of the fiber bundle. The resolution is normally 35 line pairs/mm or better at the fiber bundle. This resolution is an average taken over about 25 equal rectangular areas of the object field.

---

[16] H. Lamm, *Z. Instrumentenk.* **50**, 579–581 (1930).

[17] A. C. S. van Heel, *Tech. Wetenschap. Onderzook* **24**, 25–27 (1953).

[18] H. H. Hopkins and N. S. Kapany, *Nature* **173**, 39–41 (1954).

[19] L. E. Curtiss, B. I. Hirschowitz, and C. W. Peters, *J. Opt. Soc. Am.* **47**, 117 (1957).

[20] B. I. Hirschowitz, L. E. Curtiss, C. W. Peters, and H. M. Pollard, *Gastroenterology* **35**, 1 (1958).

[21] J. H. Hett and L. E. Curtiss, *J. Opt. Soc. Am.* **51**, 581–582 (1961).

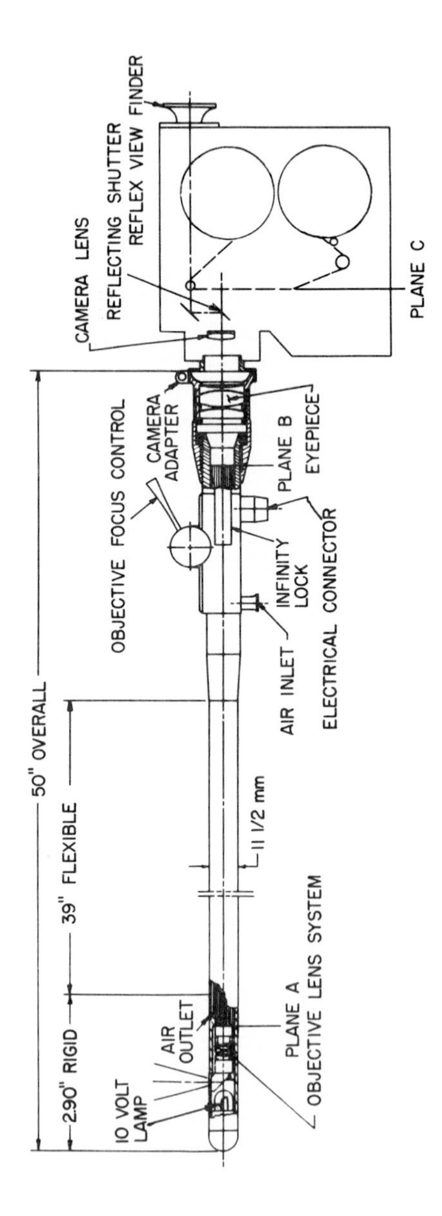

FIG. 4. The flexible fiber-optic gastroscope with camera.

When photographing through the instrument, the eyepiece is locked at infinity and the camera lens likewise set at infinity, the mucosa being imaged on the film at plane *C*. The end of the bundle may thus be viewed through the reflex finder. The mucosa is focused through the rack-and-pinion drive of the objective focus control. An infinity position lock is provided on the eyepiece.

Using the 10-V lamp, many motion pictures have been taken with an 8 mm camera at 8–16 frames/sec using Kodak High-Speed Ektachrome Type B film. If the objects are kept fairly close to the instrument, good results may also be obtained with Kodachrome II. The camera objective should have a focal length in the range 20–30 mm. It is also entirely feasible to use a small 16 mm motion picture camera, such as the Beaulieu.

Recently, fiber gastroscopes have been built which use a fiber bundle for an illumination channel and also contain a channel for a biopsy forceps. The advantages of this system are (1) the elimination of the bulb gives a much cooler distal head, (2) the distal head is smaller than with the bulb instrument, (3) the object field illumination may be well over 2000 ft-c at 1 in. working distance, allowing full-field 16 mm motion pictures at 16 frames/sec with Ektachrome EF film.

Fiber gastroscopes with a mechanically controlled flexible distal tip were introduced in Japan by Machida in 1966. In the United States, a gastroscope with distal-tip control was offered by American Cystoscope Makers in 1967. Also from Japan is a gastroscope with a built-in camera at the distal end. This instrument is an improvement over the early Uji system, since the field of view is now visually monitered with a wide-angle objective.

B. Fiber Bundles for Medical Instruments

Image-transmitting fiber bundles for medical instruments require: (1) adequate definition; (2) good color rendition over the visual spectrum; (3) maximum light transmittance; and (4) a visual field as free as possible from distortion and extraneous structures such as dots, blank spots, and grids.

As mentioned earlier, the development of the glass-on-glass fiber-drawing technique made fiber medical instruments possible. The core-glass index may be chosen at $n_D = 1.69$ and the cladding index $n_D = 1.52$. This gives a numerical aperture of 0.72, which is sufficiently fast for the lens input and output systems generally used in medical work.

Fiber diameters are approximately 13 μ (0.0005 in.) and the cladding thickness is about 1 μ. Flexible bundles about 1 m long built of these fibers may be expected to resolve 35 lines/mm or more on a production basis. If sintered rectangles are used to build up a fiber bundle, higher resolution may be expected, but a grid structure formed by the boundaries of the small rectangles will remain.

A certain amount of decollimation will occur in light carriers. If a parallel light beam enters a 3-ft-long bundle, the half-angle of the emerging cone may be 10°. This effect arises partly because of diffraction and partly because the internal reflections at the interface between core and cladding are not perfect but are subject to reflection variations not yet clearly understood. The transmission properties of individual fibers and of certain fiber bundles have been studied by Potter.[22]

V. THE ESOPHAGOSCOPE

Observational esophagoscopy, like gastroscopy, began with the open tube of Kussmaul[3] in 1868.

The open-tube type, about 48 cm long and about 8 mm in diameter, is used extensively today. However, in introducing an open tube, there is always the great danger of tearing the laryngopharynx. Apparently, obturators were not used until suggested by Schindler in 1948.[23] Through the open tube, various types of telescope have been used, chiefly with forward and foroblique vision. Optically, these telescopes are similar to the cystoscope, except for a longer relay system and the use of air in the object field.

A fiber-optic esophagoscope with direct forward-vision head has been reported by Hirschowitz,[24] but this instrument has been largely superseded by a foroblique system. The foroblique system, with the line of vision directed at an angle of 25° with the principal mechanical axis, has been evaluated by LoPresti.[25]

The diameter of this instrument is about 13 mm. The tip is made of soft neoprene to facilitate ease of introduction. A focusing objective is controlled by rack and pinion at the proximal end. Provision is made for an air channel and nozzle over the objective to keep this area clear of mucous. Illumination is provided by a round fiber bundle which, in turn, is illuminated by a high-intensity external light source. An additional channel is provided for a suction tube or for a biopsy forceps. Figure 5 shows a fiber esophagoscope with a biopsy forceps in position and with fiber light

---

[22] R. J. Potter, *J. Opt. Soc. Am.* **51**, 1079–1089 (1961).
[23] R. Schindler, *J. Am. Med. Assoc.* **138**, 885 (1947).
[24] B. I. Hirschowitz, *Lancet* **7304**, 388 (1963).
[25] P. A. LoPresti, A. Hilmi, and P. Cifarelli, *Am. J. Gastroenterol.* **47**, 11 (1967).

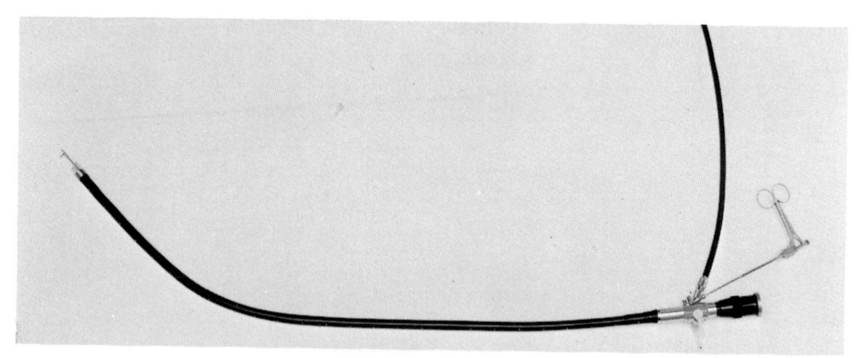

FIG. 5. Fiber esophagoscope. In addition to the objective system, the head contains a light-carrying bundle, water channel for clearing objective, and a biopsy channel.

input. LoPresti has taken excellent motion pictures of esophageal conditions, such as varices and tumors, using 8 and 16 mm cameras with Kodachrome II film.

VI. THE CYSTOSCOPE[26]

The early cystoscope optical systems consisted of a right-angled distal prism followed by a series of positive lenses which formed an image and relayed it to the eyepiece focal plane.

A simple way of improving the field angle and of obtaining a better water seal was found by Wappler,[27] who replaced the right-angled prism and simple lens with a hemisphere. The flat side of the hemisphere was silvered and set at 45° to the principal optical axis of the instrument. This system gave fields of view of over 60°. A correction for the mirror reflection was obtained by using a Dove prism, usually placed after the eye lens, since this part of the instrument was not restricted in diameter. The spherical and color aberrations of the hemisphere could be tolerated because of the small entrance pupil, usually less than 0.1 mm in diameter. However, even these aberrations were reduced by von Rohr,[28] who used the Amici prism as a first reflecting element, followed by a corrected objective system. Both the von Rohr- and the Wappler-type objective systems are still used extensively today.

---

[26] A. Gleichen, "The Theory of Modern Optical Instruments," 2nd ed., pp. 225–231. Stationery Office, London, 1921.

[27] R. Wappler, U.S. Patent 849, 344 (1907).

[28] M. von Rohr, U.S. Patent, 940,894 (1909).

Cystoscopes are available in many different designs, as examining, as catheterizing, and as operating instruments.[29] All designs have in common the use of small-diameter elements and the immersion of the object field in water. The latter is done to distend the bladder wall for examination and to avoid the danger of air inclusion in the patient's circulatory system.

The usual cystoscope has a working length of about 210 mm and diameters ranging from 9 mm down to 3 mm. These diameters are often expressed on the Charrière or French scale. Thus, a 9-mm-diameter instrument is called "27 French," each French unit being $\frac{1}{3}$ mm of diameter.

1. *Some Optical Characteristics*

These instruments are designed to give a visual magnification M_v of unity at a working distance of about 40 mm. With an exit pupil radius ρ' of 1 mm in the larger instruments, one may compute the entrance pupil from Eq. (4) by setting $M_v = 1$, $P = 40$ mm, $D = 200$ mm and $n = 4/3$. Then $\rho'/\rho = 20/3$ and since ρ' is 1.0, $\rho = 0.15$ mm (radius).

We also note that from Eq. (2), the magnification of the instrument itself, M_i, is readily determinable as being less than 1.0. One has directly

$$M_i = n\rho/\rho' = 3/20 \times 4/3 = 1/5$$

From Eq. (5), one may determine the numerical aperture of this example to be 0.005. For some large cystoscopes, this value may be as high as 0.01.

From the small size of the entrance pupil of this example, it is clear that there is a large depth of object field. Nevertheless, it has been found useful to employ a focusing eyepiece to obtain maximum resolution and to allow for the frequently low accommodation of the observer's eye.

The dependence of the visual magnification M_v on the working distance P is also apparent from Eq. (4). Since ρ'/ρ is essentially constant for a given instrument, one finds, for the example given above, that $M_v = 40/P$. Thus, if

$$P = 10 \text{ mm}, \qquad M_v = 4$$
$$P = 100 \text{ mm}, \qquad M_v = 0.4$$

or a reduction of ten times.

In 1959, Hopkins[30] achieved considerable improvement in the light transmission of cystoscopes by reducing the number of air-glass surfaces through the use of very thick lenses.

---

[29] R. W. Barnes, R. T. Bergman, and H. L. Hadley, "Encyclopedia of Urology," Vol. 4. Springer, Göttingen, 1959 (This book reviews from the clinical angle all cystoscopes available at that time).

[30] H. H. Hopkins, U.S. Patent 3,257,902 (filed 1959).

2. *Field of View*

The true field of view of the cystoscope is about 60°. This field may be set parallel to the principal axis of the instrument, as in the urethrascope, or at nearly right angles, as in the examining systems. The field may also make an angle of 28° with the principal axis, as in the foroblique system, or it may be directed back toward the observer axis, as in the rectrograde system.

The foroblique system is used extensively with resectoscopes and other endoscopes. This direction of vision is accomplished by using two successive lens wedges of flint glass at the distal end.

Experimental cystoscopes have been built with fields of view much larger than 60°, but the increase in angular field size is always at the expense of the magnification. Cystoscopes have also been built with small fields of view and magnifications M_v of about 40. Such instruments are very difficult to use, since the bladder is always in motion, even in the anesthetized patient.

3. *Using Cystoscopes*

A sheath with obturator locked in it is first introduced into the patient. After removal of the obturator, a cystoscopic telescope is locked into the sheath. Figure 6 shows two sheaths, both beaked, and each containing a lamp at the distal end and two stopcocks and the lock at the proximal end. Below the sheaths is shown a typical obturator. Water may be introduced through one stopcock. When the bladder is partially filled, the visual examination may begin. If an operative procedure is called for, the examining telescope is removed from the sheath, and a single operating channel combined with telescope, as shown in the figure, is introduced. This instrument also has a controllable lid at the distal end, which enables the operator to deflect and keep in the field of vision an instrument such as a catheter or a biopsy forceps.

An important application[14] is the use of a fiber-optic light carrier to replace the conventional bulb in the sheath. The light output may be many times that of the incandescent bulb. This illumination system employs a Sylvania 150-W DLS lamp with an elliptical reflector. The light output curve for a Brown-Buerger 21 French sheath is shown in Fig. 7.

Quartz-rod illumination systems for cystoscopes and other endoscopes were built in France in 1951, and later in this country.[11,31] The average level of illumination with these systems is much lower than that of a comparable fiber-optic system. It is, of course, possible to use flash lamps at high light levels with the quartz-rod systems as with fiber systems.

---

[31] J. F. McCarthy and J. S. Ritter, *J. Urol.* **78**, 674 (1957).

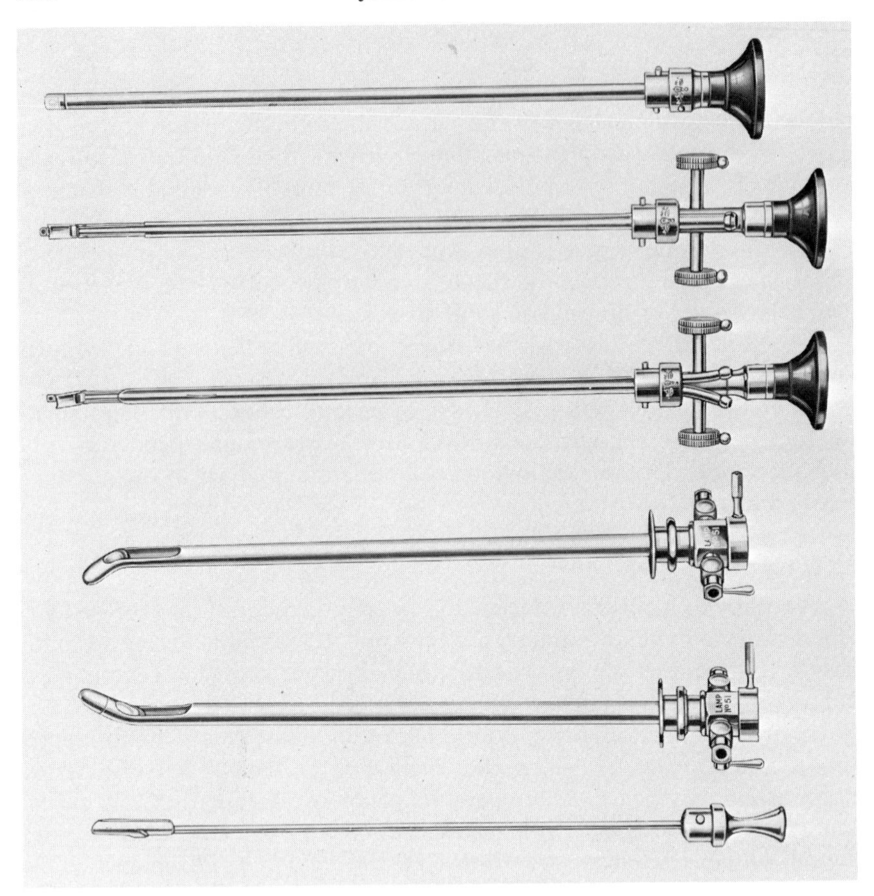

FIG. 6. Brown-Buerger universal cystoscope, 24 French. At the top is a right-angled-vision examining telescope. Below this is a right-angled-vision operating telescope, and, below this, in turn, is a double catheterising telescope, a convex sheath, a concave sheath, and the obturator.

4. *Resectoscopes*

These instruments were devised for the transurethral electrosurgical resection of the lobes of the prostate gland. Figure 8 shows a Stern–McCarthy-type instrument. The telescope used is the foroblique instrument. The sheath is straight and made of bakelite or fiberglass. In the field of view of the telescope is the cutting loop, which may be moved about 1 in. by the rack-and-pinion control handle.

Using a quartz-rod lighting system, the first motion picture of a resection was shown by B. Fey in March 1955, at Paris. In this country, McCarthy and Ritter used quartz-rod equipment to take color motion pic-

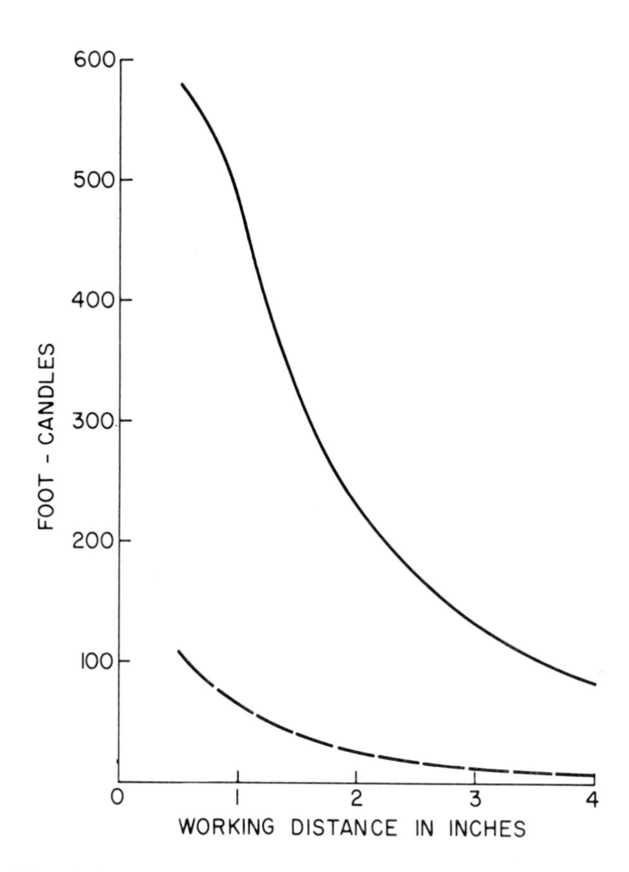

FIG. 7. The dashed line represents the measured output from a standard Brown-Buerger 21 French sheath, 1.8-V, 260-mA lamp. The solid line is from the fiber sheath of a 21 French instrument, with external light source, 150-W DLS lamp. (Curve measured by L. Scrivo.).

tures of the bladder in 1957.[31] Using quartz-rod equipment, M. Jaupitre of Paris showed motion pictures of a prostatic resection in 1957.

When the standard sheath is replaced by a fiber-optic sheath, the light output obtained is about five times greater. Using this type of sheath, a color motion picture of a transurethral resection was made in 1963 by Barnes and Mims.[32]

---

[32] R. W. Barnes and M. M. Mims, motion picture "Transurethral Resection." Am. Urol. Assoc., St. Louis, Missouri, May 13–16, 1963.

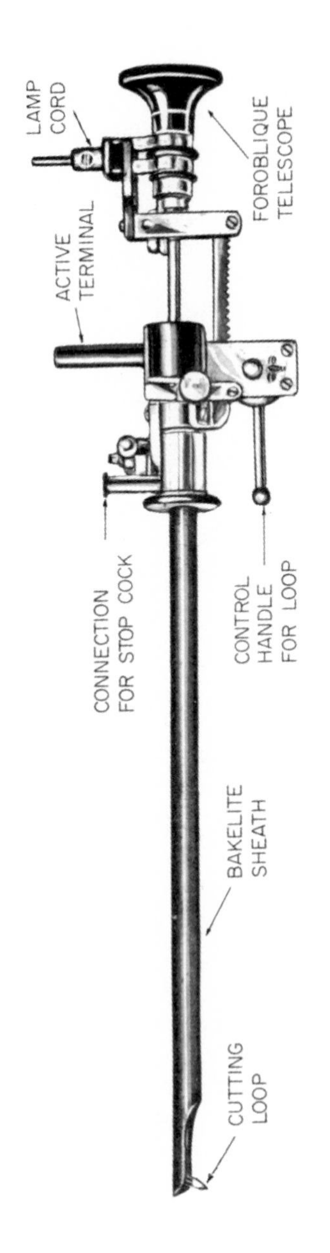

FIG. 8. Stern–McCarthy, 28 French resectoscope, showing foroblique telescope in position and with cutting loop at maximum extension.

5. *Urethrascopes*

A forward-vision telescope with a field of about 55° is often inserted through the sheath of a Stern–McCarthy electrotome after resection, to examine the total prostatic lumen. Color motion pictures of the urethra using quartz rod equipment were taken by M. Jaupitre at the Hospital Necker, Paris, in May 1955. The urethrascope has been lately improved by using an annulus of fiber optics around its objective. Motion picture studies of the urethra have also been made using this instrument.

6. *Ureterscopes*

The visualization of the lumen of the ureter presented a problem not solvable with regular lens optics. Using fiber optics with annular distal illumination, American Cystoscope Makers, Inc., have produced an instrument of 7 French diameter which covers a field of over 20°.[21,33]

VII. FLUORESCENCE ENDOSCOPY

In 1944 Figge *et al.* [34] indicated that if certain porphyrins are injected into a patient, red fluorescence will occur in neoplastic tissue if the tissue is illuminated with ultraviolet light at a wavelength of about 3660 Å. Figge and his associates at the University of Maryland used an ultraviolet lamp in open surgery to delineate the extent of a malignancy.

A cystoscope transmitting ultraviolet light has been described by Whitmore *et al.*[35] This instrument was developed in 1962. It contained a quartz rod in a straight endoscope sheath and a 100-W mercury lamp built directly on the proximal end of the instrument. This instrument clearly demonstrated tetracycline fluorescence in malignant tumors of the bladder. However, the usefulness of this original instrument was limited by its size, 28 French (9.3 mm), and the heat of the lamp. In 1964, cystoscopes using ultraviolet-transmitting fiber optics were developed by American Cystoscope Makers, Inc. and clinically used by Whitmore and Bush.[36] These were beaked, convex or concave examining instruments of 24 French (8 mm) diameter. A fiber-optic connecting light bundle was used between the cystoscope and a 200-W mercury arc lamp. A series of Corning filters placed at the junction between light bundle and instrument allow either

---

[33] J. H. Hett, U.S. Patent 3,089,484 (1963).

[34] F. H. J. Figge, G. S. Weiland, and L. O. J. Manganiello, *Proc. Soc. Expl. Bio. Med.* **68**, 640 (1948).

[35] W. F. Whitmore, I. M. Bush, and E. Esquivel, *Cancer* **17**, 1528 (1964).

[36] W. F. Whitmore and I. M. Bush, *J. Urol.* **95**, 201–207 (1966).

various intensities of ultraviolet light to be used, or a clear white light for examination. These instruments have been discussed by Quint *et al.*[37]

Fluorescent systems have also been designed to find cancers in the bronchus, esophagus, stomach, cervix, and rectum.

VIII. BRONCHOSCOPES

Modern bronchoscopy began with G. Killian in 1898 and was subsequently developed intensively in this country by Chevalier Jackson.

The simplest form of the bronchoscope is a rigid open tube up to 9 mm in diameter and up to 40 cm long. A small bulb is provided for illumination at the distal end. As an aid to surgery through such a tube, a Galilean telescope of about two power, designed for a relatively short front conjugate distance, may be mounted on a hinge at the proximal end.

An operating telescope may be mounted through the bronchoscopic tube. This is essentially a biopsy forceps with the objective of a foroblique system arranged to keep the forceps always in view. In another form, a right-angle telescope may be provided with a channel and a distal movable deflector to guide a biopsy forceps. This instrument, suggested by Holinger, is particularly useful for work in the upper lobe.

Many different types of telescope objective are used by the bronchoscopist in an attempt to see as many as possible of the branching bronchial tubes. The author has devised a fiber-optic telescope for this purpose which, because of its small diameter (3 mm) and double deflectors, allows visual exploration of some of the smaller branches.[38]

In 1948, Holinger[39] succeeded in taking motion pictures through a bronchoscope by using an external bulb, such as an aircraft landing lamp, and reflecting the light down the instrument. In 1956, Dubois de Montreynaud *et al.*[40] using a quartz-rod illuminating system, succeeded in taking endobronchial motion pictures. Later on, electronic flash lamps were used with the quartz system. Recently, Broyles[41] has discussed a bronchoscopic tube with a built-in annular fiber light carrier. Such a system gives an increase of illumination of over five times compared to the older distal lamp.

---

[37] R. H. Quint, J. H. Hett, and F. J. Wallace, *J. Urol.* **95**, 208 (1966).

[38] J. H. Hett, U.S. Patent 3,081,767 (1963).

[39] P. H. Holinger, F. C. Ainson, and K. C. Johnston, *J. Thoracic. Surg.* **17**, 178 (1948).

[40] J. M. Dubois de Montreynaud, R. J. Edward, and A. J. Gladu, *Laryngoscope* **66**, 637 (1956).

[41] E. N. Broyles, Paper accepting Chevalier Jackson award, meeting of the Am. Broncho-Esophagological Soc., Hollywood, Florida, April 1963.

IX. MISCELLANEOUS ENDOSCOPES

A. Choledochoscope

The detection of stones in the common and hepatic ducts at surgery has always presented a problem. Such methods as palpation, X rays, and electroacoustic transducers with metallic dilator probes have not been entirely successful. Direct visualization of the duct interior is required. In 1923, Bakes[42] is reported to have examined the common duct of forty patients by inserting a laryngoscope.

Cystoscopes were often used for this type of observation until 1941, when the McIver Choledochoscope was developed. This instrument had a thin shaft 70 mm long attached at right angles to a heavier shaft about 20 cm in length. A lamp was provided at the extreme distal end to illuminate the field for a direct-vision objective. A channel was also provided for flushing with water.

An improvement was offered by Wildegans[43] in Germany in 1953. The 70-mm distal tip was tilted to include a 60° angle with the main axis of the instrument. Recently, a fiber-optic, completely flexible instrument of 17 French was developed by American Cystoscope Makers (Fig. 9). The flexibility and high light levels available have enabled Shore and Lippman[44] to take excellent photos with this instrument.

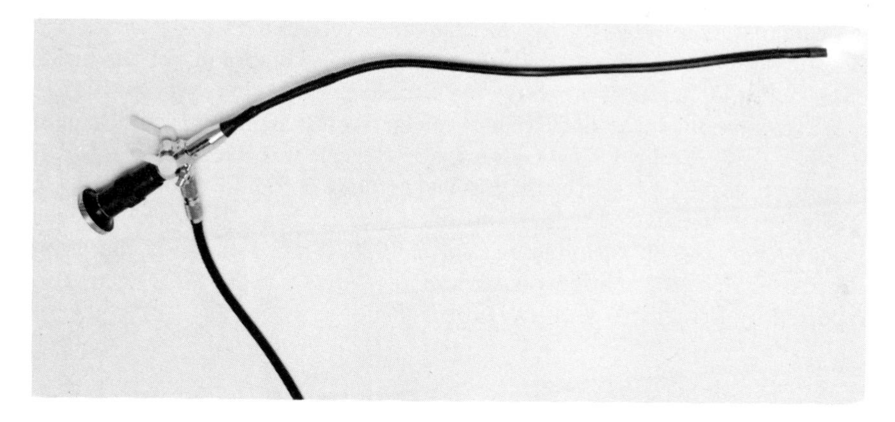

Fig. 9. Fiber-optic flexible choledochoscope. Diameter, 5.2 mm. Contains water channel and fiber illumination giving 500 ft-c.

---

[42] J. Bakes, *Arch. Klin. Chir.* **126**, 473 (1923).
[43] H. Wildegans, *Med. Klin.* **48**, 1270 (1953).
[44] J. M. Shore and H. N. Lippman, *Lancet* **7397**, 1200 (1965).

B. Pharyngoscopes and Laryngoscopes

The smallest nasopharyngoscope used for inspection of the small cavities and canals of the nose has a length of 4 in. and a diameter of only 2.16 mm or 6½ French. The field of view is about 30° at right angles to the main axis. At this diameter, the resolution of the instrument is diffraction-limited.

A larger instrument, the Broyles nasopharyngoscope, of diameter 8½ French, may be made as a foroblique vision instrument as well as a right-angle system. Still larger is the Broyles pharyngeal telescope. This instrument is 33 French and is provided with a built-in fiber-optic light carrier yielding over 2000 ft-c at 1-in. working distance. The vision is at right angles or foroblique. It is possible to take color motion pictures through this instrument. The smaller Broyles nasopharyngoscopes are also available with fiber-optic carriers giving greatly improved illumination.

The Broyles laryngoscope is made with an open tube about 16 cm in working length with a handle whose axis is parallel to the tube. At the distal end of the tube is a lamp, and at the proximal end may be attached a two-power Galilean telescope designed for the short working distance. The illumination of this instrument has recently been greatly improved by building into the tube an annular fiber-optic light carrier.

Cinematography of the larynx had been reported as early as 1930.[45] A good summary of the procedures used in the photography of the bronchus and esophagus was given by Brubaker and Holinger in 1947.[46]

A stroboscope for examination of the larynx was reported by Gutzmann in 1939 at the Berlin Charity Hospital. After the war, General Electric of Germany developed especially for laryngology a xenon strobe lamp of 260,000 ft-c with a light duration of 5 μsec. In 1953, Winckel[47] used a Fastax camera at 8000 frames/sec and a 4000-W lamp to study the vocal chords. High-speed motion pictures of the human vocal cords were also reported by Pfister[48,49] and by Dunker.[50]

---

[45] G. O. Russell and C. Tuttle, *J. Soc. Motion Picture Engrs.* **15**, 171 (1930). Also C. A. Morrison, *ibid.* **16**, 356 (1931).

[46] J. D. Brubaker and P. Holinger, *J. Soc. Motion Picture Engrs.* **49**, 248 (1947).

[47] F. Winckel, *Med. Markt.* **3**, 80 (1954).

[48] K. Pfister, "Motion Pictures of the Human Vocal Chords." *Actes 2ème Congr. Intern. Phot. Cinematog. Ultra-rapides, Paris. 1954*, p. 225. Dunod, Paris, 1956.

[49] K. Pfister, *Proc. Intern. Conf. High Speed Photography, 4th*, p. 290. Helwich, Darmstadt (1959).

[50] E. Dunker, *Med. Markt.* **4**, 159 (1961).

C. Peritoneoscopes and Laparoscopes

Peritoneoscopy of humans began with Kelling in 1910.[51] Working at the same time, Jacobaeus of Stockholm,[52] also using cystoscopes, published several papers and established the procedure in Europe. About 1934, special instruments were developed for Ruddock.[53]

The telescope is of the foroblique examining type, 11 in. long, of larger diameter, and with a larger lamp than the foroblique cystoscope. This instrument is passed through a cannula 28 French in diameter. The cannula has been placed in position by running a trocar through it into the already inflated abdomen. Inflation is by needle. With the cannula in position it is also possible to pass a biopsy forceps whose jaws may be kept in the field of vision of a thinner foroblique telescope.

In 1953, R. Wolf developed a laparoscope which used two lamps; one for observation, the other for overvolting for purposes of photography. In 1961, he developed a laparoscope using an electronic flash lamp at the distal end. Henning and Muller,[54] of the City Hospital at Berlin-Spandau, report enough illumination to allow the use, with color film, of camera focal lengths from 70 to 110 mm.

Recently, a peritoneoscope with fiber-optic illumination has been developed by American Cystoscope Makers, Inc.

D. Culdoscopes

The culdoscope was devised for the examination of the female pelvic organs *via* the vagina.[55–58] A puncture is made into the cul-de-sac with a trocar through a cannula, usually 27 French in diameter.

The photographic telescope, usually with a right-angle objective, is inserted through the cannula. The angular field of view, depending on the model, ranges from 55° to 70°. Because of the range of object distances, a strong source of illumination is required if photographs are to be taken.

---

[51] G. Kelling, *Muench. Med. Wochschr.* **57**, 2358 (1910).

[52] H. C. Jacobaeus, The use of laparo-thoracoscopy from a practical point of view. *Intern. Congr. of Med.* (1913), *London*, Sect. 6, *Medicine* Pt. 2, p. 565 (1914).

[53] J. C. Ruddock, *Western J. Surg.* **42**, 392 (1934).

[54] H. Henning and K. Muller, *Med. Markt.* **2**, 61 (1963).

[55] A. Decker and T. H. Cherry, *Am. J. Surg.* **64**, 40 (1944). Also A. Decker, *Am. J. Obstet. Gynecol.* **50**, 227 (1945).

[56] R. W. TeLinde and F. Rutledge, *Am. J. Obstet. and Gynecol.* **55**, 102, 115, (1948).

[57] A. Decker, *J. Am. Med. Assoc.* **140**, 378 (1949).

[58] N. Aresin, *Zbl. Gynak.* **74**, 401 (1952).

Excellent photographs and motion pictures have been taken by Decker using the wide-angle instrument and a fiber-optic illumination system which had an integral bundle going from the lamp to the distal end of the instrument.[59]

An operating culdoscope has also been devised to work through the standard cannula. The telescope proper is of smaller diameter than the examining instrument to allow room for an operating channel and an adjustable distal deflector. A review of this subject in book form containing references has been written by Decker.[60]

E. Colposcopes

For generations, gynecologists have examined the os uteri without visual magnification. Occasionally, some would use a cystoscope to obtain a magnification of about four. In 1924, Hinselmann developed a colposcope with a magnification of 10–20. This instrument, made by the firm of J. D. Moller, has a field of view of 22 mm at 10× and 11 mm at 20×. A camera with a flash system is available.

Several firms offer colposcopes with powers up to about 50. In the binocular stereoscopic form, excellent photographs may be obtained. Illumination may be either with electronic flash or fiber-optic systems. The binocular features also allow the use of a television camera viewing through one side of the lens system.

In Germany, Hinselmann aided the development of a colpomicroscope with powers up to 1200 using oil immersion,[61] but this technique was not used widely owing to the distress to the patient.

F. Proctoscopes and Sigmoidoscopes

The use of the open tube for examination of the rectum and sigmoid colon has been known for at least two millenia. Improvements in modern times include the use of an incandescent lamp at the distal or proximal ends. In addition, a telescope of about two power, which is removable, is often used at the proximal end.

---

[59] A. Decker, *in* "Gynecology and Obstetrics" (C. H. Davis and B. Carter, eds.), Vol. 3, Chap. 15N, p. 1. Prior, Hagerstown, Maryland, 1965.

[60] A. Decker, "Culdoscopy." Davis, Philadelphia, Pennsylvania, 1967.

[61] H. Hinselmann, "Kolposkopieschen Studien," Books I, II, and III. Thieme, Leipzig, 1954–1959. Also *Med. Markt.* **3**, 52 (1959).

A recent significant improvement is the use of an annular fiber light carrier to place a strong illumination at the distal end.[62] Thus, high-intensity illumination is achieved and the troublesome reflections down the tube are avoided.

Many proctoscopists have tried using endoscopes through the observing tubes. The magnification thus available is often useful in studying mucosal details. Motion pictures have been taken, e.g., by Berci,[63] using a forward-vision telescope and quartz-rod illumination.

An interesting development in proctoscopic cameras is that made by Benndorf *et al.*[64] Their instrument had essentially a conical reflector at its distal end and a hole through this reflector so that forward vision, as well as right-angle vision many degrees wide and 360° around, was obtained. Illumination was obtained by a doughnut-shaped flash lamp placed below the proximal end of the objective. For study, the picture is reconstituted by projecting the image obtained back on to a film wound inside a cylinder. The working length of this instrument was not over 10 in.

Photographic telescopes of large entrance pupil have been built by American Cystoscope Makers to allow the taking of large-scale pictures on 35 mm film.

X. ENDOSCOPIC PHOTOGRAPHY

Medical photography clearly serves many useful purposes: (1) A record is established and the effects of time may be observed. (2) Medical teaching is aided. (3) Many functions of the body, e.g., esophageal peristalsis, can best be recorded using motion pictures.

Endoscopic photography began with Nitze, who developed the first lens cystoscope. In 1894, he also published the first atlas of photos taken through a cystoscope.

The illumination problems of endoscopic photography have already been considered in this chapter. The informational sensitivity of an endoscope combined with a camera is a subject requiring much additional research. For relatively simple systems, such as a camera lens and a film, an excellent review has been published by Perrin.[65] For an endoscopic system, as a first approximation, one may use Perrin's curves by considering the endoscope plus camera objective as the lens input system.

---

[62] R. Turell, *Am. J. Surg.* **105**, 133 (1963).

[63] G. Berci, *Med. J. Australia* **2**, 653–658 (1962).

[64] J. H. Benndorf, H. Brachvogel, and J. N. Stavenhagen, *Med. Markt.* **1**, 23 (1961).

[65] F. H. Perrin, *J. Soc. Motion Picture Television Engrs.* **69**, 157, 239 (1960).

A much simpler optical problem is to transfer the image formed in the last focal plane of the endoscope to the film plane of the camera. This may be done by using an optical system imaging at these conjugate planes replacing the eyepiece and camera lens, but in practice, this is undesirable, since eyepieces of endoscopes, being waterproof, are usually not removable. Usually, the eyepiece is set at its infinity point or zero diopters. Then, if the camera lens is set at infinity and positioned as closely as possible to the exit pupil of the endoscope, the last focal plane of the instrument will be imaged at the film plane. The size of the image at the film plane is then given as:

$$D = Fd/f \tag{8}$$

where D is the diameter of the image in the film plane of the camera, F is the focal length of the camera lens, f is the focal length of the eyepiece of the endoscope, and d is the diameter of the image in the last focal plane of the endoscope.

The required speed (f-number) of the camera lens depends on the angles of the rim rays of the eyepiece and the diameter of the exit pupil of the endoscope. In view of the practice of placing a Dove inverting prism on the eye side of the eye lens and consequently vignetting many of the rim rays, the diameter of the exit pupil alone usually determines the area of the camera lens being utilized. For example: Suppose a camera lens of 25-mm focal length, $f/2.5$, is used with an endoscope of 2-mm-diameter exit pupil. Then the camera lens is being effectively used at $f/12.5$.

An important exception to the above occurs when the eyepiece is used to view the end of a fiber-optic bundle. In this case, the exit pupil is at the eyepiece and the rim ray for an axial point is parallel at the edges of the eyepiece. Thus the camera lens diameter, or, more strictly, the entrance window of the camera lens, may effectively be as large as the diameter of the eye lens of the eyepiece.

1. *Still Cameras*

From about 1889, still pictures have been taken through endoscopes. Later, special adapters were built with a tilting mirror, so that with a special eyepiece, the operator could view the object before taking the picture. Cameras with a spring-driven film advance, such as the Robot, were used extensively.

Since about 1948, excellent through-the-lens reflex cameras using 35 mm film have been developed. These function well with endoscopes, provided the proper focal-length camera lens is chosen. While from a projection viewpoint, it is desirable to fill the 35 mm frame, the amount of light available from the endoscope with incandescent illumination is often low and the exposure time required would be too long. Living human tissue is always in motion, so that exposure should be in the range of fractions of a second. One may infer from Eq. (8) that for a given endoscope illumination and film, the exposure time required varies directly as the square of the focal length of the camera lens. One should note that with electronic flash systems, these time problems disappear whether the flash lamp is at the distal end of the instrument or the illumination is conveyed down the instrument by quartz rod or fiber optics. In this case, the flash time is of the order of 1 msec or less. The light energy available, depending upon lamp and power supply, may vary from 25 to 800 W-sec. A discussion of flash lamps for medical purposes as developed in Germany has been given by Breuer.[66]

2. *Motion Pictures*

Since medical phenomena are often transient, it is frequently much easier to take motion pictures of an event rather than stills. The limitation on the amount of light available at the image in the film plane and the economy of operation initially led to the use of 8 mm motion picture cameras of light weight with a slow framing rate. We have already noted the work of Hirschowitz in gastroscopy and Mims and Barnes in urology. Extensive work has also been done in France and Germany with such 8 mm cameras as the Nizo, Bauer, or Leicina when used with flash lamps.

The introduction of the Super 8 camera in 1964 offered the advantages for endoscopic use of a motor drive, the film being available n icartridges, and 50% more film area in use than in the older 8 mm cameras.

The film used by many observers has often been Eastman Ektachrome ER II, Type B. If enough light is available, Kodachrome II (ASA 25) is used. Daylight-type film is used with electronic flash systems. The recent development of Ektachrome EF film, designed Type 7242 (16 mm), for tungsten balance, with an ASA speed of 125, improved granularity, and better color balance, has already yielded better endoscopic pictures at much shorter exposure times.

Infrared film has been used successfully in delineating areas of polycythemia, such as appear in Cushing's syndrome. The color balance of Ektachrome is excellent when observing tissue, even at the low light levels existing in the exit pupil of the endoscope. According to Kodak, no reciprocity failure or loss in speed occurs as long as the exposure time is 0.1 sec. or faster.[67]

Until recently, small lightweight 16 mm cameras were not commercially available. Special 16 mm cameras were occasionally developed, such as that

---

[66] U. Breuer, *Med. Markt.* **3**, 103 (1962).

[67] Eastman Kodak "Tech. Bits" #3, Oct. 1963.

of Berci.[68] In this system, the film magazine and motor are behind the examiner and the film travels in the flexible tube up to and back from the endoscope. Recently available is a small, lightweight 16 mm camera by Beaulieu. This has been used successfully with the fiber esophoscope by LoPresti,[25] and, in Europe, is used by many endoscopists. Among cameras available are the Pathe-Webo M-60 16 mm reflex camera, the Arriflex, and cameras by Eastman Kodak Co.

XI. TELEVISION

The use of television as a teaching aid in medical and dental schools has become widespread in the last decade.[69] Medical color television even spanned the Atlantic Ocean *via* Telstar I in September 1962. The use of television with endoscopes, however, is still limited.[70]

It is a simple matter with the larger-diameter endoscopes to attach a vidicon-type camera to the eyepiece in a similar manner to the application of a film camera, making sure to choose the camera-lens focal length long enough to fill most of the cathode with the image. Thus one obtains a satisfactory kinescope image provided there is enough light. This may be done with cystoscopes, as first demonstrated by Jaupitre at the Centre Bichat, Paris, in 1956.

The teaching of some aspects of gynecology by television is routine in some medical schools. The binocular colposcope is well adapted to television, since the television camera can be attached to one eyepiece while the doctor monitors through the other. Furthermore, the light source, being external to the patient, may be of very high flash intensity. The resolution on the viewing screen is limited by both the television line scan and the optical limitations of the endoscope. In addition, the smaller-diameter endoscopes, because of their small entrance pupils, show diffraction limitations in resolution.

Ideally, endoscopic television should be presented in color, particularly if inflammation of tissues is involved. Color television through endoscopes has occasionally been demonstrated using an image orthicon camera system. These cameras, however, are so massive that the patient must be maneuvered to the camera.

The most recent development in color television through endoscopes is the use of the Philips Plumbicon tube in conjunction with the rotating color disk system of the Columbia Broadcasting Co. This system has great

---

[68] G. Berci, *J. Soc. Motion Picture Television Engrs.* **72**, 715 (1963) (Abstr.).

[69] Council on Med. Television Newsletter, and Health Sci. TV Bull. (S. A. Angello, ed.). Duke Univ. Med. Center, Durham, North Carolina.

[70] G. Berci and J. Davids, *Brit. Med. J.* #**5292**, 1610 (1962).

advantage in that the incoming light is not divided by beam splitters; instead the images are formed sequentially on the tube cathode. The Plumbicon not only has a high sensitivity, but also has a very fast response time, so that the spinning color disk is now entirely practical. This system was first demonstrated successfully at CBS Laboratories in Stamford in May 13, 1966. The CBS camera was used successfully behind a fiber-optic cystoscope and culdoscope from American Cystoscope Makers to examine the interior of a phantom bladder. Using very-high-intensity light sources, I. Bush has succeeded in recording on video tape, various bladder pathologies using a Cohu-type three-vidicon camera.

XII. PHOTOFLUOROGRAPHY

Direct photography of the X-ray sensitive fluorescent screen was first used to reduce film costs in large-scale population screening. Instead of a 14 in. × 17 in. film for each chest X ray, a camera was placed about 4 ft away from a blue phosphor screen and a 70 or 100 mm film used. This procedure was improved by Bouwers, who designed a large-aperture mirror system with a concentric correcting plate plus a Schmidt-type plate for the camera objective. This system had a curved field and used 70- or 100-mm square cut orthochromatic film. At least four times as much light was obtained at the film with the mirror system as with the best refractive objective.

The need to take motion pictures and the desire to reduce the X ray dosage to the patient led to the development of the image converter intensifier tube, initially by Philips of Eindhoven in 1934.

The image intensifier tube consists of a large-diameter fluorescent screen to which a photocathode is closely applied. A high potential is applied between the photocathode and a much smaller anodic output screen. Thus the high accelerating voltage plus the size reduction may yield a brightness gain of up to 6000. The intensified image may then be observed by the eye and much more detail may be seen.

Image intensifiers have made tv fluoroscopy, cinefluorography, and spot-film work feasible. An example of an extensively developed system is shown in Fig. 10. Here, *A* is the patient, *B* is a 1-cm fine grid in front of a circular fluorescent screen, *C* of diameter $12\frac{1}{2}$ in. A 45° mirror *D* is in front of a Bouwers catadioptric mirror system *E* ($f/0.68$) which produces an image of the screen on the cathode of the image converting tube *F* at a reduction of about $2\frac{1}{2}$.

The anode screen of the converter-tube image is, in turn, reproduced on the cathode of an image orthicon *G* by means of a Rayxar lens *L* ($f/0.75$) of focal length 65 mm, and lens *K* ($f/1.4$). A 35 mm cine camera also uses a

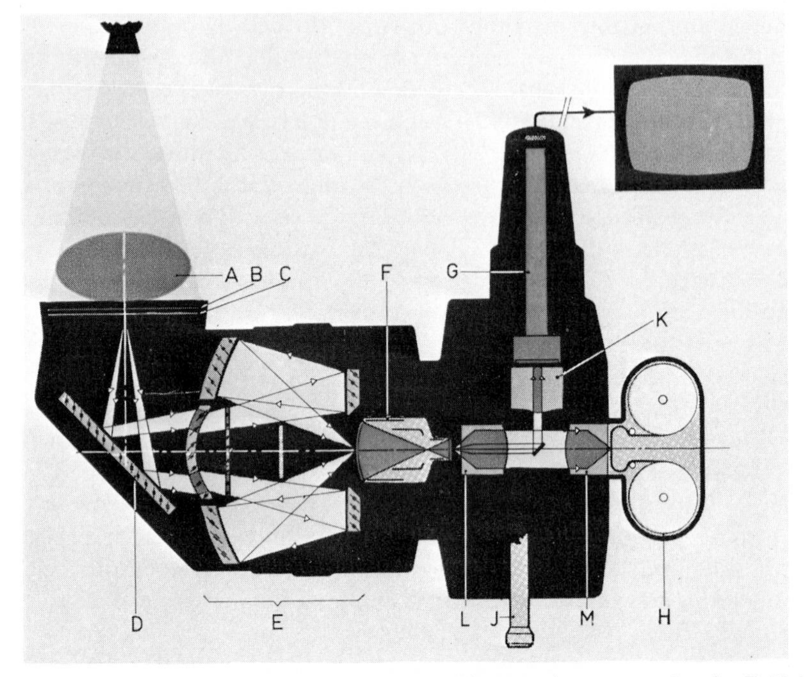

FIG. 10. X-ray image intensification system. (Photograph courtesy Aerojet-Delft.)

Rayxar (f/0.75) lens. Here, J is an optical viewing device. With this system, television fluoroscopy and cinefluoroscopy may be carried on simultaneously. The camera may be specially designed for work at very low levels of illumination. The shutter has a 270° open sector allowing, at a 16 frames/sec frequency, an exposure time of 3/64 sec and a transport time of 1/64 sec. It is clear that the motion picture camera may be used interchangeably with a single-shot camera using 100 × 100 mm sheet films.

In cinefluorography, the X-ray tube itself is sometimes pulsed to a maximum output to coincide with the camera exposure.

Most manufacturers of X-ray equipment in this country do not use an optical transfer system between the fluorescent screen and the image intensifier tube. Instead, the cathode of the intensifier tube, which may be 9 in. in diameter, is placed close to the patient to allow direct pick up of the X-ray radiation. In addition, the output end of the intensifier may simply be viewed with mirrors.

An introduction to this subject has been given by Ramsey *et al.*[71]

[71] G. H. Ramsey, J. S. Watson, T. A. Tristan, S. Weinberg, and W. S. Cornwell, *Cinefluorography, Proc. First Ann. Symposium Cinefluorography*, Univ. of Rochester, New York, Nov. 1958. Thomas, Springfield, Ill., 1959.

CHAPTER 10

Ophthalmic Instruments

HENRY A. KNOLL
Bausch and Lomb, Inc. Rochester, New York

I. INTRODUCTION

Ophthalmic instruments are used by a variety of individuals who are concerned with the health and performance of the eyes as organs of the body cooperating binocularly as optical systems. Ophthalmologists, opticians, optometrists, and orthoptists are trained to serve the visual health of the public in specific sectors. The public is often understandably confused by these terms.

Dr. Charles Sheard succinctly classified the first three groups by stating that "the ophthalmologist treats the eye, the optometrist treats the vision, and the optician treats the glasses." The orthoptist, who has come to the fore since Dr. Sheard's quotation, treats the muscles of the eyes.

The ophthalmologist, a medical specialist, is trained primarily to recognize and treat, medically or surgically, diseases and disorders of the eye. He also prescribes glasses. The optometrist is licensed by the individual states to diagnose and treat, by the use of optical appliances, optical errors of the eyes, and ocular muscle anomalies. The optician is fully versed in the fabrication of spectacles and contact lenses to the prescription of the ophthalmologist or the optometrist. He may fit contact lenses under the supervision of an ophthalmologist. The orthoptist is trained to diagnose and treat by nonmedical means binocular muscle anomalies. This work is usually done in cooperation with ophthalmologists or optometrists.

The instruments described in this chapter are used to examine the tissues of the eye, to determine its optical state, to test its sensory capacities (i.e., to light, form, and color), and to test its muscular functions. A few instruments used in treatment are described. Also included are instruments to assist in the fabrication and testing of optical appliances.

II. EXAMINATION OF EYE TISSUES

A. Ophthalmoscopes

The optical system of the eye (see Vol. II, Chapter 1 of this series) is perfectly suited to act as a magnifier for viewing the retinal surface. All that is required is an illumination system which can project light through the patient's pupil and also allow the viewer to align his own pupil with that of the patient. Such a method of illumination was devised by Helmholtz over one hundred years ago. His invention launched a new medical specialty as well as providing a tool for the entire medical profession. The condition of the blood vessels and other tissues of the retina not only indicates the health of the eye, but also reflects the condition of the entire body, providing information concerning diabetes, liver disfunctions, and other systemic diseases.

The most commonly used ophthalmoscope is a hand-held instrument using a 2.5-V lamp powered by two dry cells or a rechargeable unit. Six-volt ophthalmoscopes powered *via* a transformer are becoming more common. An example of the latter is shown in Fig. 1.

The optical arrangement of the ophthalmoscope is very simple. The lamp filament is imaged at the top of a small diagonal mirror or prism, the light diverging from that point toward the pupil of the eye to be viewed. The light entering the pupil illuminates an area of the retina. The doctor looks into the patient's pupil through a viewing aperture over the top of the mirror while holding the instrument close to his own eye as well as being close to the patient's eye. The patient's pupil acts as the aperture stop of the viewing system, whereas the viewer's pupil acts as the field stop.

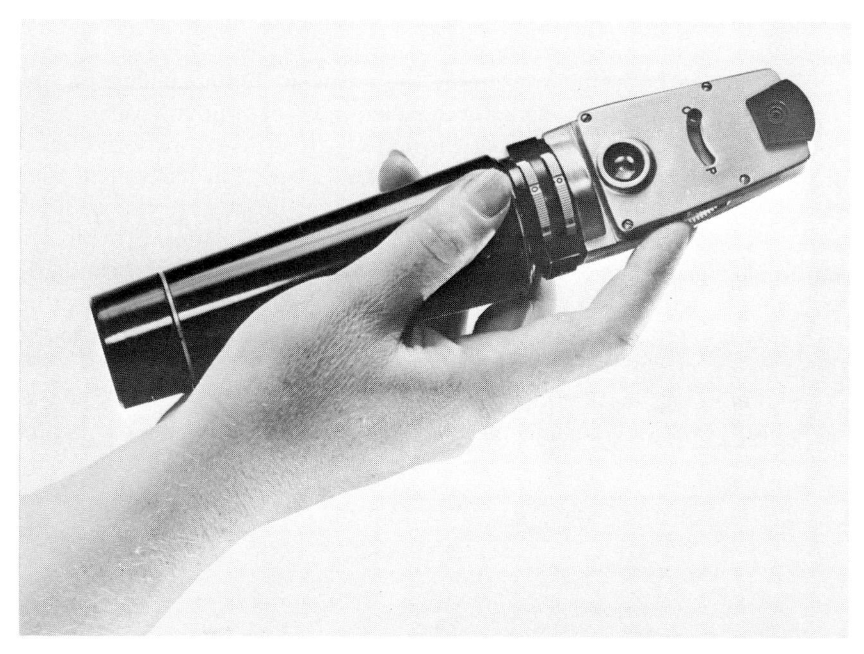

FIG. 1. A 6-V ophthalmoscope. The power cord is normally plugged into the lower end of the handle. (Photograph courtesy of Bausch and Lomb, Inc.)

Magnification is about 15 ×, the field about 5°. The total field coverage can, of course, be increased by having the doctor move the ophthalmoscope off axis.

The patient's retina will be seen clearly if both he and the doctor are emmetropic (i.e., the secondary focal points of the eyes fall on the retina), and they are both able to relax their accommodation fully. These conditions are not always satisfied. In order to compensate ametropias and/or accommodation, lenses of varying power are introduced between the diagonal mirror and the doctor's eye. These lenses are mounted on a disk which is rotated by fingertip control. Usually, some converging lenses of rather high power are included so that the viewer may examine structures in the anterior segment of the eye as well as the retina.

Figure 1 illustrates an ophthalmoscope as seen by the doctor. The small aperture at the top is the viewing aperture. His index finger is shown on the control for the lenses mentioned above. The power of these lenses is read through the magnifying lens above his thumb.

Immediately above his thumb, are two disks which contain apertures and filters of various sizes. The aperture plane is focused at infinity by a lens in the illumination system. The edges of the apertures are thus seen sharply on the retina of an emmetropic patient. They will be slightly out of focus in cases of ametropic patients.

At times, it is useful to examine the retina at less magnification, but with a larger field. This is done by so-called indirect ophthalmoscopy. The doctor moves away from the patient and observes a real, inverted image of the retina formed by an auxiliary lens held in the left hand. The auxiliary lens is also used to focus the illumination beam onto the patient's pupil. Because the alignment is critical and the illumination provided by hand-held instruments is rather low, indirect ophthalmoscopy never became very widely used until a need was created by the requirement for careful wide-field retinal viewing in the diagnosis and treatment of retinal detachment.

The need was met by mounting the illumination system on the examiner's head, providing a binocular viewing system, and having the patient

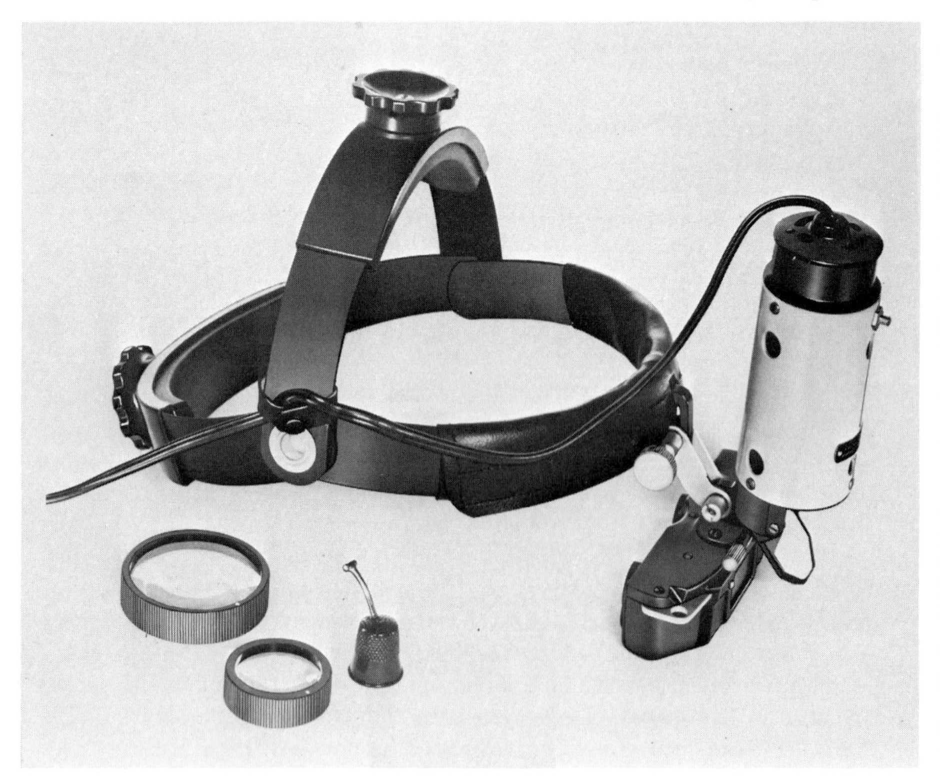

FIG. 2. A head-mounted binocular, indirect ophthalmoscope. One of the auxiliary lenses is hand-held close to the patient's eye. The thimble depressor is used to indent portions of the eye to enable the doctor to see parts of the retina close to the anterior portion of the eye. (Photograph courtesy of American Optical Co.)

recline, thereby making it possible to hold the eye very still. The auxiliary lens, which now often has one aspheric surface, is still hand-held, but the examiner can rest his hand on the patient's cheek or forehead. One such unit is shown in Fig. 2. The fact that the image is inverted is somewhat disturbing, but the examiner, with some experience, is soon able to make the conversion from one instrument to the other—as does the astronomer and the microscopist.

B. Fundus Camera

Photography of the retina is often useful, and the optics of indirect ophthalmoscopy are used in the fundus camera, the camera and film replacing the doctor's eye. Figure 3 shows the optics of a modern fundus

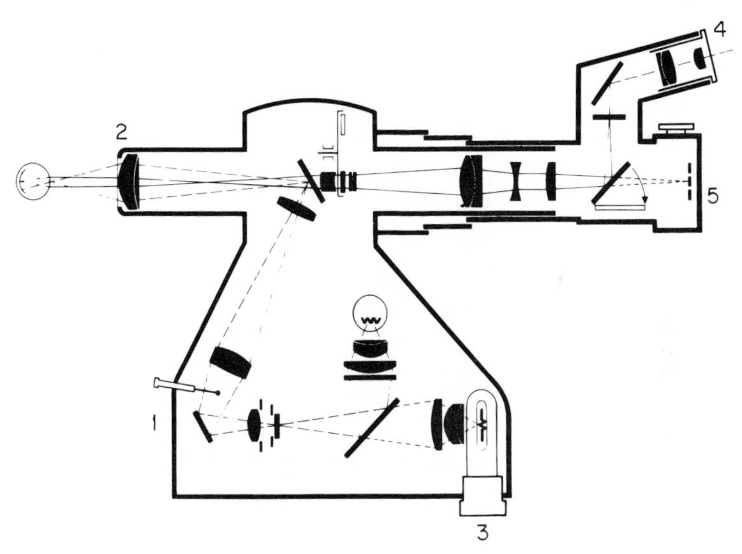

Fig. 3. Diagrammatic sketch of a modern fundus camera. (Courtesy Carl Zeiss, Inc., New York.)

camera which uses an incandescent source for the viewing system and an electronic flash source for the photographic system. Used as a fixation target (1) is a small metal ball on a moveable lever positioned in the beam path of the camera. The patient sees it as a tiny point of light in the instrument's objective lens (2). An electronic flash (3) produces light of high luminous density for photographing even the darkest fundus. A double-filament bulb provides light for observation and, in effect, converts the instrument into a high-quality ophthalmoscope. The focusing eyepiece (4) is located directly above the camera (5).

In all fundus cameras, the light source is imaged in the patient's pupil (a drug is needed to dilate the pupil because of the intense illumination), providing maximum retinal illuminance. The projection lens also serves as the objective for the imaging system. Field coverage is limited, but different areas of the retina can be covered by having the patient fixate in various directions.

Ease of positioning the instrument is essential, since patients tire easily when subjected to prolonged examination. The X–Y motion of the base

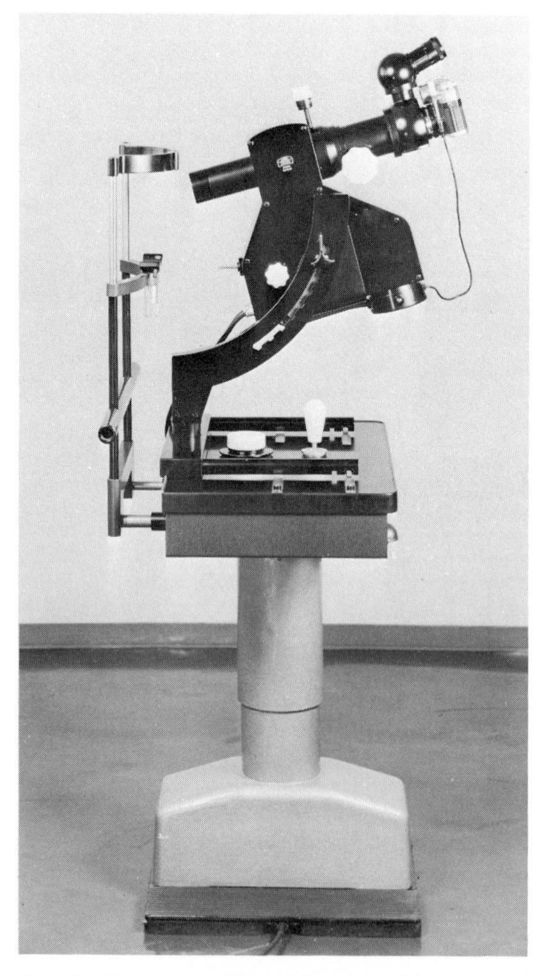

FIG. 4. A modern fundus camera. The patient is positioned by means of the head-and-chin rest shown at the left. The joystick provides smooth X–Y adjustment. The knurled knob, to the left of the joystick, provides Z adjustment. (Photograph courtesy of Carl Zeiss, Inc., New York.)

table is provided by means of a joystick arrangement (X is transverse and Y longitudinal) which makes possible smooth, fine adjustment (see Fig. 4). Adjustment in the Z direction (vertical) is accomplished by means of a knurled knob. The patient is held in a fixed position by means of an adjustable head-and-chin rest.

It is necessary to avoid having reflected light from the patient's cornea returned into the instrument, since this will seriously degrade the contrast of the image. For this reason, the light is generally allowed to enter through one side of the pupil, the reflected light then being so steeply inclined as to miss the aperture of the instrument.

C. Slit Lamp

Whereas the ophthalmoscope is used primarily to examine the retina, the slit lamp is designed to make possible microscopic examination of the anterior structures of the eye—cornea, aqueous humor, iris, lens, and the anterior portions of the vitreous. Attachments make retinal examination possible, but this is by no means the purpose of the slit lamp.

The slit lamp consists of two main optical components—the illumination system and a stereoscopic microscope viewing system. A modern version of the slit lamp is shown in Fig. 5. The two systems are mounted on a common vertical axis, which passes through the slit image formed by the illumination system and the focal plane of the stereomicroscope. This makes it possible to change the angle of illumination and the angle of viewing for maximum visibility of structures.

The illumination system, starting with the lamp housing above, is a conventional projection system, imaging a slit in the focal plane of the stereomicroscope. The slit image, normally vertical, is projected onto the optical structures of the eye. Since these are cellular structures, light will be scattered, and the scattered-light sections are viewed through the stereomicroscope. Since the slit image and the stereomicroscope are parfocal, the sections of the eye illuminated by the slit illumination will always be in focus.

As in the fundus camera, ease of positioning is important, and this is again provided by joystick X–Y positioning and simple Z positioning by means of a knurled knob seen just in front of the joystick. The adjustable head-and-chin rest make it possible to hold the patient steady during the examination. The obliquely positioned tubular assembly between the illumination system and the microscope is an adjustable fixation device. Slit controls permit its width and length to be altered as well as rotated. Various filters can also be brought into the beam.

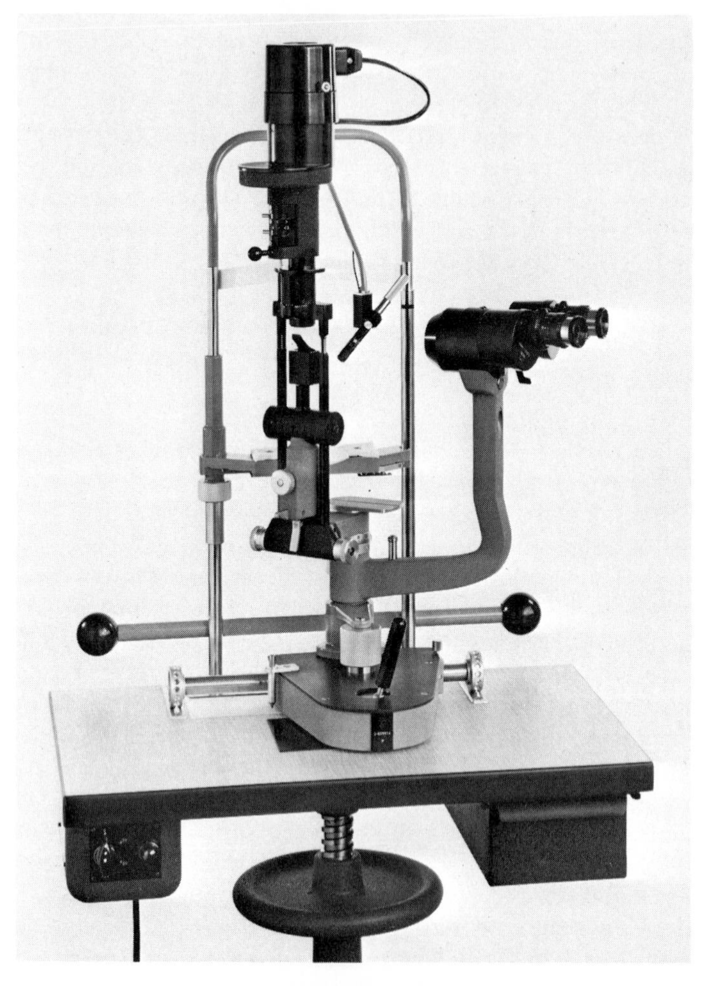

FIG. 5. A modern slit lamp. The table-mounted instrument is seen here from the doctor's side. The common vertical rotational axis is just beyond the knurled elevation knob. The joystick allows accurate *X–Y* adjustment. The two black spheres right and left are hand grips for the patient. (Photograph courtesy of Haag-Streit, Bern.)

Two sets of microscope objectives are usually provided, allowing a selection of low and high magnification, approximately 1.2 × to 2.5 ×, a 10 × eyepiece being used. Some models provide for additional powers, and, most recently, zoom microscopes have been introduced. Usually, tissues are examined at approximately 15 × total power.

III. DETERMINATION OF OPTICAL CONSTANTS OF THE EYE

A. Retinoscope

Retinoscopy is a form of Foucault knife-edge test used to determine the point in space conjugate to the retina. The principle was discovered by accident around the turn of the century by a user of an ophthalmoscope. As he approached a patient's eye, he noted the movement of lighted and dark areas in the subject's pupil. Retinoscopy has developed into quite an art, since the play of light and shadow are used to determine additional features of the eye other than the point conjugate to the retina. It is widely used by both ophthalmologists and optometrists to objectively determine the refractive state of the eye. The prescription is written only after certain subjective tests have been made, but it does offer the only practical clinical objective method. There have been objective optometers used clinically, but the retinoscope is by far and away the most widely used.

The retinoscope is a hand-held instrument utilizing a 2.5-V lamp powered either by batteries or a transformer. A battery version is shown in Fig. 6. This is a patient's view of the instrument. The bulb is mounted in

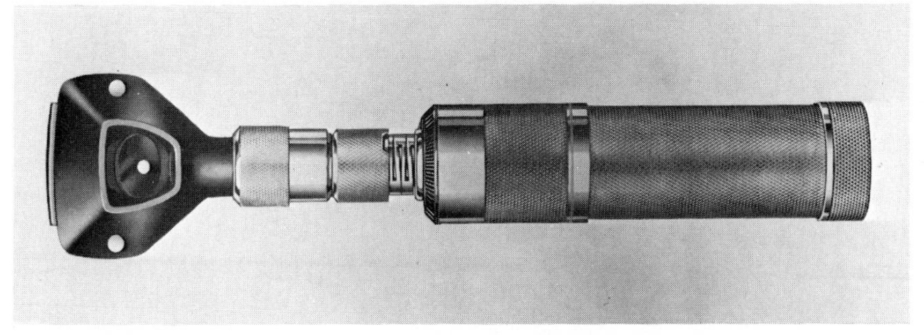

Fig. 6. A battery-powered retinoscope as seen by the patient. (Photograph courtesy of Welch Allyn.)

the neck of the instrument, base down. The filament is located inside the focal point of a converging lens mounted just below the partially reflecting diagonal mirror which directs the diverging beam towards the patient's eye. The patient's pupil is viewed by the doctor through the peephole seen in this view behind the diagonal mirror. The patient's accommodation must be relaxed.

The diverging rays enter the subject's pupil and are converged to a small patch of light on the retina. The retinoscope is rotated about an axis passing through the point where the folded optical axis touches the diagonal mirror. As the beam is "brushed" across the patient's eye, the spot of light on his retina moves in the same direction as the diverging beam. The reflected light emerges from the patient's pupil and a portion of it will enter the doctor's pupil through the peephole. The doctor's pupillary margin or the peephole, whichever is smaller, acts as the "knife-edge." If the patient's retinal conjugate lies in front of the "knife-edge," the pupil reflex (as it is called) will be seen to move opposite the direction of the "brushed" beam; if the pupil reflex is seen to move in the same direction, this denotes that the retinal conjugate lies behind the knife-edge. The former is referred to as "against" motion, the latter as "with" motion. If the knife-edge (doctor's pupil or peephole) is conjugate to the patient's retina, the movement is said to be stopped or "neutral."

If there is reflex movement, i.e., no neutrality, lenses of appropriate power are introduced to cause the rays to converge at the "knife-edge." These lenses are picked by hand from a trial set or positioned by use of a refractor (see Fig. 7). Since the doctor works at arm's length, he ends up with the retinal conjugate at about 50 cm from the patient's spectacle

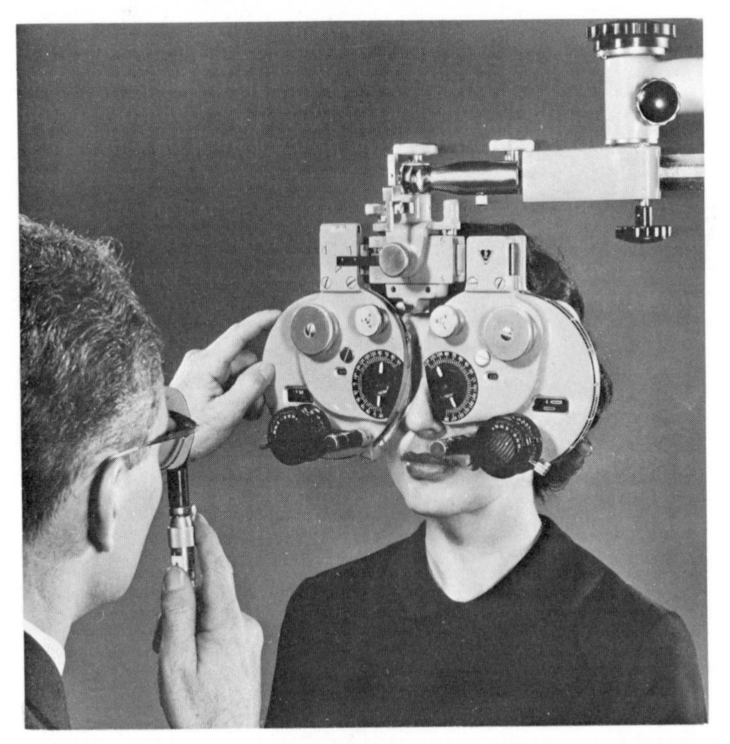

FIG. 7. A retinoscope in use in conjunction with a refractor. The doctor is changing the lens power by means of a disk controlled by his left index finger. (Photograph courtesy of Bausch and Lomb, Inc.)

plane. Since a patient's refractive state is defined in terms of infinity, 2 diopters have to be subtracted from the test-lens finding. The 50 cm is referred to as the "working distance," the 2-diopter lens as the "working lens."

The procedure described above will give the result for one meridian of the eye (the meridian parallel to the direction of motion of the "brushing" beam). If the eye has a spherical refractive error, this will end the test. If the eye has astigmatism (determined by trial "brushing" in several directions or recognized by the experienced observer), the refractive state of the two principal meridians will have to be determined separately.

So-called "spot" retinoscopes use a light source having a very compact filament. "Streak" retinoscopes use bulbs having a linear filament. In the streak retinoscope, the image of the filament must be kept normal to the direction of "brushing." This is accomplished by mounting the bulb in a rotating sleeve. The spot retinoscope utilizes a diverging beam, whereas the steak retinoscope sometimes uses a converging beam, and hence the rotating sleeve is also made to move up and down.

The circular spots seen on each side of the mirror are used as patient's fixation points during so-called dynamic retinoscopy. During static retinoscopy, a distant fixation point is used. Dynamic and static refer to the accommodative state of the patient's eye. During the determination of the refractive state, accommodation should be relaxed, and hence static retinoscopy is used.

B. Trial Sets and Refractors

A trial set consists of a trial frame, a selection of spherical and cylindrical lenses, a selection of ophthalmic prisms, a selection of accessories, and a case to store all the above.

The convex and concave spherical lenses are stored in matched pairs at the extreme right and left sides of the trial set, and positive and negative plano cylinders are stored in matched pairs in the upper central portion. Since the dioptric range required for cylinders is smaller, fewer paired lenses are needed. The accessories are stored in the small compartment to the left of the trial frame in front.

The trial frame has numerous adjustments which make it possible to adapt it to a variety of subjects' heads. An example is shown in Fig. 8, adjusted for a child's refraction. There are clips to hold three to four lenses or prisms or accessories before each eye. The inner portion of the eye ring can be rotated making it possible to orient cylindrical lenses, prisms, and accessories. The lenses supplied in trial sets were formerly made equiconvex and equiconcave, but modern ones are plano, and some are even made meniscus in an attempt to minimize marginal errors.

FIG. 8. A trail frame adjusted to a willing young man. (Photograph courtesy of Bausch and Lomb, Inc.)

The trial set is used during retinoscopy and throughout the subjective examination. During the examination known as phorometry (determination of muscle balance), various accessories in the trial set are used, i.e., prisms, Maddox rods. The Maddox rod is, in fact, a number of small, stacked cylinders used to cause a spot of light to be seen as a streak of light perpendicular to the axes of the cylinders. The "lens" is pressed from white or red glass.

A refractor is, in fact, the optomechanical equivalent of the trial set. Fortunately for the patient, no one has tried to mount one on his nose and ears. They are generally suspended from a stand or wall bracket. The spherical and cylindrical lenses are mounted in circular disks which are stacked inside a housing. The two housings can be positioned before the patient's eyes. The spherical range is generally ± 25.00 diopters in 0.25-diopter steps, the cylindrical power (+ or − by doctor selection) to about 7 diopters in 0.25-diopter steps (see Fig. 7).

On the front of the housings are a number of accessories mounted on arms to be swung into place before the patient's eyes. A variable-power prism known as a Risley prism is one of these accessories. The prism power is produced by two counter-rotating prisms mounted in tandem. Prism power can be varied continuously from 0 to 30 prism diopters. A second accessory is a pair of Maddox rods, mounted so that they can be rotated about the line of sight of the patient. A third accessory is a pair of Jackson Cross Cylinders. The Jackson Cross Cylinder is a spherocylinder lens whose principal meridians have equal powers of opposite signs, e.g., ± 0.25 diopters. This accessory is used to refine the power and axis position of the cylindrical component of the patient's refractive correction. These are also mounted to be rotated about the patient's line of sight.

A box of "plug-in" accessories is provided which permit extension of the cylinder power, changes of the cross cylinder power, and the addition of red-green filters used for various tests.

An "upside-down optical bench," known as a near-point rod, can be attached to the refractor. On this rod is mounted a holder for reading material to be used for making measurements of visual performance at near distances. Among other tests, it is used to determine the dioptric power of the bifocal and/or trifocal segment.

C. Optometers

An optometer is another instrument for locating the conjugate to the patient's retina. It consists of a lens with its posterior focus at the patient's spectacle plane. A test object is mounted on a small optical bench beyond the optometer lens. The test object can be moved from contact with the optometer lens to some finite distance beyond. In some research versions, a point source is used, and this version has been given the name of stigmatoscope.

By applying Newton's formula relating object and image distances to the focal points, it can be shown that the dioptric scale is linear, with optical infinity corresponding to the anterior focal point of the optometer lens.

The retinal conjugate can be located by moving the target until it appears sharply imaged to the patient. It is important that he maintain his accommodation relaxed. Because one cannot be sure of this, the instrument has never been used clinically, but it is often used in research studies.

D. Ophthalmometers

Since the cornea contributes the largest portion of the refractive power of the eye and it is the surface upon which contact lenses rest, its radii of curvature are of importance. The ophthalmometer, invented by Helmholtz, utilizes the measurement of reflected image size as the means of determining the radii of curvature of the cornea.

The human eye is constantly in motion, making it impossible to measure the reflected image size by a scale method. Helmholtz borrowed from the astronomers the idea of variable doubling to measure the image size. Doubling is accomplished in a variety of ways—calcite prisms, biprisms, and longitudinally moving prisms.

Figure 9 shows a cross-sectional view of an instrument using the movable prisms. The patient is positioned by means of the chin rest *C* and the forehead rest *H*. A movable occluder *T* is used to shield the eye not being measured. The patient's head is adjusted vertically by means of a knob *A*. The instrument can be rotated about the vertical axis *P*, adjusted vertically by means of knob *E*. Both of these adjustments are locked by means of knob *D*. Knob *F* is used to translate the instrument longitudinally for focusing.

The target is mounted on the plano surface of the condensing lens *R*. It consists of a thin metal plate which has etched in it a focusing circle 70 mm in diameter with four index marks just outside its outer diameter. These consist of plus symbols on the exterior of the horizontal diameter and minus symbols on the exterior of the vertical diameter. It is these symbols which are brought into coincidence by the doubling prisms *V*.

The patient's cornea is illuminated by means of the bulb *I*, the mirror *Z*, and the condensing lens *R*. An average cornea has a radius of about 8 mm, hence the reflected image is located approximately 4 mm behind the cornea.

The corneal image is reimaged in the focal plane of the eyepiece E.P. by means of the paired achromats L_1 and L_2. In the small mirror *Q* the patient sees a reflection of his own eye, serving as the fixation point. At the left of L_1 is an aperture plate containing four circular apertures. Two apertures, vertically arranged, serve as focusing apertures. The sum of their areas is equal to the area of each of two horizontally arranged apertures. The rays passing through these apertures pass, respectively, through two achromatic prism assemblies, one of which has its base-apex line oriented vertically, the other horizontally. These prisms provide the two secondary images used for measuring the size of the corneal image, all three images being equally bright. Two external drums, *M* and *N*, control the longitudinal position of the doubling prisms *via* a rack-and-pinion

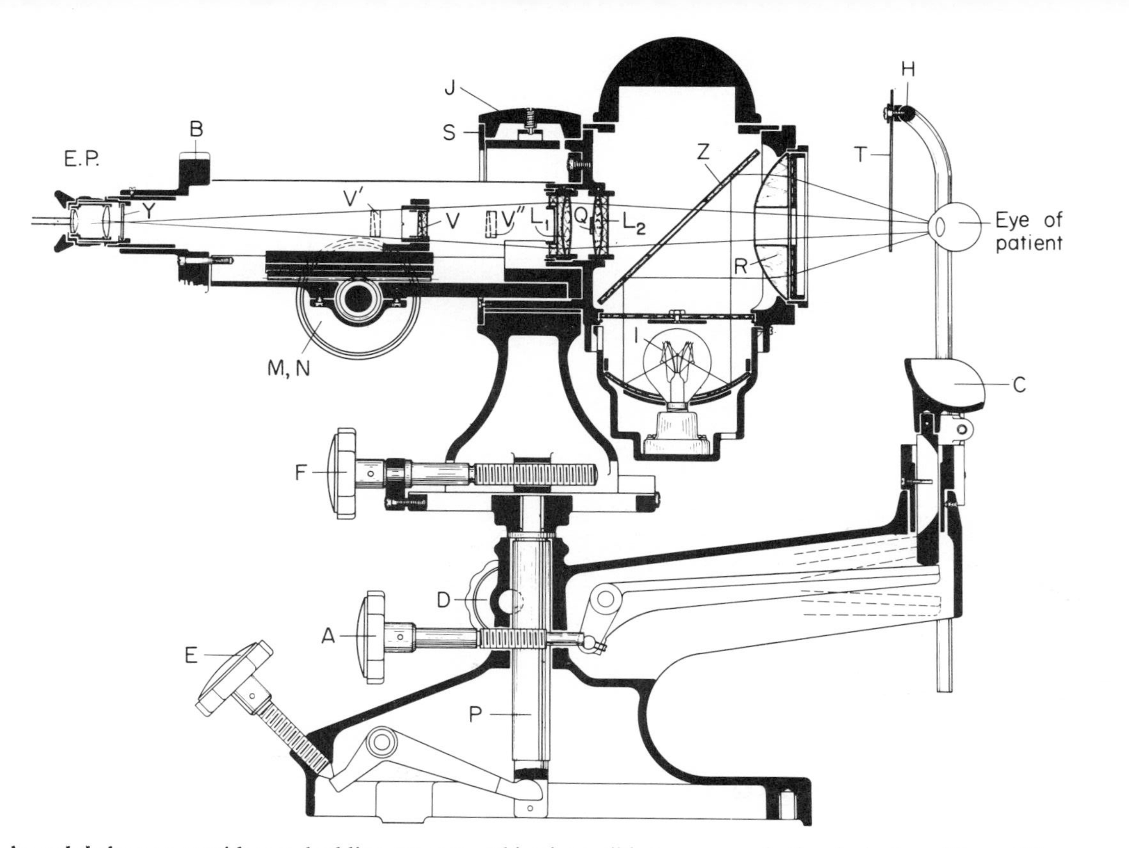

FIG. 9. An ophthalmometer with two doubling systems making it possible to measure both principal meridians of the cornea at a single setting. (Courtesy of Bausch and Lomb, Inc.)

arrangement. The scale on these drums is given in diopters of refraction of the cornea.

The four apertures act as the entrance pupils of the viewing system; hence there are four exit pupils at the eye point. These are small enough to fit into the pupil of the doctor's eye.

The entire optical assembly is rotatable about its axis, making it possible to align the doubling system with the principal meridians of the cornea. All other ophthalmometers have only one doubling system, so that it is necessary to make two settings.

E. Keratoscopes

A qualitative instrument of superb simplicity is the keratoscope, some times called Placido's disk. It consists of a metal disk carrying a series of black and white concentric circles, held before the patient's eye. The doctor looks through a lens mounted at the center of the disk at the reflected image of the circles. The image is interpreted as if it were a contour map of the corneal surface. Perfect circles evenly spaced, represent the surface of a sphere. Regular ellipses tell the doctor that the surface is a torus. Irregular images reveal irregular astigmatism.

The keratoscope is used to quickly diagnose irregular astigmatism, such as that arising from a condition known as keratoconus. This is a pathological condition wherein the center of the corneal tissue thins out and the corneal apex protrudes as a cone.

IV. TESTING VISUAL FUNCTIONS

A. Visual Acuity

Clinical testing of the acuity of the eye is done using alphanumeric characters printed on a chart, either front or rear illuminated, or by means of a projected matrix of characters. Each character is constructed in a 5×5 block, the stroke width of the characters being one unit. Acuity is expressed as a "Snellen fraction," the numerator expressing the testing distance, the denominator expressing the distance at which the "normal" eye could resolve the character. In a 20/20 Snellen character, the block subtends $5'$ of arc, the stroke width subtending $1'$ of arc. Hence 20/20 vision implies the ability to recognise a character subtending $5'$ of arc; 20/10, $2\frac{1}{2}'$ of arc; 20/40, $10'$ of arc; etc.

Acuity projectors are classical projection systems with relatively long-focal-length objectives and large F-numbers. The magnification between slide and screen is of the order of $20\times$. The slides are glass mounted in a metal holder which moves vertically in the slide plane by means of a rack-and-pinion arrangement. A horizontal aperture plate is also provided, making it possible to expose the entire matrix of characters, a vertical row of characters, or a horizontal row of characters. In addition to alphanumeric characters, other material is used, such as a radial pattern for testing astigmatism, and special symbols for testing children, illiterates, and for special vocational applications.

B. Visual Field Studies

The clinical science of testing the completeness of the visual field is called scotometry (see Vol. II, Chapter 1 of this series). One instrument for performing scotometry is known as a "perimeter," which usually consists of an arc of 33-cm radius, mounted at its center and capable of rotating about a horizontal axis to cover various radial orientations. Clinically, the

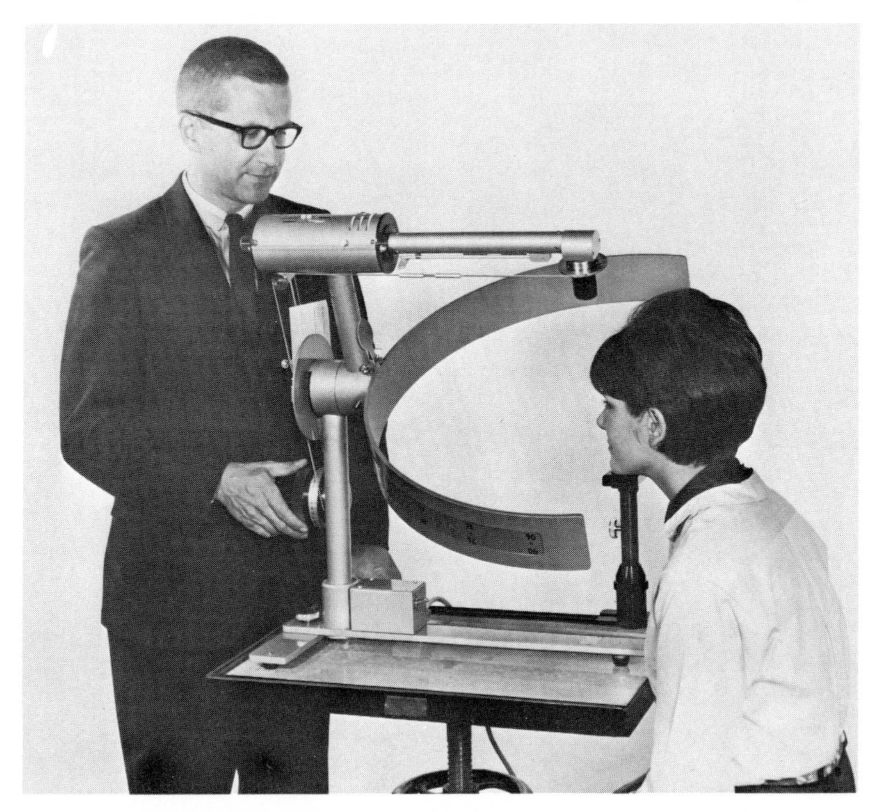

FIG. 10. A projection perimeter used for testing visual fields. The recording system is synchronized *via* the cable shown, allowing the doctor to record the limits of the field on the chart just above his right hand. (Photograph courtesy of Bausch and Lomb, Inc.)

targets have been hand-held, being moved from beyond the area of recognition into the "edge" of the field where the target is first seen by the patient. However, in recent years, projection perimeters have reached the market, wherein the target is projected on the arc (Fig. 10), and, most recently, a hemispherical perimeter has been introduced (see Fig. 11). Recording is done on polar coordinates using the fixation point as the center of the coordinate system.

FIG. 11. A hemispherical perimeter showing the doctor's side of the instrument. The pantographic linkage synchronizes the stylus with the projection system. The telescope permits the doctor to monitor the patient's fixation. (Photograph courtesy of Haag-Streit, Bern.)

The central visual field (25° outward from the fixation point) is especially sensitive to changes in many pathological conditions. Glaucoma diagnosis is one of the most important. Central field testing is done using a tangent screen. Here, the patient is located 1 m in front of a plane screen. Test targets are introduced by hand at the end of a slender rod. Recently, a projection tangent screen has been introduced commercially.

C. Color Vision

Color vision is most commonly tested using so-called pseudoisochromatic plates. These are cards covered with dots of various colors, in which recognizable patterns of dots of a particular color are visible to a person with normal color vision but are invisible in the presence of color anomalies.

An instrument using a variable mixture of red and green to match a given yellow was introduced by Nagel. The three colors are chosen to lie on the straight-line portion of the spectrum locus of the CIE Color Diagram, making it possible for a normal individual to obtain a perfect match. The proportions of red and green are varied by the movable plate shown diagrammatically in Fig. 12, the yellow intensity being independently

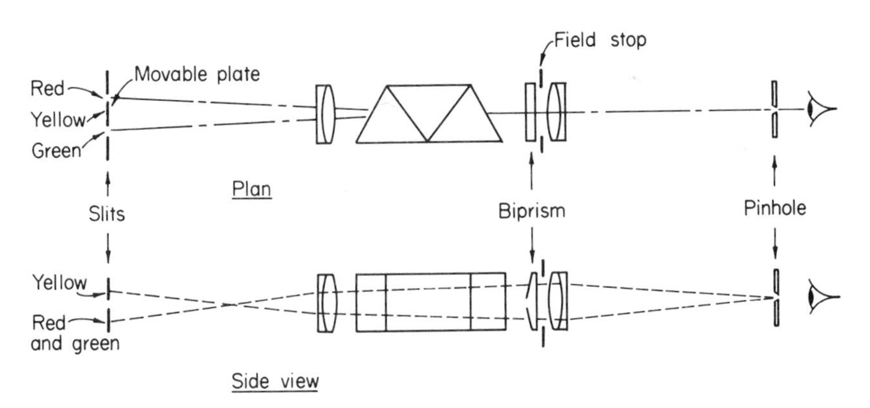

FIG. 12. Diagram of a Nagel anomaloscope. The eye sees a divided circular field, the upper half being formed by a mikture of red and green, the lower half being yellow. The proportions of red and green can be adjusted by the movable plate, and the intensity of the yellow by variable slit width.

variable by adjusting the "yellow" slit width. The instrument is known as the Nagel anomaloscope. Four types of color anomalies can be distinguished with this instrument from the normal, i.e., protanomaly, deuteranomaly, protanopia, and deuteranopia (see Vol. II, Chapter 1 of this series). Because of their high cost, these instruments have not been used clinically, the pseudoisochromatic plates being more popular.

V. TREATMENT

A. Orthoptics

When the binocular muscle function is disturbed, the patient is said to have a tropia. The two eyes are not able to fixate a single fixation point simultaneously. The measurement of deviation is performed by means of a modified mirror stereoscope. The arms of the stereoscope are made to coincide with the lines of sight of the two eyes, and the deviation can be measured on scales provided on the instrument. A representative instrument is shown in Fig. 13. The optics are simple. The target is set in the

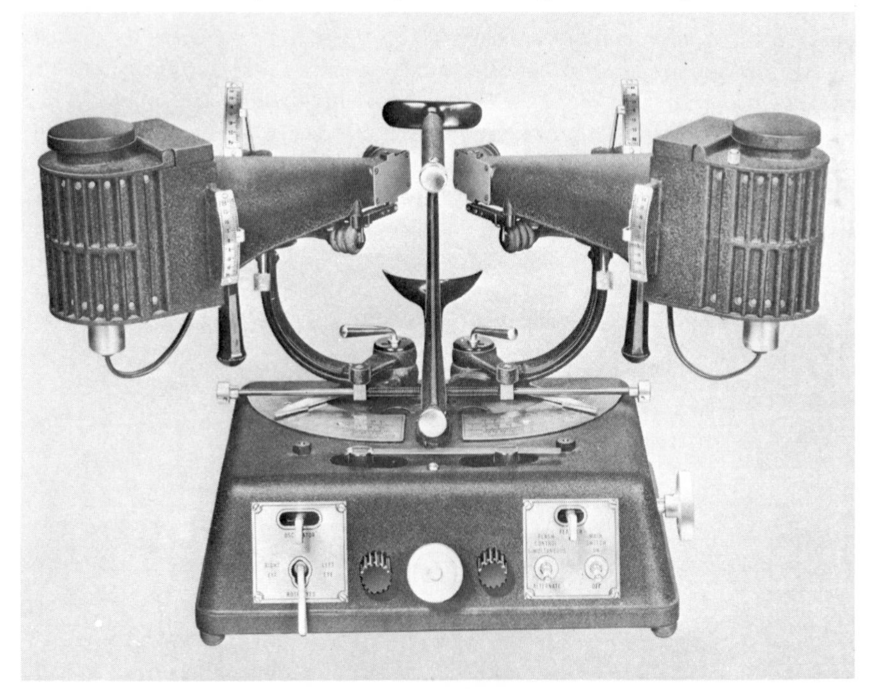

Fig. 13. A mirror stereoscope designed to diagnose and treat binocular muscle imbalances. (Photograph courtesy of American Optical Co.)

focal plane of a single lens, and hence the image is seen at infinity. Other lenses can be used to induce various amounts of accommodation. By alternating the illumination of the two targets, the patient will report that the two images are seen in the same visual direction, i.e.; there is no visual parallax.

These same instruments can be used in the treatment of the tropia by training the eye muscles to keep the images fused as the arms of the stereoscope are brought into their normal geometrical relationship. All tropias cannot be "cured" by this procedure, and surgery is often needed to correct muscle anomalies, which can then be stabilized by training.

B. Retinal Photocoagulators

The retina is a very thin tissue which has two zones of close adherence to the underlying tissue, the choroid. One zone surrounds the optic nerve, and the other is at the outer margin near the ciliary body. In certain conditions, the retina develops tears and becomes detached from the choroid, and the treatment is to literally "spot weld" the retina back into place. This is done by inducing injury to the retina and the choroid in selected areas, and the healing process unites the two tissues.

At one time, radio-frequency currents were used to induce these injuries. A small probe was surgically introduced behind the eye and contact made at the point requiring treatment. Another method of inducing injury is to "burn" the retina with an intense beam of light. Originally, this was done using sunlight. Next, a Xenon light source was used, and, most recently, ruby pulsed lasers have been used (Fig. 14).

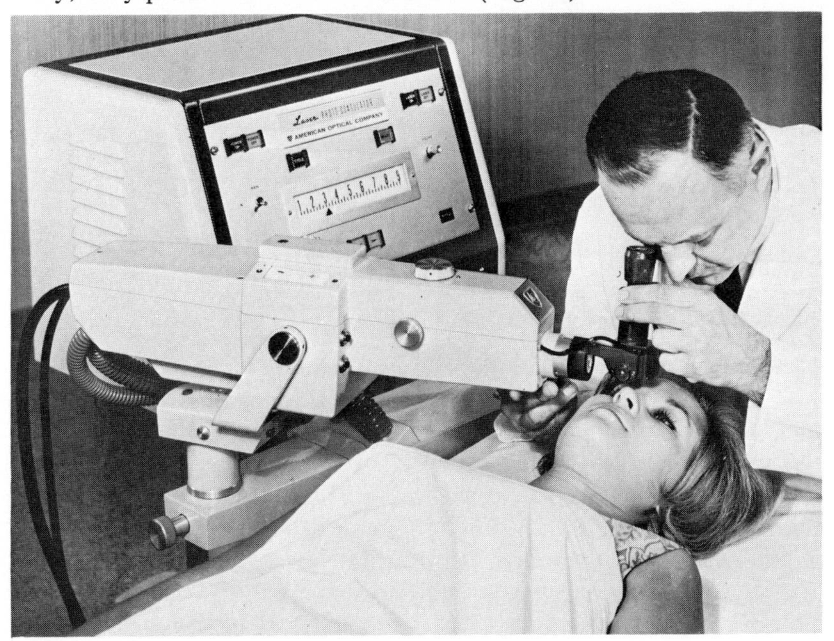

Fig. 14. A ruby laser photocoagulator for retinal detachment work. (Photograph courtesy of American Optical Co.)

All of these devices are really ophthalmoscopes with dual light sources. A low-power source is provided which the doctor uses to locate the lesion and aim the beam. When the lesion is centered in the ophthalmoscope beam, the high-power source is discharged. The lower-power source is again used to evaluate the treatment.

VI. AUXILIARY INSTRUMENTS

A. Pupillometers and Interpupillometers

Measurement of pupil size and interpupillary distance is of importance in contact-lens and spectacle-lens fitting. In the type of instrument shown in Fig. 15, a scale is optically superimposed on the patient's eye so that the

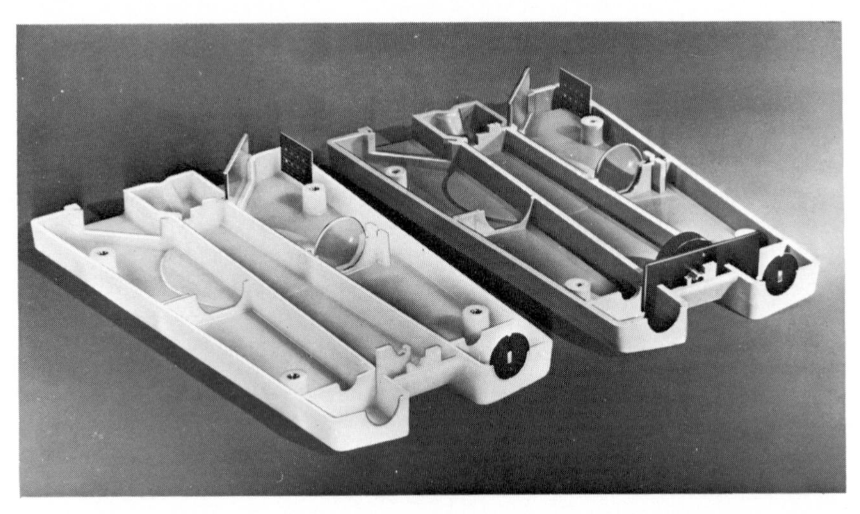

Fig. 15. A combination pupillometer and interpupillometer. (Photograph courtesy of Bausch and Lomb, Inc.)

doctor can directly compare the scale with the pupil. The two separated halves shown in Fig. 15 are identical. The patient is at the far end, and the doctor views the patient's pupil through the rectangular aperture shown. The scales, which are illuminated by light diffusing through the translucent plastic housing, are seen reflected in partially reflecting diagonal mirrors. The knurled knob shown in the right half of the instrument shifts an occluder from eye to eye. The instrument is placed on the bridge of the patient's nose, and hence the doctor can also measure the distance to the center of each pupil from the nose. Another type of interpupillary distance gauge uses a doubling principle which allows the doctor to obtain the total interpupillary distance.

B. FOCIMETERS

The measurement of the dioptric power of spectacle lenses and contact lenses is done by means of an instrument based on the telecentric system described in the discussion of optometers. The test lens (sunglasses shown in Fig. 16) is placed against a stop which lies in the anterior focal plane of

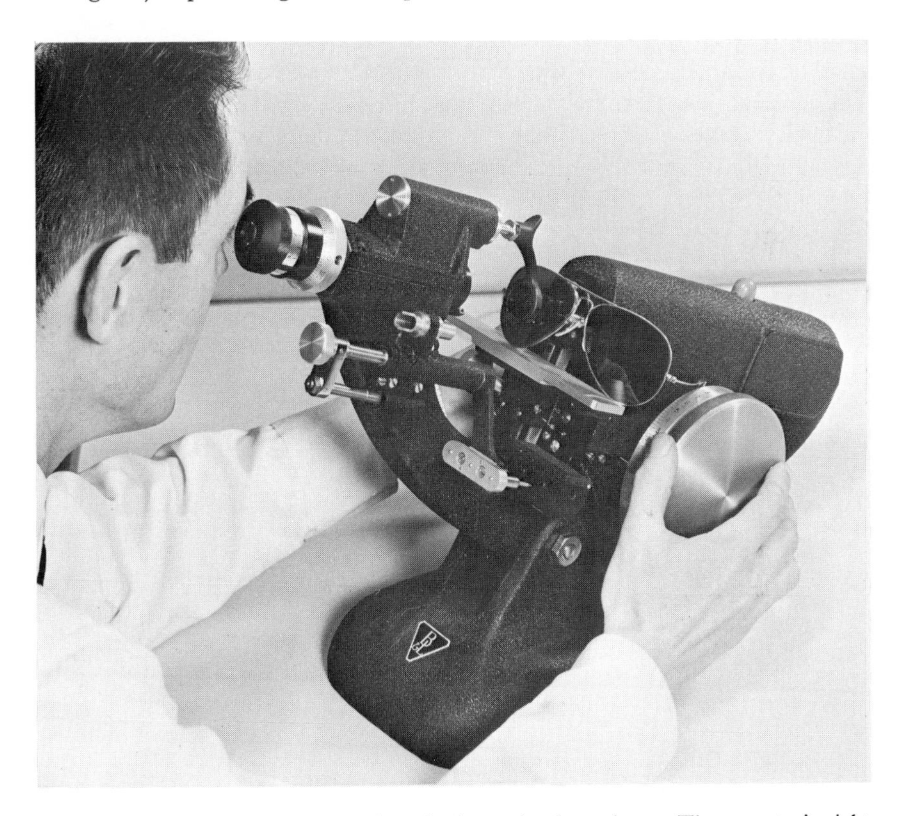

FIG. 16. A focimeter being used to check a pair of sunglasses. The operator's right hand is on the power drum. His left hand is on the axis drum on the far side of the instrument. (Photograph courtesy of Bausch and Lomb, Inc.)

the standard lens. The observer views the target image formed by the standard and test lenses as seen through the afocal telescope. The target is moved by rotating the power drum until the image is seen clearly. The vertex dioptric power is read directly from a linear scale on the drum. The usual range is from +20.00 to −20.00 diopters. Because the drum reads directly in terms of vertex diopters, the instrument has been called a vertometer.

The target is designed to allow the measurement of astigmatic (cylindrical and spherocylindrical) lenses as well as spherical lenses. The target is rotatable and the axis of the cylinder component can be determined. The reticle in the eyepiece of the viewing telescope has a cross hair which is used for centering of the lens being tested. The reticle also contains a series of concentric circles calibrated to measure prismatic power.

A marking assembly can be seen mounted on the right side of the instrument. It is in position to be pressed into the inking receptacle (the small box-shaped assembly with three holes in the side toward the operator). The assembly has three points arranged linearly. The central point marks the optical center of the lens, the two outboard points provide points marking the cylinder axis or some line related to the cylinder axis.

GENERAL REFERENCES

M. L. Berliner, "Biomicroscopy of the Eye," Vol. I. Harper (Hoeber), New York, 1943.

I. M. Borish, "Clinical Refraction," 2nd ed. Professional Press, Chicago, Illinois, 1954.

W. S. Duke-Elder, "Text Book of Ophthalmology," Vol. IV. Mosby, St. Louis, 1949.

H. H. Emsley "Visual Optics," Vol. I. Hatton Press, London, 1952.

G. H. Giles, "The Principles and Practice of Refraction," 2nd ed. Chilton Books, Philadelphia, Pennsylvania, 1965.

A. C. Hardy and F. H. Perrin, "The Principles of Optics," Chapter XX. McGraw-Hill, New York, 1932.

D. O. Harrington, "The Visual Fields," 2nd ed. Mosby, St. Louis, 1964.

K. N. Ogle, "Optics, An Introduction for Ophthalmologists." Thomas, Springfield, Illinois, 1961.

CHAPTER 11

Motion Picture Equipment

ROBERT M. CORBIN
Rochester, New York

I. GENERAL DISCUSSION

Motion pictures consist of a succession of still pictures exposed along the length of a ribbon of sensitized film. When this succession of pictures is shown upon a screen at the same rate at which it was taken, an illusion of motion at normal speed is produced. The spacing and lateral location of the exposures are maintained by the use of accurately punched perforations, and, in almost all systems, the film moves downward from a supply reel at the top to a takeup reel below the gate.

When photographing a motion picture, the camera lens is covered by a shutter while each new area or "frame" of film is moved into place, and the exposure is made while the film is stationary. Upon projection, the light is cut off by a shutter while the film is being moved, persistence of vision serving to prevent the viewer from seeing a dark screen. It was early determined that the minimum acceptable rate of presentation of individual pictures is 16 per second. This rate was used in the days of silent pictures and is still used in many silent amateur cameras. With the advent of sound, the rate was raised to 24 pictures/sec as a help to the sound quality. To prevent visible flicker on the projection screen, one or two so-called "flicker blades" are added to the projector shutter to increase the flicker rate to an acceptable level.

Early in the history of motion pictures, various widths of film were used, but by the time any great amount of motion picture production took place, the width of $1\frac{3}{8}$ in. (35 mm) was established. A picture format with a height-to-width ratio of 3 : 4 was also used. These film dimensions remained standard for many years, except for a few experimental programs.

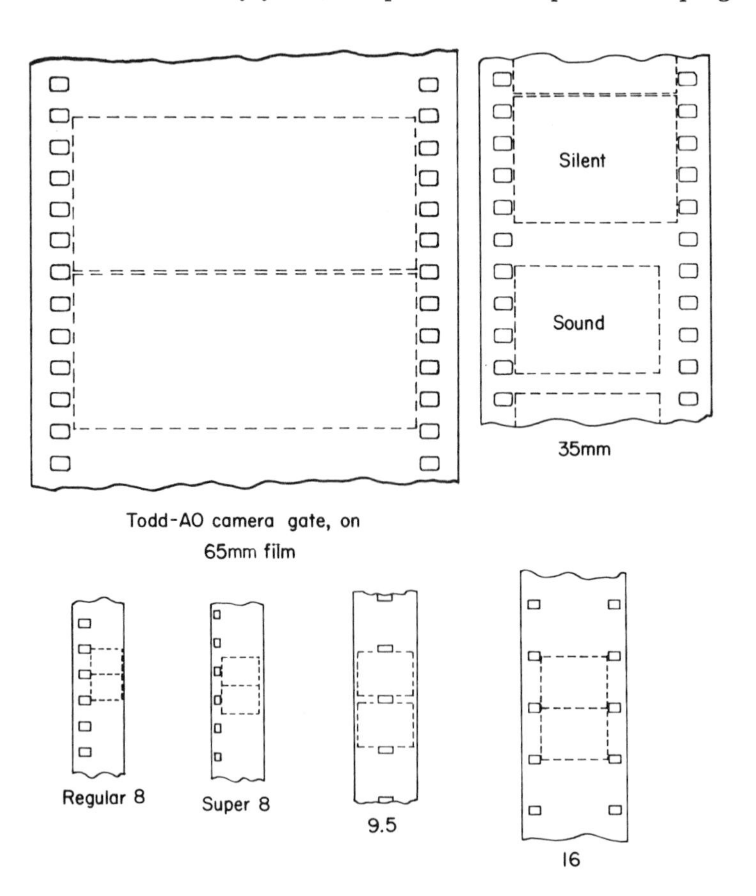

FIG. 1. Standard film and picture gate dimensions.

TABLE I

Film	Camera image area			Projector image area		
	Inches	millimeters	Standard	inches	millimeters	Standard
Normal 8	0.192 × 0.145	4.88 × 3.68	PH22-19	0.172 × 0.129	4.37 × 3.28	PH22-20
Super-8	0.226 × 0.166	5.74 × 4.22	-157	0.209 × 0.158	5.31 × 4.01	-154
16 mm	0.404 × 0.295	10.26 × 7.49	-7	0.380 × 0.284	9.65 × 7.21	-58
Sound 35 "Academy Aperture"	0.868 × 0.631	22.05 × 16.03	-59	0.825 × 0.600	20.95 × 15.25	-8
Todd AO (65 mm)	2.972 × 0.906	52.60 × 23.01	-	—	—	-
70 mm	—	—	-	1.912 × 0.870	48.56 × 22.10	-152
Silent 35 (obsolete)	0.980 × 0.735	24.89 × 18.67	-	0.906 × 0.680	23.01 × 17.27	(JSMPE May 1930)

More recently, other film widths have come into use, as well as different aspect ratios.

There were early attempts to market narrower films for amateur use, but the only successful widths were $9\frac{1}{2}$ mm, 16 mm, and, later, 8 mm. At present, the $9\frac{1}{2}$ mm film has a very restricted use, and 16 mm is employed for the most part in professional work, such as in industrial, educational, and journalistic applications. Most amateur motion pictures are now made on 8 mm film, using either the standard format or the recently introduced super-8 format, which is 50% larger in area and has different perforations.[1–3]

Table I and Fig. 1 give the dimensions as currently standardized for various ordinary film sizes. For other dimensions, such as location of the frame on the film, size and location of perforations, position and width of sound track, etc., the reader is referred to the relevant American Standards.

By far the greatest part of motion pictures for entertainment purposes are now photographed on film 35 mm wide, exposed at a rate of 24 pictures/sec. Of those not so photographed, the major part are made on 65 mm negative film, exposed sometimes at 30 pictures/sec.

A. Use of Reversal Film

The film used in 8 mm cameras, and the greatest share of the film used in 16 mm cameras, is of the reversal type.

The reason for using reversal film is that in the case of amateurs, one copy only is usually all that is required. The use of reversal film obviates the necessity for making a print, saving cost and trouble. In professional work, the use of reversal film reduces the number of printing steps, which is highly desirable from the standpoint of quality. In addition, any dirt spots on the original show as black spots on the projection screen, which are less objectionable to the viewer than white spots.

The reversal process has not been used with the 35 mm width, mainly because of the difficulty of obtaining pictures that are uniform from frame to frame and because of cost. These two factors are much less serious with the smaller films.

1. *16 mm prints* are made by one of the following procedures:

1. Direct exposure in a 16 mm camera on reversal film.

2. Printing onto positive film from a 16 mm negative exposed in a 16 mm camera.

---

[1] E. A. Edwards and J. S. Chandler, *J. Soc. Motion Picture Television Engrs.* **73**, 537–543 (1964).

[2] C. L. Graham and W. L. Stockdale, *J. Soc. Motion Picture Television Engrs.* **73**, 934–936 (1964).

[3] R. A. Colburn, *J. Soc. Motion Picture Television Engrs.* **70**, 603–606 (1961).

3. Printing from a 16 mm reversal original onto 16 mm reversal print film.

4. Direct reduction from a 35 mm negative.

5. Printing from a 16 mm duplicate negative made by reduction from a 35 mm original.

2. *8 mm prints*, regular or super-8, are very often made by:

1. Direct exposure in the camera; this is the method used when extra prints are not required.

2. Contact printing from a special 8 mm negative produced from a 16 mm or 35 mm original in such a way as to get the ultimate in sharpness.

3. Direct reduction from a 16 mm original, reversal or negative (either duplicate or original).

4. Contact printing onto a special reversal print stock.

B. Frame Rates

For the most part, 16 mm film is exposed at a rate of 24 pictures/sec, although occasionally, a rate of 16 pictures/sec is used. Eight millimeter film is exposed at either 16 (18)[4] or 24 frames/sec.

For special purposes, various slower and faster frame rates are employed. For instance, pictures are sometimes made at very slow rates, such as one per second or one per hour or even longer. When the finished picture is projected at normal speeds the action is greatly speeded up. An example is the photography of the growth of a plant; by the above process, the plant can be made to appear to grow and bloom in a few seconds.

Pictures made for the purpose of studying athletic events are often photographed at a slightly higher than normal speed. When projected at normal speed, the action can be better followed. Photography of small-scale models used in making entertainment pictures is done at higher than normal speed, the increase in speed being proportional to the reduction in size of the model. If this is not done, the final action appears much too fast.

Then there are high-speed motion pictures made at rates from several hundred to many thousand pictures per second. Such photography is done to study motion that takes place too rapidly for the eye to follow.[5] High-speed motion picture photography is defined as exposure at a rate in excess of 128 frames/sec.

---

[4] Eighteen frames/sec is often used in amateur cameras as a matter of convenience in design.

[5] T. E. Holland, High-speed photography, This series, Vol. IV, Chapter 6.

C. Wide-Film Systems

Greatly enhanced definition and greatly increased screen illumination result from an increase in the frame size used in motion picture production. Some experimental applications of 65 mm and 70 mm films were made in the 1930's and 1940's, but extended use did not commence until the introduction of the Todd-AO system in October 1955[6]. The camera gate was 0.906×2.072 in. on 65 mm film, which could be projected directly from contact prints on 70 mm film (the additional width allowed for several sound tracks), or from reduced prints on 35 mm film.

Another system known as "Vistavision" employs two ordinary frames of 35 mm film, but the film is run sideways instead of downwards. The picture area is 0.911×1.485 in., which is either projected full-size or reduction-printed on 35 mm film run vertically.

Very complete analyses of the gains and losses resulting from the use of a larger-than-normal camera gate have been given by Hill[7] and by Wolfe and Perrin.[8]

D. Anamorphic Compression

The French scientist Henri Chrétien demonstrated in 1929[9] a system for motion pictures in which a very wide picture was compressed laterally in the camera by means of a cylindrical lens arrangement, and decompressed or expanded laterally by a similar cylindrical lens on projection. Thus an original scene with an aspect ratio of 2.6 to 1 could be recorded in a normal motion picture camera equipped with a 2 : 1 compression lens. This arrangement was first used commercially by Twentieth-Century Fox in 1954 in their "CinemaScope" system.

For various reasons, the use of a 2 : 1 compression in the original camera is undesirable. Furthermore, for the best possible definition, the use of a camera film wider than 35 mm is to be preferred. These two features have been combined in the MGM "Camera 65" process.[6] Here, a partially compressed (1.3 : 1) image is recorded on 65 mm film, the height of the frame being five perforations instead of the usual four. A contact print on 65 mm or 70 mm can be projected with a 1.3 : 1 anamorphic expansion. Alternatively, reduction prints can be made on 35 mm film with a further 1.5 : 1 compression, for projection using ordinary Cinema-

---

[6] Wide-screen motion picture systems. Soc. of Motion Picture and Television Engrs., New York, (1959).

[7] A. J. Hill, *J. Opt. Soc. Am.* **46**, 691–698 (1956).

[8] R. N. Wolfe and F. H. Perrin, *J. Soc. Motion Picture Television Engrs.* **65**, 37–42 (1956).

[9] H. Chrétien, *J. Opt. Soc. Am.* **18**, 174–175 (1929).

Scope 2 : 1 lenses. By the use of prisms giving variable degrees of compression in camera and printer, a number of different alternative arrangements are possible under the general name of "Superscope."

Yet another arrangement has recently been introduced by Technicolor which is known as "Techniscope." In this system the original camera image covers the full film width but is only two sprocket holes high. This is expanded vertically in the printer to give a "CinemaScope" print which can be shown by use of an ordinary 2 : 1 horizontal expansion. In the "Technirama" system, partially compressed negatives are made on horizontally moving double-frame film, the remainder of the compression being performed in the printer.

There have been a number of other systems introduced for the purpose of producing a large picture, but any of those in use at present involve principles similar to those described above.

E. American Standards for Motion Picture Photography

A number of different types of film, cut and perforated to various dimensions, are required for use in the processes described above. The dimensional characteristics are quite well covered by American Standards. There are also American Standards for 16 mm and 8 mm widths of film as well as for camera and projector apertures, sound track placement, etc. These standards are published in the *Journal of the Society of Motion-Picture and Television Engineers*, or they can be obtained from the American Standards Association.

These standards do not cover the physical behavior of film, such as thickness, stiffness, curl, and frictional characteristics,[10–12] nor do they cover the effect of shrinkage. Variation in the physical behavior of film may cause serious trouble in motion picture equipment.

II. MOTION PICTURE CAMERAS

A. Professional

All motion-picture cameras used for making entertainment, educational, industrial, documentary, or other pictures operate on an intermittent principle.[13] The film is pulled into place while the light from the lens is

[10] R. H. Talbot, *J. Soc. Motion Picture Television Engrs.* **45**, 209–217 (1945).

[11] C. R. Fordyce, J. M. Calhoun, and E. E. Moyer, *J. Soc. Motion Picture Television Engrs.* **64**, 62–66 (1955).

[12] A. J. Miller and A. C. Robertson, *J. Soc. Motion Picture Television Engrs.* **74**, 3–11 (1965).

[13] H. Weise, Die kinematographische Kamera, *in* "Die wissenschaftlichen und angewandten Photographie" (K. Michel, ed.), Vol. III, Springer, Wien, 1955.

cut off by a shutter of some type, and is held at rest while the shutter is open, allowing the exposure to be made for that picture or frame.

The film is pulled down by a claw mechanism, a sprocket, or a beater device.[14,15] Nearly all intermittent cameras used today have a claw pull-down,[16,17] and many professional cameras are equipped with a set of registration pins to hold the film in a definite position in relation to the perforations during exposure. The pins are, of course, withdrawn while the film is in motion.

A large number of nonintermittent camera mechanisms[18] have been invented, some of which have found use in high-speed photography, where rapid movement is more important than perfect steadiness.[5]

1. *Shutters*

Most normal motion picture cameras use a rotating blade shutter. Other types of shutter have been used experimentally and for some special purposes.

Professional cameras generally have a shutter with an adjustable opening. The maximum aperture is ordinarily 170°, and those that are adjustable can be closed down completely. Many can be slowly closed while the camera is running, producing what is known as a fadeout; conversely, if the shutter is gradually opened during operation, the result is a fadein. This method of producing a fadeout is, however, not normally used when taking an original negative, as the desired location for a fadeout is not known until the film is edited.

Recently, some so-called "reflex" cameras, such as the Arriflex, have been developed in which the shutter is actually formed from a rotating mirror set at 45°, so arranged that during the closed portion of the cycle, the shutter reflects an image into the viewfinder (Fig. 2 and 3). The operator can thus view the scene while the camera is running, although his view will suffer from a very marked flicker. This also requires the use of an eyepiece which is closed when not in use, for otherwise, the film would be exposed to light coming in through the eyepiece. Some typical cameras are shown in Figs. 2–7.

---

[14] E. M. DiGiulio, E. C. Manderfeld, and G. A. Mitchell, *J. Soc. Motion Picture Television Engrs.* **76**, 665–670 (1967).

[15] F. T. O'Grady, *J. Soc. Motion Picture Television Engrs.* **67**, 385–388 (1958).

[16] J. Behrend, *J. Soc. Motion Picture Television Engrs.* **73**, 12–17 (1964).

[17] G. A. Mitchell, *J. Soc. Motion Picture Television Engr.* **11**, 267–269 (1927).

[18] F. Tuttle and C. D. Reid, *J. Soc. Motion Picture Television Engr.* **20**, 3–30 (1933).

FIG. 2. The Arriflex 16 reflex camera (16 mm), made in West Germany by Arnold & Richter. (Photograph courtesy Arriflex Corporation of America.)

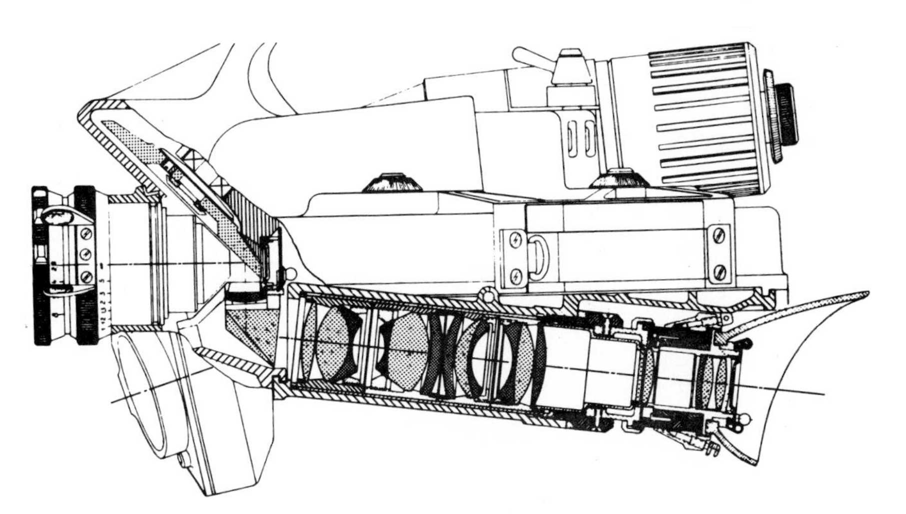

FIG. 3. Cross section of Arriflex viewfinder system. (Courtesy Arriflex Corporation of America.)

FIG. 4. The Mitchell BNC reflex camera (35 mm). (Photograph courtesy Mitchell Camera Corp.)

FIG. 5. Bolex H-16 Rex-5 16 mm camera with 400-ft magazine. (Photograph courtesy Paillard, Inc.)

FIG. 6. The Bach-Auricon Pro-600 16 mm sound camera, equipped with zoom lens. (Photograph courtesy Bach-Auricon, Inc.)

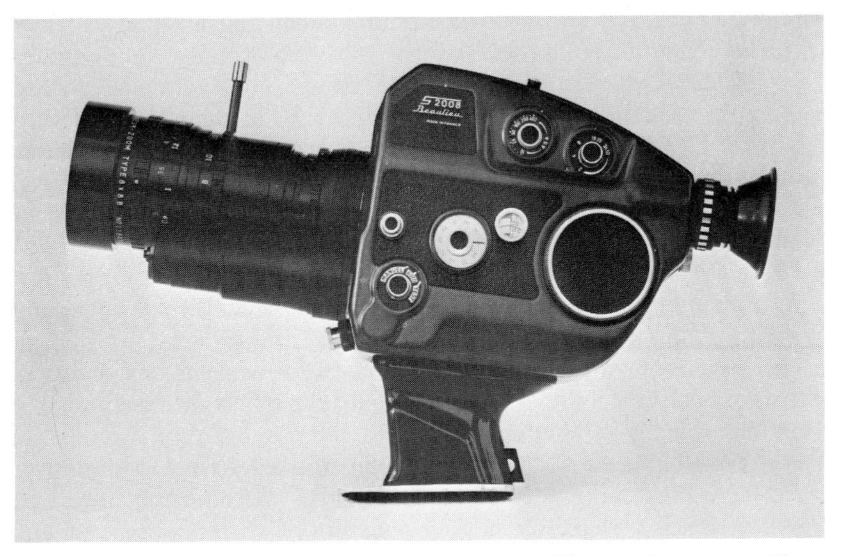

FIG. 7. The Beaulieu 2008S super-eight camera. (Photograph courtesy Cinema Beaulieu, Inc.)

2. *Viewfinders*

Nonreflex cameras require some sort of viewfinder, and many professional cameras are so designed that everything except the lens can be moved sideways ("racked over") to allow the scene to fall on a ground glass in place of the film at the gate. Before "shooting," the camera is, of course, racked back again into position. During shooting, a separate finder is normally used. This produces a large upright image on a ground-glass screen. It is mounted on a level with the camera lens and is swung out at the rear to take care of parallax. On cameras for newsreel photography, a sinple gun-sight type of finder is often used, an open wire indicating the area being covered.

3. *Lenses*

A variety of focal lengths are used in motion picture photography, the range for the standard "academy" gate being from about 18 to 200 mm or even longer. The standard focal length is 50 mm, the shorter foci representing wide-angle lenses, and those with longer foci being known as long-focus lenses. The same angular fields will be covered on 16 mm film if the focal length is about half the corresponding focal length for 35 mm film. It is common practice to mount several lenses on a revolving turret so that the desired lens can be quickly brought into operating position. Lenses intended to be used on a reflex-type camera must have a sufficiently long back-focus to clear the shutter.

Zoom lenses have come into common use in amateur motion picture photography, which is now almost entirely on 8 mm film. They are also used in news photography (Figs. 6 and 7), but very little in entertainment pictures. The change in focal length during the shooting of a scene does not give the desired effect, and since there is sufficient time to select and change lenses, it is thought that the best results are obtained by using a fixed lens of the desired focal length.

When it is desired to appear to be moving closer to the subject during the shooting of a scene, the camera is actually moved forward, a special wheeled stand known as a "dolly" being used for the purpose. It may be necessary to continually refocus the camera if the object is very close, or if a long-focus lens is being used.

When making professional motion pictures, it is desirable to have lenses with as large an opening as possible. This permits the use of the least light indoors, which is desirable from the standpoint of cost and of the comfort of the actors. For most closeups and semicloseups, a short depth of field is wanted so as to separate the subjects from the background. A great share of the long shots, where considerable depth of field is required, are made outside where sufficient light is available to allow stopping down the lens.

To reduce the exposure outdoors, when using a lens with a large opening, a neutral density filter is used. This is normally placed in the matte box in front of the lens. In black-and-white photography particularly, colored filters are often used to produce various effects. In closeup work, diffusion disks are sometimes used to soften the lines in the actors' faces. Special soft-focus objectives are sometimes also used for this purpose.[19]

In the early systems of color photography, it was necessary to produce separation negatives, which was often accomplished with the use of beam splitters.[20] The Technicolor camera was an example of this type of procedure. Modern color photography involves the use of a monopack negative, and so standard optical systems can be used. Care must be taken, however, that the color of the light coming through the lens systems stays the same with all the various objectives used. It has been necessary to standardize the color of lens coatings for this reason.

4. *Dollies and Cranes*

There are a number of miscellaneous items that are important in the production of motion pictures. These are chiefly mechanical in nature.

As mentioned previously, the camera is often moved toward or away from the subject during shooting, and this is done either by mounting the camera on a special dolly or setting the tripod on a dolly which is run on a track laid for the purpose.[21] Sometimes, the camera and crew are placed on a platform at the end of a crane, particularly, where shooting is to be done from above or where part of the action takes place at a higher level. The camera is then free to be moved in any direction in relation to the subject.

Often, the camera is mounted on a crane which is in turn mounted on a dolly. There are many versions of such equipment, some devised by the director of photography to handle a particular situation. These pieces of equipment have beome very sophisticated, with controls that allow moving the camera smoothly in almost any fashion.

---

[19] This series, Vol. III, Chapter 3, p.118

[20] A Cornwell-Clyne, "Color Cinematography," 3rd ed., pp. 509–520. Chapman and Hall, London, 1951.

[21] E. A. Hunter, *Am. Cinematographer* **29**, 234 (1948); F. E. Lyon, *Ibid*, **30**, 242 (1949); J. duValle, *Ibid*, **31**, 270 (1950); L. Garmes, *Ibid*. **31**, 307 (1950);

5. *Lighting Equipment*[22]

The earliest motion pictures were photographed by daylight, but mercury arcs soon came into use for inside photography, and these were followed by carbon arcs. In those days, the negative film was sensitive only to blue and blue-green light.

With the coming of panchromatic negatives in the 1920's, tungsten light was introduced and soon came into general use. Carbon arcs were still used where an intense light source was required. The Technicolor Three strip and monopack process required a daylight color balance, so arcs were used in making Technicolor pictures and even for fill light outdoors. In recent years, a number of different sources have been introduced, such as tungsten–halogen lamps.[23]

6. *Exposure-Control Devices*

For many years, the professional cameraman or director of photography depended upon his experience to judge the proper exposure level for any given scene. In the earliest days, the problem of making a correct judgment was complicated by the fact that there was not much control in the processing of the negative. The negative was developed by inspection, and the operator would watch the image appear and take the negative out of the developer when he thought it was developed to the proper density. For years, an attempt was made to get better controls installed in the laboratories, and this gradually came about. With the advent of high-speed panchromatic negatives, and later with color, it was no longer possible for the operator to process by inspection, and laboratory controls became a necessity. Under these conditions, the cameraman came into more nearly complete control of the final results on the negative and began to use light-measuring devices to determine the proper amount of light to be used and the camera settings. These measurements are now made with an exposure meter at the plane of the subject, and, in some cases, with a spot photometer, where the measurement must be made at a distance and where a certain part of the picture has special significance.

7. *Animation Equipment*

The technique of animation is used in producing motion picture cartoons for entertainment purposes, for making certain advertising pictures, and in producing some educational subjects.[24] Most animation involves the

---

[22] M. A. Hankins, *J. Soc. Motion Picture Television Engrs.* **76**, 671–674 (1967).

[23] S. C. Peak, *J. Soc. Motion Picture Television Engrs.* **71**, 667–669 (1962).

[24] E. L. Levitan, "Animation Techniques and Commercial Film Production." Reinhold, New York, 1962.

photography of drawings, usually transparencies, wherein some change is introduced in each frame before it is photographed by additions to the drawing or by the substitution of another. Special equipment is used in this operation (Fig. 8), of which there are a number of types and modifications.[25,26]

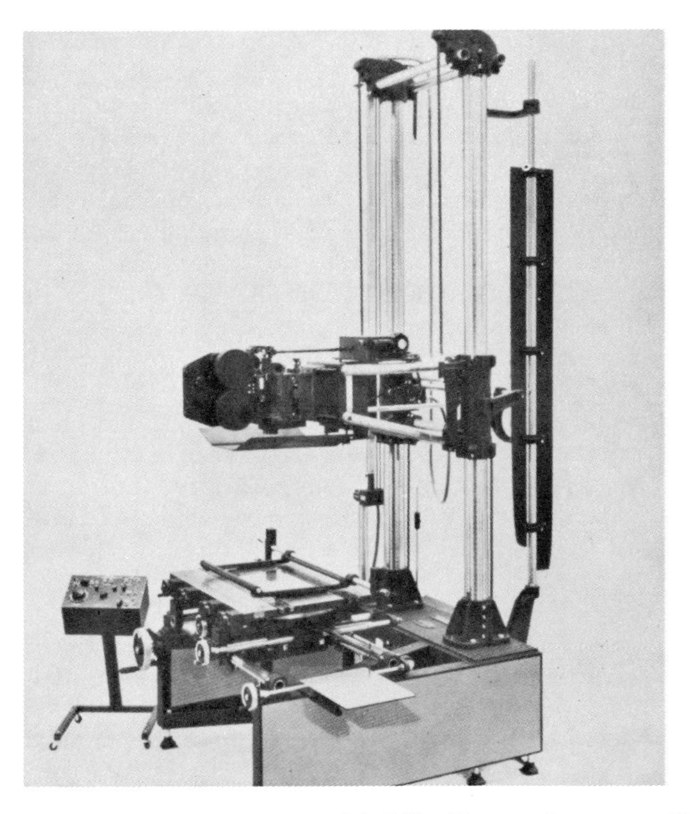

FIG. 8. Oxberry animation stand model 4200. (Photograph courtesy Technical Division, Berkey Photo, Inc.)

Some animation is done with dolls or models, movements being produced by a slight change in position of certain elements of the subject or subjects between successive frames of the picture.

The sound for animated pictures is recorded first and the pictures made to conform to the sound.

---

[25] W. E. Garity and W. C. McFadden, *J. Soc. Motion Picture Television Engrs.* **31**, 144–156 (1938).

[26] E. H. Bowlds, *J. Soc. Motion Picture Television Engrs.* **60**, 58–65 (1953).

B. Amateur

At the present time, amateur motion picture cameras almost invariably use 8 mm or super-8 film, although 16 mm was once considered to be an amateur size. The simplest possible camera is one equipped with a single lens, commonly $f/2.7$ or $f/1.9$, a spring drive, only one frame rate, and no exposure control other than a manual adjustment for the lens aperture. More expensive cameras feature several film rates and interchangeable lenses, or interchangeable afocal attachments used in front of a fixed lens, and, more recently, an electric drive operated by penlight cells. At the next level, we find some form of automatic exposure control, usually a swinging vane with a wedge-shaped slot driven across the light path by a tiny electric motor, the current for which comes either directly from a photovoltaic cell or from a battery *via* a cadmium sulphide cell which acts as a light meter. The most expensive cameras have in addition a zoom lens, with a focal range lying between 2 : 1 in the simplest case to 8 : 1 or even 12 : 1 in the most expensive. Modern zoom lenses are virtually as good as a single lens of the same focal length and aperture.

The viewfinder of an 8 mm camera is generally of the telescopic type, lying along the top of the camera, with an objective in front, a central relay lens, and an eyepiece at the back. Many zoom lenses have a small mirror close to the aperture stop, which reflects some light upward into the viewfinder so that the observer is able to see at all times exactly the scene that will be photographed.

To simplify the problem of loading a roll of film and forming the necessary loops, some recent cameras have film magazines which are loaded either at the factory or by the user. Super-8 film always comes preloaded in some form of magazine. Summaries. of currently available amateur movie cameras are given frequently in the various camera magazines.[27]

III. SOUND RECORDING

A. General

Although some earlier experiments were made using sound with motion pictures, the real beginning of sound pictures was "The Jazz Singer," released in October 1927. The early pictures were supplied with sound on disk records, but the difficulty of editing and shipping, and of keeping the

[27] See, e.g., the current "Master Buying Guide" published by The Photographic Trade News, 41 East 28 Street, New York, 10016. Also the current "Photography Directory and Buying Guide" published by Ziff-Davis, New York. English readers can refer to *Amateur Photographer* **135** April 3, 1968.

pictures and sound in synchronization, made this method obsolete as soon as it could be replaced by a sound track on the picture film itself.[28]

The sound is recorded on a film separate from the picture negative film. There are several reasons for this. Usually, in making a movie several cameras are used to record closeups, long shots, and different views of the actors, but there is only one sound record, and this must provide the continuity in the cutting and editing of the final picture. Second, the photographic requirements for a good sound negative are quite different from those of a good picture negative. Third, the sound may be recorded as a single variable-density track, a push-pull track, or even as a magnetic record, while the final release prints may carry only a single variable-area track. Finally, background music and other sound effects must be added to the sound track before it is finally made into a negative for printing.

Because the picture is shown intermittently and the sound must be reproduced from a uniformly moving film, the sound record is displaced 21 frames ($15\frac{3}{4}$ in.) ahead of the corresponding picture, the sound reproducer being situated below the picture gate on the projector. In 16 mm film, this distance is 26 frames (7.8 in.) for optical sound and 28 frames (8.4 in.) for magnetic sound. It is thus impossible to cut and splice a film when both sound and picture are in their final positions. Occasionally, the sound and picture are recorded on the same film as a matter of convenience in news or documentary photography, but in such cases, it is necessary to rerecord the sound to allow for editing. Almost all original sound is now recorded magnetically, but optical (i.e., photographic) sound tracks are made for final editing and making the production prints.

For much news work, and for documentary pictures using 16 mm film, a magnetic stripe is placed on the film used for taking the picture. Here again, the sound is rerecorded onto a negative film for editing and printing. As with 35 mm, most pictures made on 16 mm film have the sound recorded on a separate sound negative, but today, practically all 16 mm sound recording is done magnetically. The sound track on 16 mm film replaces the perforations along one edge of the film.

Magnetic sound tracks are used to a limited extent on theater prints, particularly on 70 mm film. Photographic prints are more economical, and in most cases, there has not been enough advantage in the quality of magnetic sound, as projected in the theater, to justify the extra expense of striping the film and recording the sound on each separate release print.

Most 8 mm sound films have a magnetic track,[29,30] partly because of

---

[28] E. A. Kellogg, The ABC of photographic sound recording. *J. Soc. Motion Picture Television Engr.* **44**, 151–194 (1945).

[29] L. Thompson, *J. Soc. Motion Picture Television Engrs.* **70**, 588–589 (1961).

[30] R. G. Hennessey, *J. Soc. Motion Picture Television Engrs.* **70**, 590–592 (1961).

the availability of 8 mm magnetic projectors, and partly because of the very narrow space available on 8 mm prints for the sound track. In regular 8 mm prints, the standard distance between picture and sound is 56 frames (8.4 in.), and on super-8 prints, it is 18 frames (3.0 in.). For future photographic sound on super-8 film, the proposed distance is 22 frames (3.7 in.). In all cases, there is a tolerance of ± 1 frame.

Many articles and books[31, 32] have been written on the numerous problems, and the great variety of equipment used, in recording and reproducing sound on film, and the different types of sound track that can be obtained.[33, 34] The reader is referred to these sources for details. Briefly, the advantage of a variable-density track (Fig. 9a) is that it can be made very narrow if necessary without running the risk of overloading loud sounds. A variable-area track (Fig. 9b), on the other hand, permits a larger amplitude range in the reproduced sound, but it is liable to overload, resulting in clipping of the tops of the sound waves on the film. Present-day standards permit a track width on 35 mm film of 100 mils. (0.100 in.) for variable density, and 76 mils for variable area, the reproducer scanning line being 84 mils long. For 16 mm film, the standard widths of sound track are 80 mils for variable density and 60 mils for variable area, with a scanning line 71 mils long.

Variable-area tracks are possibly more sensitive photographically than variable-density tracks, since even a small error in the density of a variable-area track can cause serious distortion, whereas a variable-density track that is so thin as to be barely visible may still yield acceptable sound. The general adoption of magnetic recording for the original sound has been a godsend, as the photographic rerecording of the sound track can then be done under proper working conditions in a laboratory.

Optical sound tracks can be printed at the same time as the picture if a continuous printing machine is used. Sound reduction prints can also be made from 35 mm to 16 mm film,[35] at an optical reduction ratio along the film length equal to the velocity ratio of the two films, namely, 2.493:1. The lateral magnification, however, must be equal to the ratio of the

---

[31] Academy of Motion Pictures Arts and Sciences, "Recording Sound for Motion Pictures."'McGraw-Hill, New York, 1931.

[32] Academy of Motion Picture Arts and Sciences, "Motion Picture Sound Engineering." Van–Nostrand, New York, 1938.

[33] W. H. Offenhauser, "16mm Sound Motion Pictures." Wiley (Interscience), New York, 1949.

[34] J. G. Frayne and H. Wolfe, "Elements of Sound Recording." Wiley, New York, 1949.

[35] C. W. Clutz, F. E. Altman, and J. G. Streiffert, *J. Soc. Motion Picture Television Engrs.* **52**, 669–675 (1949).

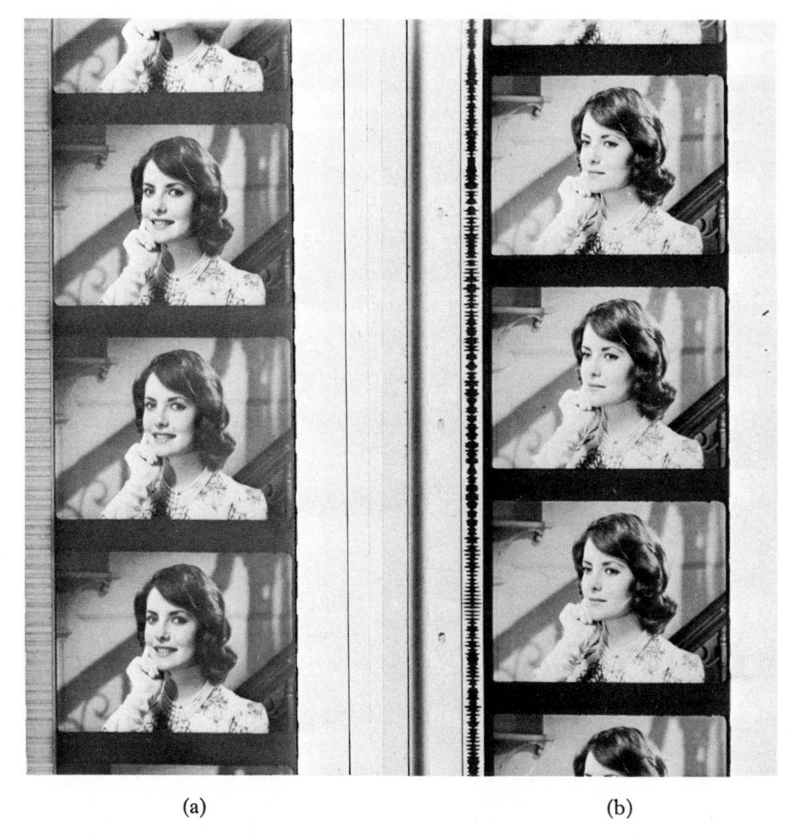

FIG. 9. Typical sound tracks on 35 mm film. (a) Variable density. (b) Variable area.

widths of the sound tracks, namely, 1.26 : 1. Hence, some form of anamorphic expansion must be included in the optical system used in a sound reduction printer.

B. SOUND RECORDERS

The purpose of the optical system in a sound recorder is to convert the fluctuating current from a microphone, or from a magnetic tape recorder, into a varying light beam that can be used to form an optical sound track. Suppose by some means we project a thin, transverse line of light onto the film in the space assigned to the sound track. Then our problem is to modulate this line of light in accordance with the amplitude variations in the microphone current, while the film is moving. The modulation may be in the intensity or in the breadth of the line, in which case, a variable-density track will result, or it may be in the length of the line, in which case, a variable-area track will result.

This problem has been the subject of an enormous amount of research and development,[36] especially between about 1925 and 1950. After trying many different types of modulator, the industry has now settled upon two. For a variable-density track, the light valve[37] is the principal modulating device. This was invented and originally sponsored by the Bell Telephone Laboratories, but today it is seldom used, as variable-density tracks are disappearing. A variable-area track is formed by a complicated optical system indicated diagrammatically in Fig. 10. This was designed by engineers at RCA and is still supplied by that company.[38] In this system, the

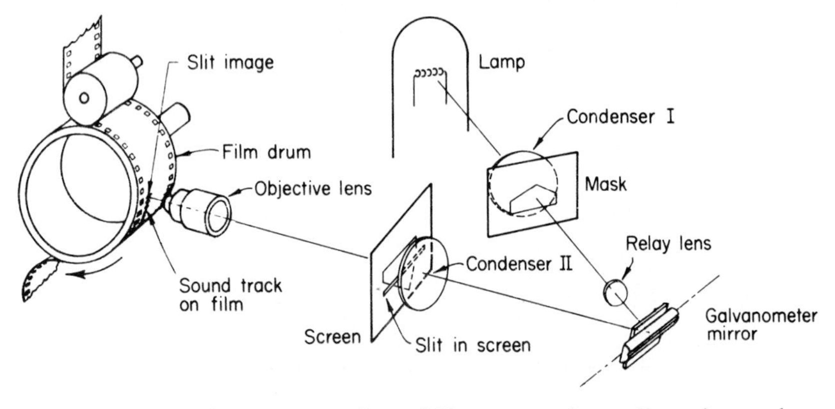

FIG. 10. Schematic arrangement for variable-area sound recording using a galvanometer (RCA).

image of a triangular mask is formed on a transverse slit, the light beam first being reflected by the mirror of a galvanometer. As the mirror oscillates about a horizontal axis in response to the microphone current, the mask image oscillates up and down over the slit, thus varying the effective length of the slit. The slit and mask are in turn reimaged on the moving film, causing a variable-area track to be formed. Many variations of this basic design are described in the literature, including those in which the mask shape is such as to generate push-pull tracks of various types, or even variable-density tracks if required.

[36] E. W. Kellogg, History of sound motion pictures, *J. Soc. Motion Picture Television Engrs.* **64**, 291–302, 356–374, 422–437, with 406 references (1955).

[37] T. E. Shea, W. Herriott, and W. R. Goehner, *J. Soc. Motion Picture Television Engrs.* **18**, 697–731 (1932).

[38] G. L. Dimmick, *J. Soc. Motion Picture Television Engr.* **29**, 258–273 (1937).

Background noise may be eliminated by a bias current proportional to the loudness of the received sound. This moves the mask image so that the tip of the triangular opening is on the slit when no sound signal is coming in, but it allows the image to move to a more normal working position as the sound becomes louder. Alternatively, two small, magnetically-operated shutters may be made to shield the ends of the slit when the sound is quiet, the shutters moving out of the way when a loud sound is being recorded. The clear area on the sound negative becomes solid black on the print, thus preventing light from reaching the photocell of the sound reproducer during quiet periods.

C. Sound Reproducers

The simplest type of sound reproducer consists of a narrow mechanical slit which is imaged on the film at a considerable reduction by a small $f/2$ lens similar to a microscope objective. To compensate the field curvature of the lens, the slit is often curved in toward the lens at the ends by forming a mechanical slit in a hemispherical metal shell.[39] A condenser is used to image a lamp filament, either on the slit itself (Abbe illumination) or into the aperture of the objective lens (Köhler illumination). A photocell is either mounted immediately behind the film, or, preferably, an optical collector system is used to image the aperture of the objective lens on the photocell surface. As the lens aperture is generally large ($f/2$), depth of focus is correspondingly small, and if the emulsion side of the film is reversed, it is necessary to refocus the sound reproducer system.

The length of the reproducing line of light has already been discussed. Its width, however, is not standardized, and is generally about 0.3 mils for 16 mm and 0.7 mils for 35 mm film. If the line is too narrow, there will be very little light and the reproduced sound will have a low signal-to-noise ratio. If it is too wide, the high-frequency reproduction will suffer. At the standard 35 mm film velocity of 90 ft/min, a 10,000-cycle note will have a wavelength of 1.8 mils, and the scanning line must obviously be a small fraction of this for good reproduction.

The most important factor, however, is the angular orientation of the scanning line, which must be exactly perpendicular to the sound track. A very small error in this regard will cause a serious drop in high-frequency response.

In lower-cost projectors for 16 mm film, a slitless reproducer is often used (Fig. 11), in which a straight, single-coil lamp filament is imaged directly on the film at a considerable reduction by a small cylindrical lens E.

---

[39] L. V. Foster, U.S. Patent 1,833,073 (filed Feb. 1929).

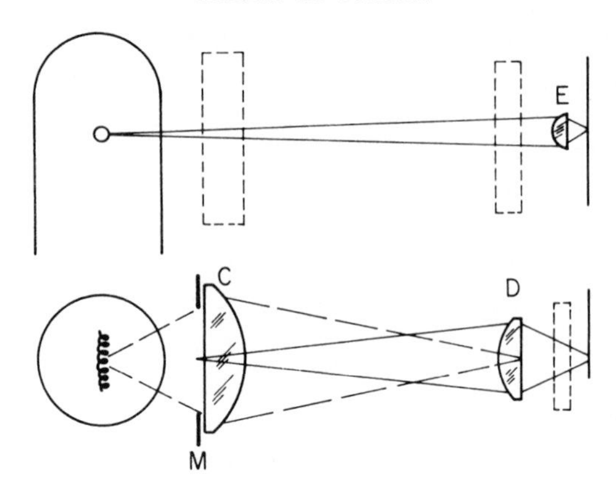

FIG. 11. Slitless sound reproducer for 16 mm film.

By using a cylindrical rather than a spherical lens, the individual turns of the filament are smoothed out, giving a uniform line of light on the film. The ends of the image formed by a cylindrical lens would normally fade out indefinitely, and the scanning line would spill over into either the picture area or the sprocket holes of the film. To avoid this, another system of cylindrical lenses (C, D) set perpendicular to the main lens is used to image a rectangular mask M on the film, which forms sharp ends to the filament image.[40]

A cylindrical lens bent into a curved form has been used to form a curved filament image, which can be substituted for the curved mechanical slit already mentioned.[41]

IV. MOTION PICTURE PRINTERS

The printing of motion pictures originally involved simply making a contact print from the negative obtained in the camera. With the development of color, sound, anamorphism, new materials, and new techniques, printing has become a complex and highly sophisticated operation. It would require several books to cover all phases of the subject, and we can therefore discuss the processes and equipment only in a very general way.

A. PRINTING METHODS

Motion picture printing may be done either by exposing one frame of the processed negative onto the positive raw stock and moving on to the next frame (known as step printing), or the negative may be exposed

---

[40] J. H. McLeod, U.S. Patent 2,161,368 (filed July 1937).

[41] J. H. McLeod and F. E. Altman, U.S. Patent 2,146,905 (filed Oct. 1936).

onto the raw stock continuously. With either type of printing, the two films may be held in contact during exposure, or the negative image may be focused onto the raw stock optically.

1. *Contact Printing*

The greatest amount of release printing is done continuously by contact, because this can normally be done at a higher rate than any other method, and the sound track can be printed at the same time if desired (Fig. 12). However, contact printing requires that the two emulsion surfaces be pressed into close contact, which may eventually cause damage to the valuable negative film.

The practice has normally been to hold the original material and the raw stock in register by means of teeth on a sprocket. The film base which

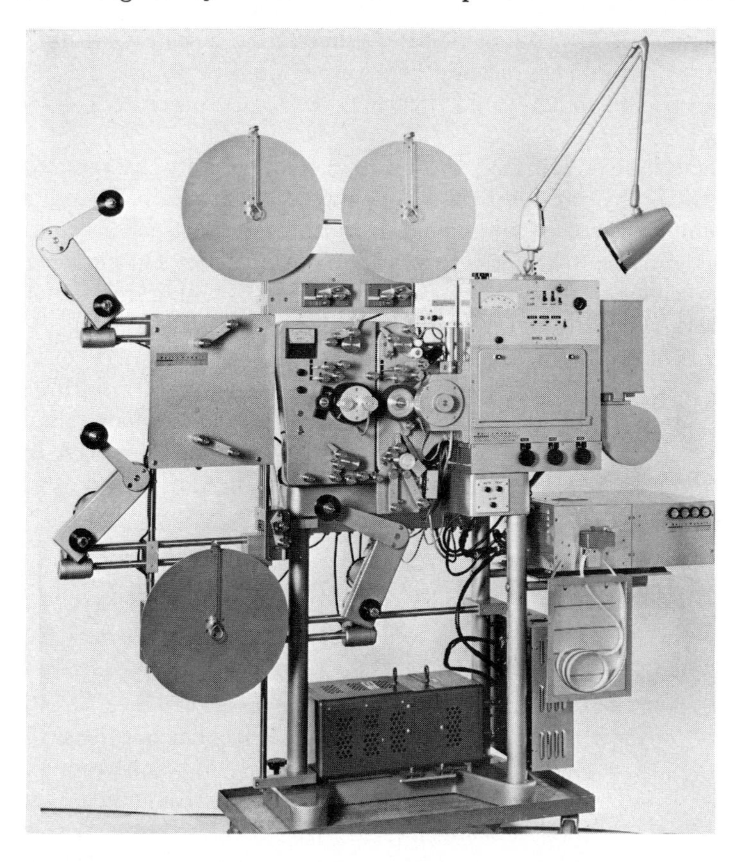

FIG. 12. Bell and Howell model C high-speed contact printer (35 mm). (Photograph courtesy Bell and Howell Co.)

was used until the late 1940's or early 1950's had a shrinkage of as much as 1% between the time of shooting and the final printing. This required that a compensation be made for the difference in pitch between the edited original and the raw stock, to avoid slippage during printing. A continuous printer was therefore equipped with a sprocket of such a diameter that the path of the original film on the inside was enough shorter than the path of the raw stock on the outside to reduce the slippage to a minimum. Generally, a 42-tooth sprocket was found to be satisfactory. With the introduction of lower-shrinkage material, it was necessary to consider altering existing equipment. This was avoided by slightly shortening the pitch of the negative material, thus producing a final negative that had a pitch similar to that of the earlier shrunk material.

In addition to this so-called permanent shrinkage, there is a smaller swelling and shrinking that takes place with humidity changes, and it is the normal practice in printing rooms to control the humidity within certain limits; the humidity is usually kept rather high to reduce the effects of static electricity, which would attract dirt and cause other difficulties in the handling of the film.

In continuous printing, the original material and the raw stock are pulled in front of an illuminated opening. If this opening is too large, it is difficult to keep the illumination uniform over the whole area of the opening, and also to keep the two films in good contact. On the other hand, if the opening is too small, there is a problem of amplifying the effect of nonuniform motion and the production of streaks by any small piece of dirt that may collect on the edge of the opening.

With step printing, it is difficult to obtain economical printing speeds and, because of the size of the aperture, it is difficult to secure sufficiently even illumination over the entire area. It is also necessary to have uniform pressure, for if the pressure is too high at one point, it may cause a pressure mark on each frame of the print.

2. *Optical Printing*

Besides reducing the risk of damage to the negative emulsion, optical printing is used particularly when it is desired to introduce some special effects, or if there is a size change between the original and the print (Fig. 13). Examples of the latter are the making of 35 mm prints from a 65 mm negative, or 16 mm prints from a 35 mm negative. Sometimes, enlarged prints are made from a smaller negative by optical printing. It should be noted that 16 mm prints cannot be made continuously from a 35 mm negative because of the wide gap ("frame line") between frames on 35 mm film, which is not present on 16 mm prints. The picture magnification in

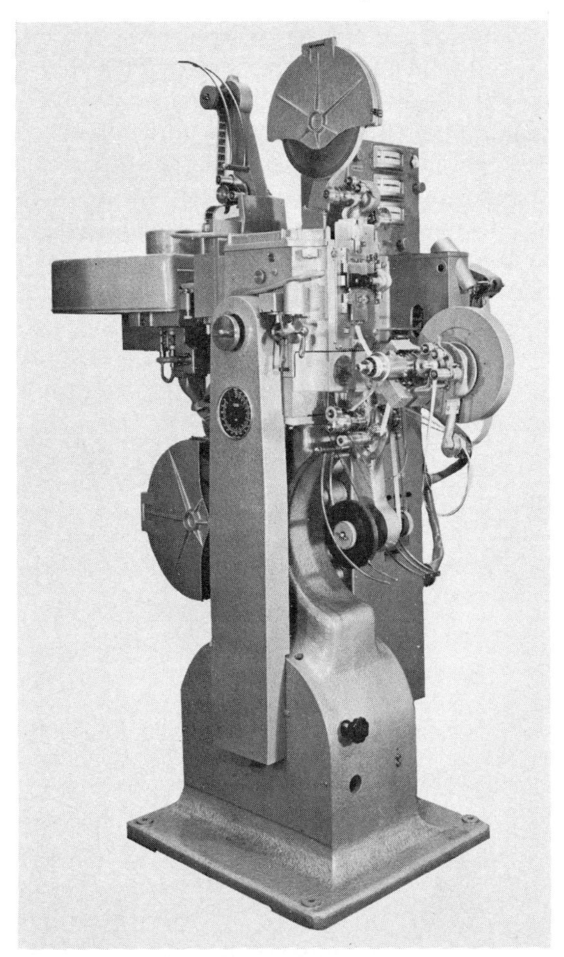

FIG. 13. Debrie Tipro optical printer 35–16 mm with sound. (Photograph courtesy Andre Debrie of New York.)

this case is 2.14 : 1, whereas the ratio of film speeds is 2.493 : 1. An optical step printer is therefore necessary for this operation. However, 8 mm prints can be made from a 16 mm negative by continuous printing because the magnification is the same as the ratio of film speeds in this case.

Optical printing can be and is used to some extent to reverse the image on 16 mm prints. This is desirable in order to keep the emulsion side of the print in the same relation to the lens on the projector. Reversal originals must be projected with the emulsion side facing the projection lens, whereas in prints from negatives, the emulsion side must be away from the lens. If originals and prints are spliced together, the focus of the lens would have to be changed each time there was a change from reversal to print and *vice versa*. This problem is particularly acute in the case of projection for television, since material from many sources is spliced together, and the operation of the projector is automated.

One of the big uses of optical printing is in the production of what are known as special effects.[42] These effects range from enlarging only a portion of a picture, to inserting part of one film into a limited area of another; making fadeins and fadeouts and lap dissolves; assembling a composite picture by means of a traveling matte, etc. (see Fig. 14). Formerly, printers for these purposes were built by specialists in the field and

FIG. 14. Acme special-effects optical printer model 103. (Photograph courtesy Producers Service Co.)

[42] R. Fielding, "The Technique of Special-Effects Cinematography." Hastings House, New York, 1965.

no two printers were alike. There are now, however, very versatile printers of this type regularly available.[43]

B. Light Sources for Printers

1. *Black and White*

The light used for a printer must be uniform over the printing area, sufficiently bright to permit high-speed printing, and remain constant in intensity over a long period, since making each print of a long picture may take from 1 to 2 hr. Furthermore, it must be possible to vary the illumination quickly if required. Since the negatives of different parts of a picture may not be equally dense, it is necessary to set up a schedule of variations in printer illumination from splice to splice so that the final print will be equally dense throughout its length. The variation in printer light is accomplished either by varying the lamp voltage, or by means of vanes or an iris diaphragm inserted in the light path. The required succession of light changes is determined in advance by a "Timer," who sets up a series of pegs in a resistance board or punches a tape which is advanced each time there is a notch in the negative film.

In recent years the notch has often been replaced by a small patch of an electrically conductive material such as metal foil. The light change mechanism is actuated electrically. This patch method has the advantage of permanently marking the original as does a notch.

Tungsten lamps are generally employed for printing, but occasionally, small mercury lamps have been used where a slow, fine-grain material is to be printed at high speed. Contact printing is generally performed with diffused light, which gives lower contrast and tends to eliminate scratches on the film. Optical printing, on the other hand, uses virtually collimated light, which aggravates the effect of dirt and scratches, but improves contrast and sharpness. This is particularly valuable when printing 8 mm film. One of the most effective ways for removing scratches is to use a so-called liquid gate, in which the negative is immersed in a liquid of matching refractive index during the printing cycle.[44, 45]

Although the effect of alternating current upon the intensity of lamps is not noticeable to the eye, it often shows in printing, and for this reason direct current is most commonly used. This is produced by motor generators

---

[43] L. S. Dunn, *J. Soc. Motion Picture Television Engr.* **42**, 204–210 (1944).

[44] J. G. Stott, G. E. Cummins, and H. E. Breton, *J. Soc. Motion Picture Television Engrs.* **66**, 607–612 (1957).

[45] J. R. Turner, P. A. Ripson, F. J. Kolb, and E. A. Yavitz, *J. Soc. Motion Picture Televison Engrs.* **71**, 100–105 (1962).

or by batteries. When printers are run at high speed, and lamps with heavy filaments are used, the variation in light intensity produced by alternating current is reduced to a negligible amount.

2. *Color*

With the coming of color, it became necessary to control not only the density of the print, but the color balance as well. This is done in one of two ways, known as additive and subtractive printing. Additive printing is done with three separate light sources, red, green, and blue, which are combined at the printer gate. The intensity of each light source is controlled separately. This is done by vanes actuated by rotary solenoids, or by interposing neutral densities.[46] In subtractive printing, filters are interposed in the light beam to produce a light of the proper color balance. The filters may be interposed automatically by solenoid, they may be fed from a magazine, or the filters may be inserted over holes in the timing tape if one is used.

Additive printing has the advantage over subtractive printing in that narrowband filters matching the peaks of sensitivity of the color print material may be used, giving no interference of one color with another. the absorptions of subtractive filters overlap seriously, so that unwanted absorptions are produced. Furthermore, it is usually necessary to use a number of filters in each pack, introducing many surfaces that may become dirty or damaged.

It is possible to "time" a black and white negative by merely looking through it and using personal judgment. This has even been done with color negatives in rare instances. If a large number of prints are to be made and the loss of one or two prints in making corrections is a small matter, this system is not too impractical. Very commonly, tests are made on each scene by printing on what is known as a "scene tester," which gives the range of intensities and/or color balances for each scene at one printing. These are judged after processing, and the negatives timed accordingly. Another system involves a scanning method similar to that used for color television, a producing positive color image on a screen, the balance and intensity of which can be controlled by calibrated dials.

When printing color film, it is necessary to take into account the fact that most lenses focus the different colors in slightly different planes, and hence it has become necessary to design lenses for optical printers that give the best possible compromise between the foci of the three colors used in the particular process involved.

---

[46] J. G. Streiffert, *J. Soc. Motion Picture Television Engrs.* **59**, 410–416 (1952).

C. Editing Equipment

The equipment used in editing a motion picture[47, 48] comprises a film viewer or editing machine[49], really a small sound projector (Fig. 15), a splicer, and a synchronizer. Some editing viewers are of the nonintermittent type, using a hexagonal glass block between the film and projection

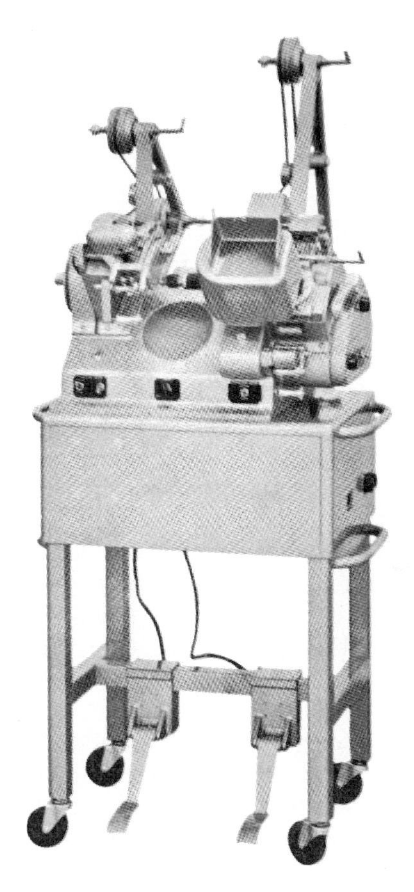

Fig. 15. Moviola film editing machine series 20, for 35 or 16 mm film. (Photograph courtesy Magnasync/Moviola Corp.)

---

[47] G. R. Crane, F. Hauser, and H. A. Manley, *J. Soc. Motion Picture Television Engrs.* **61**, 316–323 (1953).

[48] G. W. Tressel and S. J. Andrews, *J. Soc. Motion Picture Television Engrs.* **73**, 959–960 (1964).

[49] J. V. Aalberg, *J. Soc. Motion Picture Television Engr.* **31**, 426–428 (1933).

lens to form a stationary picture even though the film is moved continuously;[50] Splicers are important mechanical devices, of course. A sychronizer is a shaft having several sprockets to move the various camera films and the sound film while maintaining synchronism between them.

Editing is performed on a master positive print which, when everything has been approved, is used as a guide in cutting and splicing the original negatives. As a safety precaution, a duplicate negative is often made from the edited master positive; this can be used to make release prints if the original negative becomes worn or damaged.

V. FILM PROCESSING MACHINES

The processing of film is an exceedingly important operation in the production of a motion picture, but as it is not an optical problem we shall not discuss it in detail here.

For satisfactory results, the processing conditions must be very carefully controlled in accordance with the recommendations of the film manufacturer. This applies particularly to the composition of the solutions, the temperature and pH of the various baths, and the time spent by the film in each bath. It is essential that the processing conditions remain constant for long periods of time, and that densitometer test strips be run through the process from time to time to check that everything is in order.

Originally, motion-picture film to be processed was wound emulsion side out on a large drum which dipped into a tray of processing solution, the drum being moved from bath to bath while being rotated to expose all parts of the film to the solutions. The whole matter was under very poor control, and the long exposure to air caused the developer to oxidize very rapidly. Today, films are processed continuously in a series of tanks, the film passing over a series of rollers immersed in the liquid, the rate of film travel and the number of rollers in each tank being chosen so as to ensure the correct time of immersion in each solution. The solutions are constantly replenished at a fixed rate, and a washing stage is provided between one solution and the next to avoid contamination by chemicals being carried over by the film as it is moved along. Sometimes, an "air squeegee," or a blast of air to remove liquid from the film, is used between one tank and the next. As a thin layer of liquid is liable to move with the film in the tank, it is advisable to provide quick agitation, either by jets of fluid impinging on the film in the tank, or by bursts of nitrogen bubbles which serve to stir the fluids thoroughly.

---

[50] T. Johnke, *J. Soc. Motion Picture Television Engrs.* **60**, 253–259 (1953).

After processing is complete, as much water as possible is removed from the film, which not only reduces drying time, but prevents the formation of water spots on the emulsion. The film then passes into the drying chamber, where warm air is passed over the surface, and, in some cases, heat is applied by infrared lamps. Impingement drying has come into more frequent use in recent years. Here, air is impinged on the film at high velocity, which reduces the time and space required for drying, although the net cost is likely to remain about the same. The wet-bulb temperature of the drying air must be kept low or the film may become very brittle, and reticulation of the emulsion may even result. Impingement drying has the disadvantage of producing a very shiny negative, which may lead to the formation of Newton's rings when printing by direct contact.

Three very important items connected with the processing of film are cleaning,[51] lubricating,[52] and splicing. Since these are nonoptical problems, they will not be discussed here, but the greatest care must be taken with all of them in any actual film production.

VI. THE PROJECTION OF MOTION PICTURES

A. Professional Projectors

With a few exceptions, the finished prints are projected on equipment that operates with a shutter and intermittent pulldown motion as in the case of most cameras.[53, 54] There have been a few nonintermittent projectors made and used commercially, but they are by far in the minority. Theater projectors do not use registration pins or a pulldown claw, but instead the film is moved by an intermittent sprocket driven by a Geneva movement. As was indicated previously, one or two additional "flicker blades" are provided in the shutter in addition to the 90° capping blade which cuts off the light during pulldown, to raise the flicker frequency to an acceptable 48 flashes/sec.

Release prints in standard 35 mm size are normally handled in 2000-ft reels having a core diameter of 5 in. and an outside diameter of about 15 in., and weighing some 11 lb. These run for about 20 min at the standard

[51] D. W. Fassett, F. J. Kolb, and E. M. Weigel, *J. Soc. Motion Picture Television Engrs.* **67**, 572–589 (1958).

[52] F. J. Kolb and E. M. Weigel, *J. Soc. Motion Picture Television Engrs.* **74**, 297–307 (1965).

[53] R. A. Mitchell, "Manual of Practical Projection." International Projectionist Publ., New York, 1956.

[54] E. H. Richardson, "Bluebook of Projection." Quigley Publ., New York.

90 ft/min. This happens to be a convenient interval of time for replenishing the carbons of an arc lamp. The roll of film is inserted into a metal box at the top of the projector (Fig. 16), and the print is fed downward through a pair of fire rollers (a relic of nitrate film days!) to a continuously rotating upper sprocket. A loop of film is formed above the projection gate in order to permit the intermittent pulldown mechanism to operate freely. Below the gate, there is a second loop and another continuously moving sprocket which feeds the film to the sound head.

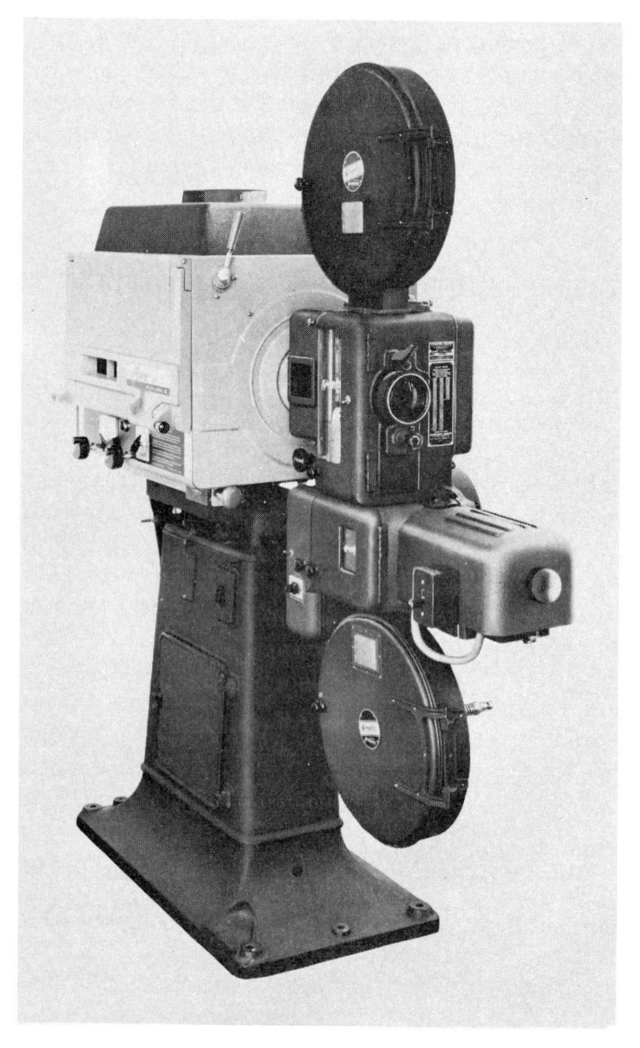

FIG. 16. Simplex 35 mm projector. (Photograph courtesy National Theater Supply Co.)

Many projectors in use today can handle either 35 mm or 70 mm film by making a few changes in the system.[55]

The sound head contains the lamp and optical sound-reproducer unit, which projects a thin line of light on the sound track, and a photocell behind the film to transform the fluctuating light beam into an electric current which is then amplified and fed to the loudspeakers behind the projection screen. Since it is essential that the film be moving absolutely uniformly at the sound pick-off point, it is pulled by a third sprocket over a sound drum at or close to the sound reproducer, the drum being driven by the film itself and coupled to a heavy flywheel which absorbs any minor fluctuations in speed. The dimensions of the equipment and the size of the lower loop must, of course, be such as to ensure the standard distance of $21 \pm \frac{1}{2}$ frames between picture and sound reproducer. Finally, the film passes over a fourth sprocket which feeds it into the takeup reel at the bottom of the projector.

Amateur 8 mm and 16 mm projectors are equipped with a claw pull-down which is satisfactory so long as the gate friction is maintained correctly. There is, in this case, nothing but friction to hold the film stationary while the shutter is open and the frame is being projected.

Those 16 mm projectors which are built for auditorium use generally have a sprocket pulldown instead of a claw, the sprocket being often, but not always, driven by a Maltese cross arrangement. Some projectors have been built with a variable-pitch worm to provide the intermittent mechanism.

Theater projection lenses generally have an aperture of $f/2$ or $f/1.9$, and today are of the double-Gauss type, although Petzval lenses were commonly used until the end of World War II. Projection lenses are made in steps of $\frac{1}{4}$ in. in focal length between about 2 in. and 6 in., with many additional focal lengths up to 10 or 12 in. The choice of focal length depends on the screen size, the film gate size, and the "throw" or projection distance. For many systems where compressed film images are used, an anamorphic attachment must be added in front of the projection lens to expand the picture to its proper format.

The lenses used on 16 mm projectors generally have an aperture of $f/1.6$ and range in focal length from $1\frac{1}{2}$ to 4 in. Anamorphic attachments are occasionally used here also. Projection lenses for 8 mm and super-8 films have an aperture between $f/1$ and $f/1.6$, with a focal length which may be as short as 22 mm ($\frac{7}{8}$ in.). Zoom projection lenses are occasionally used in 8 mm projectors for home movies. The purpose of a zoom lens on a projector is to fill the screen properly for a range of screen distances.

---

[55] W. Borberg and B. N. Plakun, *J. Soc. Motion Picture Television Engrs.* **69**, 176–178 (1960).

B. Light Sources for Projectors

1. *Theater Projectors*[56]

a. Carbon Arcs. The light source for most projectors used in the entertainment field is the carbon arc. These carbon arcs are mounted in special lamphouses,[57] which are usually vented to the outdoors so as to remove the products of combustion. In arc projectors, the highly luminous crater formed at the end of the positive carbon is imaged directly into the film gate such at a magnification as to fill the gate. At one time, very thick carbons were used and a two-element aspheric quartz condenser magnifying about $3\frac{1}{2}$ times was adequate. Today, most arc projectors use thin carbons requiring six times magnification to fill a standard 35 mm gate, the image being formed by an elliptical mirror some 10–12 in. in diameter, the two focal distances being typically 5 and 30 in. from the mirror.

Constancy of brightness is secured by slowly rotating the positive carbon and automatically feeding both carbons at the rate at which they are consumed. Since the crater is imaged into the gate, no visible shadows are formed by the carbon supports even though these items do cause some small loss of light. The positive carbon has a core of rare-earth material which produces a whiter light with less current consumption.

Direct current for operating the arc is produced by a motor generator or by the use of a rectifier. The amount of current used depends upon the size of picture to be produced. Larger carbons must be used for higher currents. There are, of course, physical limitations as to how much current can be used. One of these limitations is the amount of heat at the projector gate, since too high temperatures will damage the film. Black-and-white pictures are affected by heat much more than color pictures, since most colored images absorb very little of the infrared.

In order to reduce the amount of heat at the film gate, infrared-absorbing filters are often used in the light beam. These are cooled with an air blast. An air blast is also often used on the film at the gate itself. Another approach to reducing the amount of heat is to use a mirror that reflects only the visible light and transmits a good share of the infrared.[58]

Uniformity of the light at the gate is very important, not only from the standpoint of the picture on the screen, but also because hot spots will

---

[56] H. E. Rosenberger, *J. Soc. Motion Picture Television Engrs.* **67**, 378–384 (1958).

[57] R. J. Zavesky, C. J. Gertiser and W. W. Lozier, *J. Soc. Motion Picture Television Engrs.* **48**, 73–81 (1947).

[58] G. L. Dimmick and M. E. Widdop, *J. Soc. Motion Picture Television Engrs.* **58**, 36–42 (1952).

cause film damage even when the overall illumination is well below the safe limit.

It should be noted that the aperture stop of the projection system is the rim of the elliptical mirror, and because of the considerable distance of the mirror from the gate, the projection lens is very nearly telecentric. It is important that this fact be taken into account in the design of projection lenses.

b. Xenon Lamps. In some theaters, xenon projection lamps are used.[59–63] Since these xenon lamps operate under rather high pressure, they are installed in a special lamphouse designed to protect the operator in case of an explosion. Xenon lamps have the advantage of a long life, with no need for intermittent shut-down intervals as needed to replace carbons.

2. *Projectors for Nontheatrical and Amateur Uses*

Most 16 mm and smaller projectors employ a tungsten lamp as a light source. These lamps have been specially designed for the purpose, some even for a specific projector. It is important that the lamp and reflector be properly aligned to most nearly fill the projector gate with a uniform light. Projector lamps are all made to fit into the socket in one position only.

In 8 and 16 mm projectors, the lamp filament image is usually formed close to the film gate, but not exactly in it, as the structure of the filament would then appear in focus on the screen and cause a highly nonuniform illumination. Modern lamps equipped with a built-in elliptical mirror to image the filament into the gate are usable only because the coma in the mirror tends to blur the filament image, and a simple sagged glass reflector is imperfect enough to complete the process.

Recently, some quartz–iodine lamps have been introduced for use in 16 mm and 8 mm projectors where maximum light output is required.[64] These lamps require special optics in the lamphouse, since the source is exceedingly small; nevertheless, they result in more screen lumens with a smaller current consumption, and so are considered economical in spite of increased initial cost and higher lamp-replacement cost.

---

[59] D. V. Kloeppel, *J. Soc. Motion Picture Television Engrs.* **73**, 479–480 (1964).

[60] W. T. Anderson, *J. Soc. Motion Picture Television Engrs.* **63**, 96–97 (1954).

[61] H. Ulffers, *J. Soc. Motion Picture Television Engrs.* **67**, 389–392 (1958).

[62] W. B. Reese, *J. Soc. Motion Picture Television Engrs.* **67**, 392–396 (1958).

[63] A. T. Puder and D. Mortensen, *J. Soc. Motion Picture Television Engrs.* **74**, 594–597 (1965).

[64] R. E. Levin and T. M. Lemons, *J. Soc. Motion Picture Television Engrs.* **77**, 124–128 (1968).

C. Amateur Projectors

1. *Regular 8 and Super-8*

There are many projectors available for 8 mm films (Fig. 17), some for only one size and some able to handle both. These are mostly silent (18 frames/sec), although some have a magnetic sound head for use at 24 frames/sec. Some projectors can be run at a number of speeds, forward and reverse, with the possibility also of single-frame stationary projection. A 150-W lamp with an internal elliptical mirror is common, although some

Fig. 17. Kodak Ektagraphic MFS-8 super-8 projector. (Photograph courtesy Eastman Kodak Co.)

of the older projectors use a normal 500-W lamp and a glass condenser. Many 8 mm projectors are self-threading, with holders for 200- or 400-ft reels and a built-in rewind mechanism. A projected screen image about 50 in. wide is considered normal, although some high-powered 8 mm projectors have been built to cover auditorium screens. Lenses of $f/1$ to $f/1.6$ are available, and many excellent zoom lenses also.

2. *Sixteen-millimeter projectors*

These are now considered to be in the school or church auditorium class, although they are sometimes used in homes. Some models can be run at 16 frames in addition to the normal 24 frames/sec, and they are generally equipped with magnetic and/or optical sound reproducers. They always hold 400-ft reels, and many can take 1600-ft reels for a 45-min showing of sound film (Fig. 18). Projection lenses of *f*/1.6 aperture are available in a range of focal lengths. Anamorphic expanders can be attached for the projection of 16 mm prints from CinemaScope originals.

FIG. 18. Kodak Pageant 16 mm sound projector. (Photograph courtesy Eastman Kodak Co.)

D. REPETITIVE PROJECTORS

Besides the reel-to-reel type of projector described above, there have been a number of projectors known as "repetitive projectors"[65] in use for many years for advertising and display purposes. More recently, they have also been used for sales training and general education.

---

[65] A. S. Bradford, U.S. Patent 2,651,966 (filed July 29, 1949).
E. Busch, U.S. Patent 2,837,332 (filed May 21, 1953).

The earlier types were always left threaded. They used an endless loop of film which was drawn out from the center of the spool and fed back onto the outside, so that during the operation the film reel was continuously being tightened up. Today, most of these projectors take a film cartridge which operates on the same principle.

The picture from this equipment is normally viewed on a small translucent screen, employing mirrors to obtain a sufficiently long projection distance.

E. Projection Screens

All motion pictures are eventually viewed on some type of screen. There are two general types of screens, translucent and reflection.

1. *Translucent Screens*

Translucent screens are rarely used in theater presentation. The light distribution does not cover a very large angle and the space behind the screen would have to be long, or would at least require the use of mirrors. In addition, because of the amount of diffusion required, the apparent definition is not equal to that from a reflection-type screen.

Very large translucent screens are, however, used to some extent for projection backgrounds in doing special effects.[66]

As mentioned previously, small translucent screens are used on equipment designed for display, training, general education, etc., and for some amateur projection purposes.

2. *Reflecting Screens*

There are many varieties of reflecting screen, from matte screens, which have a large solid viewing angle, to very efficient screens that direct the reflected light to the audience.[67] Some screens are made with a good reflecting surface and with vertical lenticules which produce a wide horizontal viewing angle but a comparatively narrow vertical viewing angle, most theater audiences being distributed in this fashion. Large semicylindrical screens have been installed for the Cinerama process, and these have been made of vertical strips so that the viewing angle can be kept more or less constant, and also to prevent light on one side of the screen from being reflected to the other side of the screen.

With the advent of sound, the speakers were installed behind the screen, and small holes were put in the screen to prevent the absorption of too

---

[66] F. Edouart, *J. Soc. Motion Picture Television Engrs.* **40**, 368–373 (1943).

[67] L. A. Jones and M. F. Fillius, *Trans. Soc. Motion Picture Engrs.* **11**, 59–73 (1920).

much of the higher frequencies. A perforated screen of this type reflects only about 70% of the incident light.

It is important that a projection screen be kept taut, and this is usually done by lacing the edges of the screen to a rigid frame.

Smaller screens used with 16 mm or 8 mm projectors have been made with glass beads on the surface. These produce quite a bright screen, but have a narrow horizontal viewing angle. These smaller screens are also made of the lenticular plastic material mentioned above.

A small, metallized screen has recently been introduced which is highly reflective, the viewing angle being controlled by shaping the screen.

Author Index

Numbers in parentheses are references numbers and are included to assist in locating references in which authors' names are not mentioned in the text.

Subject Index

Cumulative Index

Roman numerals indicate volume number.

A

B

C

D

E

F

I

J

K

L

M

N

Q

R

S

T

W

X

Y

Z